Scour Technology

Scour Technology

Mechanics and Engineering Practice

George W. Annandale, D.Ing., P.E., D.WRE
President
Engineering and Hydrosystems Inc.
Denver, Colorado

McGraw-Hill

New York Chicago San Francisco Lisbon London Madrid
Mexico City Milan New Delhi San Juan Seoul
Singapore Sydney Toronto

The McGraw·Hill Companies

CIP Data is on file with the Library of Congress

1 2 3 4 5 6 7 8 9 0 DOC/DOC 0 1 0 9 8 7 6 5

ISBN 0-07-144057-7

The sponsoring editor for this book was Larry S. Hager and the production supervisor was Richard C. Ruzycka. It was set in Century Schoolbook by International Typesetting and Composition. The art director for the cover was Anthony Landi.

Printed and bound by RR Donnelley.

This book is printed on recycled, acid-free paper containing a minimum of 50% recycled, de-inked fiber.

To my lovely wife Nicolene and my children, for their unconditional love, patience, and continued support.

Contents

Preface

Engineers all over the world are responsible for maintaining existing infrastructure and building new infrastructure in a manner that will safeguard the public, and protect property and the environment. When such infrastructure interfaces with flowing water, it is necessary to investigate the potential effects of scour. Scour can lead to failure of infrastructure, with disastrous consequences.

Scour, another name for extreme erosion, occurs when the erosive capacity of water resulting from natural and manmade events exceeds the ability of earth materials to resist its effects. Scour adversely affects the integrity of earth embankment dams, levees, concrete dam foundations, plunge pools downstream of spillways, bridges, water-bearing tunnels, river banks, and pipelines crossing rivers and oceans, as well as coastlines. It is an international problem of potentially huge proportions, adversely affecting public safety, property, infrastructure, and the environment.

Scour results from the natural processes of intense precipitation, floods, hurricanes, and tsunamis, and from manmade events such as dam failures. We were recently reminded of the effects of scour during the occurrences of the 2004 Asian tsunami and the flooding that occurred in New Orleans directly after Hurricane Katrina in 2005. Scour resulting from the tsunami resulted in the destruction of infrastructure, like bridges. The most devastating effects of Hurricane Katrina resulted from the failure of multiple levees, which led to flooding of New Orleans, displaced an estimated 500,000 people, resulted in incapacitating utilities like water-supply systems, sewerage systems and electrical supply, destruction of property and infrastructure, and significant environmental damage. These events were truly extreme and reminded us of our vulnerability to nature.

Historically most of the research in the field of scour focused on prediction of scour at bridges and in plunge pools downstream of dams. The empirical nature of this research, which principally focused on predicting scour in noncohesive granular earth materials affected by flowing water, can only be applied to problem types for which they were specifically developed. For example, one cannot use an empirical equation developed to predict scour around a bridge pier to analyze scour at a pipeline crossing a river. The results of this research were empirical equations that related scour depth to, principally, a number of hydraulic parameters. The only geotechnical parameter, included in some of the equations, is a representative grain size for noncohesive granular earth material.

The need to investigate and solve scour problems that have not been addressed before and for which no ready-made empirical equations exist points to a need for deeper understanding of the fundamental nature of scour processes. This includes needs to quantify the magnitude of the erosive capacity of water for almost any flow condition that might be encountered in practice and to quantify the ability of almost any earth material, not only noncohesive soils, to resist the erosive capacity of water. Earth materials most often encountered in practice include cohesive soils like clay, sandy clay, and silty clay; cemented soils; vegetated soils; and rock of various types and descriptions. A unified approach to quantify the relative ability of these materials to resist the erosive capacity of water is useful when addressing scour problems. At the point of incipient motion, i.e., when scour is just about to occur, such relationships are known as scour or erosion thresholds. Erosion or scour thresholds serve a useful purpose for determining the potential for scour—scour extent and scour rate.

The approach in this book is to provide a framework that can be used by practicing engineers to investigate various kinds of scour problems, varying in flow conditions and material types. This decision-making framework can be used to investigate scour and develop defensible solutions to scour problems. The framework is based on objective and subjective reasoning, which are respectively supported by a solid understanding of fundamental scour processes and the experience of the individuals conducting the analysis as well as that of the profession as a whole.

A sincere attempt was therefore made to present theory in a pragmatic manner that will allow development of insight into and understanding of scour processes; theory is combined with analysis, examples, and case studies. The theory deals with the essence of hydraulic processes characterizing the erosive capacity of water and with practical methods to quantify its relative magnitude. Additionally, the inherent nature of different material types and how it affects a material's ability to resist the erosive capacity of water is discussed in detail. Practical methods are presented for implementing this understanding of material properties to quantify the ability of varying material types to resist the erosive capacity of water.

In this regard it is demonstrated that quantification of the magnitude and frequency of turbulent pressure fluctuations, the dominant process leading to erosion and scour in rough turbulent flow, is currently, in many situations, not a practical approach. Methods to quantify indicator parameters that represent the relative magnitude of turbulent pressure fluctuations, and thus the relative magnitude of the erosive capacity of water, are presented in a practical manner. In the same vein it is also demonstrated that detailed characterization of the properties of varying earth materials, representing their ability to resist the erosive capacity of water, is not currently feasible in most situations encountered in practice. Scour of earth and engineered earth materials, other than those that act like noncohesive soils, are often the result of brittle fracture or fatigue failure. Although theory exists for calculating scour due to the effects of brittle fracture or fatigue failure, it is found that practical limitations often lead to the preferred use of indicator parameters to quantify the relative ability of these materials to resist the erosive capacity of water.

The analysis procedures presented in this book, focusing on applying a cause-and-effect approach to scour investigations, rely heavily on threshold relationships to determine the potential, extent, and rate of scour. The benefit of this cause-and-effect approach is that it provides a framework that allows engineers to analyze unique scour problems and to use the results as part of an objective and subjective reasoning process that results in defensible solutions to scour problems. Application examples are presented that demonstrate identification of scour potential, calculation of scour extent, and quantification of the rate of scour. The case studies presented toward the end of the book further demonstrate application of the knowledge and methods explained in this book, and, by comparing the analysis results with observed scour provide validation of the concepts.

The book is intended for senior undergraduate and postgraduate students, and practicing engineers interested in scour. I hope that the material presented, developed over a period of almost 15 years, will provide the profession with fresh insight into scour processes and will find useful application in practice. I trust that this small contribution will be of value to man and the environment.

This book is an ongoing project. Readers are invited to communicate with the author by making suggestions on how the book can be improved. E-mails can be sent to the author at *george.annandale@enghydro.com*

Errata can be found on Engineering and Hydrosystems' website: *www.enghydro.com*

George W. Annandale, D.Ing., P.E., D.WRE
Denver, Colorado

Acknowledgments

Embarking on a project to write a book, in between all the other things that needs to be done in one's life, is like running the gauntlet, which in the *Webster Dictionary* is defined as "a double file of men facing each other and armed with clubs or other weapons with which to strike at an individual who is made to run between them." The aim for the runner is to reach the other end of the gauntlet before getting beaten down. The runner runs as fast as he can, and the men with the clubs beat upon him as hard as they can. It is a competition between two diverse goals. So it is with writing a book on a "part-time" basis, especially when the "part-time" in one's life is almost nonexistent.

Running my engineering practice—traveling extensively nationally and internationally on project assignments, and attempting to spend the occasional hour or two with my family, while writing this book—has been trying. I am deeply indebted to my wonderful wife, Nicolene, who has always been at my side supporting me in whatever I attempted, and picking up the pieces behind me. I also owe my children gratitude for being patient with a dad that always works. I have tried my best to be there for you, and will continue to do so in the future.

This book is the product of many years worth of thinking, researching, discussing, analyzing, and reflecting on the topic of scour and how it might be dealt with in a cause-and-effect manner. It is not only the result of my own thinking, but on many occasions I came upon new ideas in a purely serendipitous manner when, for example, discussing certain topics with friends and colleagues over a glass of wine or at a meal. It is therefore appropriate to recognize those that played important roles in my development as a professional engineer, and who influenced my life and way of thinking. Thanking colleagues in this regard is not easy, as one runs the risk of leaving someone out. If I have done so, I sincerely apologize.

Reflecting on the development of the erodibility index method I certainly need to thank Dr. Hendrik Kirsten, former principal of Steffen, Robertson and Kirsten, Inc., for introducing me to rock mechanics. When, as a civil engineer specialized in hydraulics and sediment transport,

I started off investigating scour of rock these geologic structures appeared to me to be no more than chaotic masses of hard material. He introduced me to the essential mechanics of rock behavior, which I then enhanced by further reading, studying, observing, and analyzing.

Additionally, I also wish to thank John Moore from the U.S. Department of Agriculture (USDA) for introducing me to the database of scour data that has been collected by the Agricultural Research Service (ARS) over a period of more than 10 years. These data are well organized and are the product of the hard work of Darrel Temple at the ARS in Stillwater, Oklahoma. The immense amount of detail in this database, comprising geologic, soils, vegetation, hydrologic, and geometric information pertaining to the performance of the USDA's dams' spillways, has been extremely valuable in the development of scour threshold relationships.

Another professional colleague I wish to thank is Dr. Hank Falvey—a grand, graceful, insightful, and world-renowned practicing hydraulic engineer who has, perhaps unknown to him, revealed many wonders of the field of hydraulics to me. His reliance on basic principles of physics and guidance on how to apply them to solve hydraulic problems have and remain to be very educational to me.

Another engineer from whom I learned a lot, who is somewhat younger than Hank but as bright, is Dr. Erik Bollaert, founder of the firm Aquavision. I got to know Erik when he was still studying for his PhD at the ETH, Lausanne. I had the privilege to be one of his advisors, but think that I most probably learned more from him than what he learned from me. The ETH, Lausanne, has wonderful testing facilities and a great approach to engineering research.

Mike Rucker, a geotechnical engineer with AMEC in Phoenix, Arizona, and a graduate from MIT has opened my eyes to see the world in a new way. Mike is an engineer's engineer; with his hands in the dirt, always in the field testing and characterizing materials, but with a mind like few I have met before. Normally one would not associate a geotechnical engineer who spends a considerable amount of his time in the field, with dirty boots, dusty trousers, and scrubby hands with someone busily studying apparently esoteric topics in the field of physics, like chaos theory, fractal geometry, and percolation theory. Mike does not only study these new advances in physics but also applies it to his day-to-day work, making him a very successful geotechnical engineer. Mike provided me with new insight and a fresh approach to characterizing earth materials and determining their erosion resistance. He explained the value of percolation theory to me, which I subsequently used to categorize the scour characteristics of different material types. It forms the basis of material characterization in this book.

Others from whom I have learned a lot as regards the behavior of streams and rivers and how to protect them in an environmentally acceptable way are Jennifer Patterson, a fluvial geomorphologist who works with me, and Linda Aberbom, a restoration ecologist with LSA in the San Francisco Bay area and a dear friend. Together we developed approaches to determine engineering specifications for plant material that can be used to protect river and stream banks and beds against erosion and scour. Jennifer and I developed procedures to calculate the required root architecture and growth habit for plants that would protect soils against the effects of scour. Linda Aberbom, nd expert in native California plants, used these specifications to select plant material that can be used to protect stream banks and beds against erosion in a natural manner. Projects we have jointly executed have been proven to be successful.

Tamara Butler, an engineer who was willing to risk her career by joining me when I started my practice, is someone I trust and often lean on to assist me in developing solutions to challenging problems. She patiently applies my "bright" ideas (which at times might not be that "bright") in developing programs and solution procedures we implement on projects. She developed the programs we use to simulate the anticipated scour of the fissures in foundation of the flood control dams of the Maricopa County Flood Control District in Arizona.

I would also like to recognize Mike George, a young geologic engineer and graduate from Colorado School of Mines who works in my office. He has enthusiastically embraced scour studies and assisted me on projects in the United States and abroad. His insight and the understanding he has developed as a young engineer is remarkable and he does not waiver when asked to tread in areas that have not been investigated before. He has been instrumental in developing some of the techniques presented in this book.

Another person, who is dear to me, is Rebecka Snell, a librarian specializing in engineering in Denver, Colorado. I met Rebecka soon after we moved to the United States in 1991. We found our way of thinking about life, in general, to be synchronized. This common understanding resulted in Rebecka not only searching and finding interesting research papers and books for me but in us often spending time together philosophizing about life. My wife and I enjoy the times we spend with Rebecka and her husband Vic Labson. Rebecka has been kind enough to review and edit some of the chapters in this book, and has also done searches for me to find information.

Other individuals who influenced my professional development are George Beckwith, a visionary and insightful engineer who passed away last year and was instrumental in providing me with the opportunity to work on the Maricopa County Flood Control District's dam safety program; Professor Albert Rooseboom, who was my advisor for my Doctor of Engineering degree many years ago; Professor Steve Abt, from the University of Colorado, Fort Collins, whose gut-feel and insight into hydraulic processes are truly amazing; and Dr. Rod Wittler, a Bureau of Reclamation engineer with whom I conducted near-prototype research at Colorado State University, Fort Collins; and, Hasan Nouri, a very close friend of mine who is not only an excellent engineer in the field of fluvial hydraulics, but a humanitarian with a sincere and true concern for the welfare of others. I also wish to thank my partner Gregory L. Morris who provided me with the opportunity to become an independent consultant and has always been a source of inspiration.

Others that contributed to this book are Bob Wark, a dam engineer with GHD, Perth Australia; the Washington Group for providing me with permission to publish the work I have done on the San Roque Dam; S. A. Iberdrola, Bilboa, Spain, who provided permission to publish the case study on Ricobayo Dam; Dave Anglin and Robert Nairn from Baird and Associates in Canada for preparing the case study on the Confederation Bridge; Dr. Jose de Melo from Portugal for sharing photographs and the findings of his research on rock scour; Ravi Murthy, currently with the Department of Water Resources in Phoenix, Arizona, with whom I previously worked on the dam safety project for Maricopa County; Jon Benoist for providing the photo of the Narrows Dam failure; Richard Humphries for providing photographs of the San Roque tunnels during a flood event; Gregory L. Morris, my partner, for providing photographs of the effects of the 2004 floods in Haiti; Professor Bruce Melville from the University of Auckland, New Zealand, for providing copies of research reports on erosion of cohesive material conducted at this university; Tamara Butler, an engineer with Engineering and Hydrosystems for taking the trouble to write up a case study which I did not use in the end.

My editor at McGraw-Hill, Larry Hager, has been very patient during the course of developing this work, and I would like to thank him for his support in this regard.

Then, last but not least, I wish to thank my neighbors, Mike and Joanie Armstrong, and Rick and Maureen Birkel, for their continued friendship and for the time we can spend together enjoying good conversation, good food and wine, and the occasional cigar.

Scour Technology

Chapter

1

Scour Management Challenges

Introduction

The objective of this book is to provide a unified approach for solving scour problems by providing a decision-making framework, presenting information that will assist the reader in developing an understanding of scour processes, and by providing an internally consistent approach that can be used in practice to solve a variety of scour problems. The intended audience is practicing engineers, senior undergraduate students, and postgraduate students with a basic knowledge and understanding of geotechnical and hydraulic engineering, and geology.

Practicing civil engineers are responsible for designing and constructing, safe and economical infrastructure systems. The development of such designs is an acquired skill, often tested to the limit by inadequate data and resources. Under such conditions, good engineering design results from extensive experience that is complimented by a solid understanding of basic theoretical principles, insight into natural processes, analytical capabilities, and engineering judgment.

The book aims at providing theoretical information in a manner that will assist the engineer in developing insight into scour processes. Guidance on how to apply such insight combined with visualization of natural flow processes when solving engineering problems in this field of expertise are offered. This is done against a formalized background of the decision-making process conventionally, albeit often subconsciously, applied in engineering design and problem solving. Case studies are presented to assist the reader in learning from another's experience.

Infrastructure in contact with flowing water such as bridges, dams, and water-bearing tunnels should be designed to resist the effects of scour. Scour is a term that is used to describe significant localized erosion of earth materials (e.g., scour around a bridge pier or in a plunge pool downstream of a dam spillway) that occurs when the erosive capacity of water exceeds the ability of the earth material to resist it. The erosive capacity of water originates with fluctuating pressures in turbulent flow, and the ability of earth materials to resist it is a function of inherent material characteristics such as mass strength, dimensions, internal friction, and shape and orientation. The terms *erosion* and *scour* are used interchangeably in this book, and are assumed to have the same meaning.

Conventional approaches to scour technology development focus on particular scour types. For example, in the past numerous researchers focused specifically on developing empirical equations to calculate the extent of bridge pier scour (e.g., Richardson and Davis 2001; Melville and Coleman 2000). Other researchers developed empirical equations to predict scour caused by plunging jets (e.g., Mason and Arumugam 1985; Yildiz and Üzücek 1994), while still others developed empirical equations to calculate scour downstream of sills (see e.g., Hoffmans and Verheij 1997), and so forth. These empirical equations represent scour for a particular earth material type, usually non-cohesive granular material like sand.

Although the conventional approach to scour technology development outlined above is useful, it does not provide practicing engineers with the capability to solve problems for which no "ready-made" empirical equations exist. For example, how would an engineer go about solving plunge pool scour in rock if the available equations were developed in a laboratory using non-cohesive sand? The material properties of rock and non-cohesive sand differ significantly.

One of the principal aims of this book is to develop insight into scour processes that will provide engineers with the capability to develop solutions to scour problems for which no "ready-made" empirical equations necessarily exist. This is accomplished by developing an understanding of scour mechanisms and by providing generalized methods for quantifying the erosive capacity of water and the ability of any earth material to resist it. The approach to solving scour problems in this book uses a cause-and-effect approach instead of empirical relationships relating scour to a number of independent variables.

Scour and Infrastructure Safety

The parts of the infrastructure mostly affected by scour include bridges, dams, pipeline crossings in rivers and oceans, and tunnels. Additionally the scour of riverbanks and shorelines, although not necessarily always considered part of the built infrastructure, can adversely affect infrastructure and property. Engineers are required to anticipate the effects of flowing water on infrastructure and property, and protect the public by designing systems that will prevent failure.

Bridges

The failure of river bridges can result from a number of factors, including overtopping, structural failure, debris accumulation, embankment erosion, and scour (Annandale 1993). Figure 1.1 summarizes the relative contribution made to bridge failure in the United States, New Zealand, and South Africa by each of these causes.

Overtopping and debris accumulation lead to functional failure. This occurs if water flows over a bridge or debris accumulates on a bridge deck. Such conditions make the bridge impassable during or immediately after a flood event. Functional failure is temporary and easily repaired at relatively low cost.

The other mechanisms lead to physical failure. Structural failure indicates that the structural components of the bridge were not strong enough to withstand the forces of floodwaters. For example, the piers may collapse if they are not strong enough to withstand the lateral forces imposed on them from the flowing water.

Embankment failure refers to erosion of the earthen approaches to a bridge. The occurrence of this type of failure is high; often resulting in the bridge structure itself remaining intact, but abandoned, in the middle of the river. Reconstruction of the approaches makes it possible to pass over the bridge again.

The term scour, in the context used in this subsection, refers to bridge pier scour, abutment scour, contraction scour, and general scour. Bridge

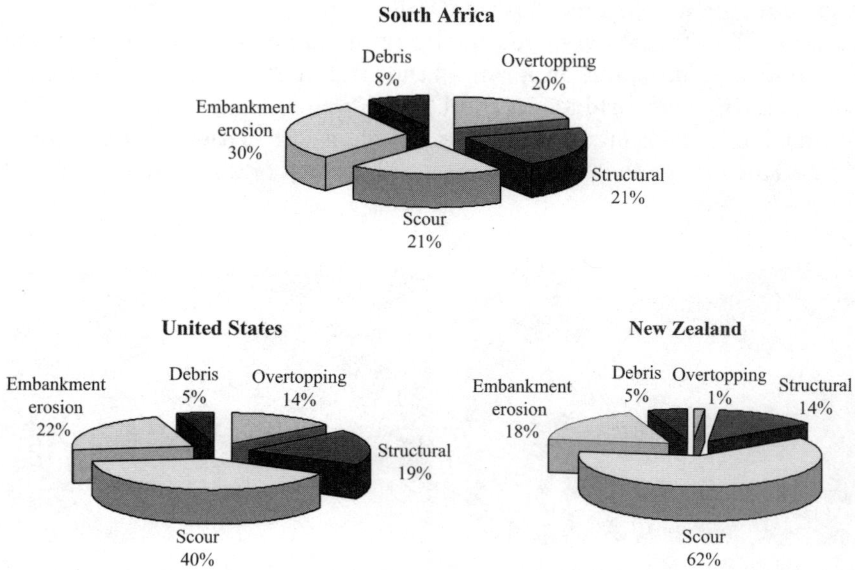

Figure 1.1 Causes of bridge failure in the United States, New Zealand, and South Africa (Annandale 1993).

pier and abutment scour occur when the flowing water in the immediate vicinity of these structural elements remove significant amounts of earth material. Removal of enough earth material can lead to failure of the bridge.

Contraction scour under bridges results from the acceleration of flow when the channel width decreases as the water flows underneath a bridge. This leads to an increase in the erosive capacity of the water, which can remove earth material from underneath and downstream of the bridge.

General scour occurs when a riverbed degrades. This is often not due to the presence of a bridge, but result from fluvial geomorphologic processes in the river. For example, a dam crossing a river might lead to accumulation of sediment behind it. If the amount of deposited sediment upstream of the dam is significant, the river downstream of the dam will be sediment-starved and degrade. Contraction and general scour can obviously exacerbate the effects of pier and abutment scour.

Scour contributes to failure of 21 percent of the bridges in South Africa, and 62 and 40 percent of the failures in New Zealand and the United States, respectively (Fig. 1.1). Scour is the biggest single cause of bridge failure in the United States and New Zealand. The statistics for other countries have not been included in this summary, but it is conceivable that it most probably follow similar trends.

The scour failure of the Schoharie Creek Bridge on the New York State Thruway in Montgomery County, New York, on April 5, 1987, was a landmark event in the United States (Fig. 1.2). When the center span and east center span of this 540-ft-long bridge over Schoharie Creek collapsed during a near record flood it killed nine people, with one person missing. This tragic event led to the implementation of a nationwide bridge scour assessment program in the United States. The study found that 141,405 river bridges in the United States either have unknown foundations, are scour susceptible, or are scour critical (Pagan-Ortiz 2002). This means that approximately 29 percent of the existing 484,246

Figure 1.2 Schoharie Creek Bridge, New York, failed by bridge pier scour in glacial till in 1987 (National Transportation Safety Board, 1988).

bridges currently crossing rivers in the United States are potentially subject to failure by scour. This percentage is reasonably close to the percentage of bridges that actually failed due to scour in the United States in the past (Fig. 1.1).

Modern bridge design recognizes the vulnerability of bridges to scour, even when founded on rock. Design of the 13-km-long Confederation Bridge across the Northumberland Strait in Eastern Canada included assessment of scour at its 65 bridge piers, mainly founded on rock. The potential for rock scour during extreme weather conditions at this bridge led to the installation of a monitoring system (Nairn and Anglin 2002).

In other cases, like the replacement design of the Woodrow Wilson Bridge across the Potomac River in the United States, the presence of clay in the foundation required a detailed assessment of its scour potential (Fig. 1.3). Cost estimates indicated several million dollars in savings for every meter of pier length reduction. Detailed assessment of the ability of cohesive clay to resist the erosive capacity of water was therefore necessary.

Dams

Dam safety concerns require assessment of foundation scour that might result from overtopping events, scour of auxiliary spillways, and the effects of fuse-plug scour. Internal erosion in embankment dams, fissure

Figure 1.3 Woodrow Wilson Bridge over the Potomac River is subject to potential scour of cohesive soils (Photo: Federal Highway Administration).

erosion in dam foundations, and scour of plunge pools can also lead to dam failure.

The erosive capacity of water flowing over dams and through spillways can be very high. For example, scour of granite at Bartlett Dam, Arizona, led to the development of a 30-m-deep plunge pool just downstream of its spillway channel. Similarly, high discharges at Kariba Dam on the border between Zambia and Zimbabwe in Africa led to scour of gneiss that formed an 80-m-deep plunge pool. This is significant when compared to the total dam height of 130 m (Mason and Arumugam 1985). Similar problems were experienced at Tarbela Dam in Pakistan, where extensive scour in its plunge pool occurred soon after commissioning, removing large masses of rock.

Reservoir inflows into Gibson Dam, Montana, reached *unimaginable* levels due to a combination of sustained upslope winds and unusually heavy moisture from the Gulf of Mexico in June 1964 (Fig. 1.4). By 1400 h on Monday, June 8, overtopping of the dam began as inflows reached an estimated maximum discharge of 1700 m^3/s and remained there for 3 h. The overtopping event, affecting the dam abutments, lasted 20 h. Although the dam did not fail, it experienced minor scour of the abutment rock.

Overtopping is the principal cause of embankment dam failure, and internal erosion and piping are the second most important (McCook

Figure 1.4 Gibson Dam, Montana, overtopping by approximately 1 m during a flood that occurred on June 8, 1964, with overflows impacting the rock abutments (Photo: Bureau of Reclamation, US Department of the Interior).

2004). A well-known example of failure by piping is Teton Dam in Idaho that failed on June 5, 1976. The dam failed over a period of several hours at first filling, killing 11 people and resulting in damages estimated at millions of U.S. dollars.

Desiccation of soils in arid regions can lead to cracking of earth embankment dams. If water flows through such cracks during filling of the dams, it could lead to failure by internal erosion. Inspection of earth embankment dams in Arizona often indicates the presence of regular cracks, both transversely and longitudinally, at a spacing of roughly 6 m center on center in both directions (Beckwith 2002). Failure of the Narrows Dam, Arizona, occurred because water seeped into and through such cracks (Benoist and Cox 1998). Figure 1.5 shows the breaches at this dam at the conclusion of the failure event in 1997.

Another scour problem that has not previously received much attention is the potential for dam failure by foundation scour due to the presence of earth fissures. The formation of earth fissures are characteristic of arid regions that are subject to groundwater abstraction. For example, regional groundwater abstraction in Arizona has led to as much as 7 m of surface subsidence over large areas, which gives rise to the development of large earth fissures (Figs. 1.6 and 1.7).

The earth fissures extend over significant depths underneath the earth's surface and are known to pass underneath dams. Internal erosion in such fissures can lead to dam failure, if water flowing through them result in significant scour.

Figure 1.5 Narrows Dam, Arizona, failed by internal erosion of desiccation cracks in the embankment in 1997 (Benoist and Cox 1998).

Figure 1.6 Earth fissure development due to regional groundwater abstraction in Arizona.

Figure 1.7 The Picacho earth fissure in Arizona resulting from groundwater abstraction (Cox 2002).

The sudden release of water flowing through a fissure, once enlarged by scour, can adversely affect public safety and lead to property and infrastructure damage. The Flood Control District of Maricopa County, Arizona, and the Natural Resource Conservation Services of the United States have invested several million dollars to investigate the scour of such fissures and to develop solutions for preventing failure of dams affected by fissures.

The Picacho fissure in Pinal County, Arizona (Fig. 1.7), passes underneath Picacho Dam, which is approximately 11 m high and 9 km long. The dam, constructed in 1889 by a private irrigation company, failed on five occasions in the past; in 1925, 1931, 1955, 1961, and 1983. The failures in 1925 and 1931 were due to overtopping, but the failures in 1955 and 1961 were due to the presence of the earth fissure passing underneath the dam and due to desiccation cracks in the embankment. The failure in 1983 is attributed to the presence of the earth fissure in the foundation (Cox 2002).

A similar problem exists at Twin Lakes Dam in Arizona, where earth fissures pass underneath a dam and affects its safety. Notice the emergence of the eroded fissure on the downstream side of the dam in Fig. 1.8.

Flow through auxiliary spillways and activation of fuse-plug spillways can also lead to scour and potential failure of storage facilities. For example, activation of a fuse-plug spillway at Silver Lake, Michigan, led to considerable scour and drainage of the lake in 2003. The depth of scour that occurred after activation of the fuse-plug was on the order of about 6 to 7 m, while the fuse-plug was only about 1.8 m high. No lives were lost, principally, due to execution of a well-organized and rehearsed emergency response plan; however, the economic loss was significant (FERC 2005).

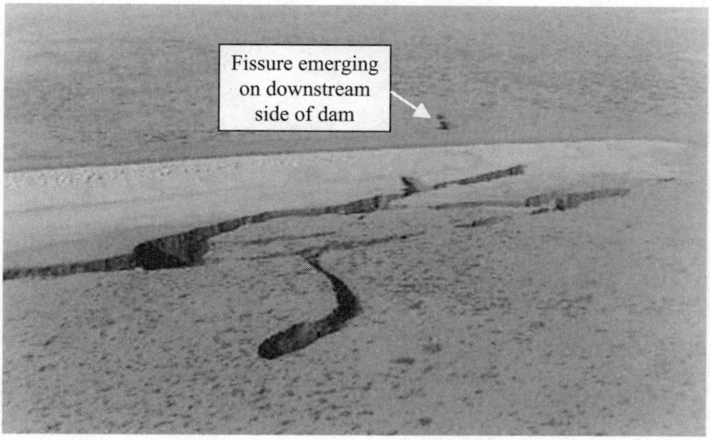

Figure 1.8 Earth fissure erosion at Twin Lakes Dam, Arizona. Notice emergence of fissure erosion on downstream side of embankment (Flood Control District of Maricopa County, 2002).

Figure 1.9 shows the auxiliary spillway at Harding Dam, Western Australia, cut in rock, releasing a flood in 2004. The 952 m^3/s flood resulted in a flow depth of 2.78 m within the spillway. Water plunged over a distance of approximately 18 m at its downstream end without resulting in scour. The reason for this is the high strength of the rock.

However, already as indicated rock is not always scour resistant. This has been demonstrated at the spillway outlet of Ricobayo Dam, Spain. In the early 1930s when this dam was built knowledge to assess the erodibility of rock was lacking. The design called for the water from the spillway to be released onto bare rock as shown in the photograph of the physical hydraulic model study (Fig. 1.10). Once built, it was found that the rock was not strong enough to resist the erosive capacity of the water. Figure 1.11 shows the scour of rock that occurred within the first 19 days of operation of the spillway at the end of 1933 and the beginning of 1934. These releases continued for about 3 months, causing considerably more damage than shown in the two photos. Chapter 10 presents a case study with more detailed discussion and analysis of the scour that occurred at the Ricobayo Dam.

Tunnels

Scour in water-bearing tunnels, such as penstocks, and diversion and water supply tunnels, is of interest in the design, construction, and operation and maintenance of these facilities. Tunnel lining can be

Figure 1.9 Harding Dam, Western Australia. Spillway cut in rock passes a flood in 2004 (Photo: Bob Wark).

Figure 1.10 Model study of Ricobayo Dam, showing spillway discharge design (Photo: Iberdrola, Spain).

expensive. Considerable saving is possible if the tunnel's rock formation can withstand the erosive capacity of water, omitting the lining.

During construction of the San Roque Dam in the Philippines it was determined that omission of the tunnel floor lining could lead to significant

Figure 1.11 First activation of spillway in 1933/1934 showing the amount of rock scour that occurred within 19 days (Photo: Iberdrola, Spain).

savings as well as allowing the contractor to make up for lost time (Fig. 1.12). An erosion assessment of the rock, conducted during construction, led to the conclusion that the rock was strong enough to resist the erosive capacity of the water. The tunnel was therefore commissioned without a floor lining. Subsequent discharges of up to 3500 m^3/s were experienced during construction, without scour or failure.

Pipelines

Scour in rivers and in the oceans can lead to damage and failure of pipeline crossings. General degradation of rivers and scour around river bends can lead to lowering of riverbeds, which can expose pipelines if they are not buried deep enough. Pipelines on ocean floors are particularly vulnerable to scour in the presence of strong ocean currents.

An example of river crossing pipelines is the extensive network of pipelines in the Mahakam Delta in Kalimantan, Indonesia (Fig. 1.13). This is one of the largest deltas in the world and is a significant natural resource in terms of the amount of oil and gas extracted from below it. The pipelines, conveying gas and oil, were designed and constructed using conventional design standards, but long-term degradation in the delta channels and scour around channel bends, exposed pipelines and led to fatalities when one of the exposed pipelines exploded.

Figure 1.12 Diversion tunnels at San Roque Dam in the Philippines, passing a discharge of approximately 3500 m^3/s during construction of the dam without significant erosion of rock on the unlined tunnel floor (Photo: Rich Humphries).

Figure 1.13 Mahakam Delta in Kalimantan, Indonesia one of the largest deltas in the world, is characterized by numerous pipeline crossings.

Bank and shoreline scour

High flows in rivers and waves in the ocean and on lakes often lead to erosion of riverbanks and shorelines. The effects of waves on shorelines were vividly illustrated in Indonesia, Thailand, Sri Lanka, India, and other coastlines affected by the tsunami of 2004. An example of the effects of riverbank erosion is also found in the aftermath of the storms that occurred in Haiti during the same year. The river flows that resulted from these storms killed numerous people and destroyed property and infrastructure (Fig. 1.14).

Approach of Book

The approach of this book is to provide information that will allow practicing engineers to develop insight into scour processes, and to use the insight thus obtained concurrently with available technology and a comprehensive decision-making process to investigate and solve scour problems. The book does not provide "recipes" for solving scour problems, but focuses on developing insight and generalized procedures that allow solution of unique problems not investigated before.

The methodology offered in this book differs from conventional scour technology. It is based on cause and effect, and does not rely on empirical equations that relate scour depth to other parameters for particular situations like bridge pier scour or plunge pool scour. By following

Figure 1.14 Damage resulting from scour and floods in Haiti in 2004 (Photo: Gregory L. Morris).

a cause-and-effect approach, it is possible to use the essential understanding developed in this book to solve most scour problems without the need for empirical equations.

Considerable effort is therefore spent on developing understanding and generalized procedures for quantifying the relative magnitude of the erosive capacity of water and for quantifying the relative ability of any earth material to resist scour. Threshold relationships that relate the relative magnitude of the erosive capacity of water and the relative ability of earth materials to resist scour at the point of incipient motion forms the basis for calculating the extent of scour and its time-related behavior.

In summary, the approach followed in this book is to emphasize conceptual understanding of scour processes, and using this understanding concurrently with available technology, experience, and subjective and objective reasoning to solve scour problems in practice. The generalized methods are used to quantify the erosive capacity of water for most flow situations and to investigate the resistance offered by any earth material to scour. The results of near-prototype scale physical hydraulic model studies and case studies of scour events validate the new proposed methods.

How to Use This Book

The book consists of 10 chapters. The second chapter deals with engineering decision-making and the third chapter deals with the essential elements of scour. The rest of the book applies these concepts to develop insight into the use of pragmatic, generalized procedures for quantifying scour in engineering practice. The reader is encouraged to work through the book in a systematic manner and make a sincere attempt to comprehend the material. A good way to accomplish this is to reflect on the contents and discuss them with others. The practice of scour entails more than just application of equations. The characterization of earth materials and quantification of the erosive capacity of water requires insight, experience, and understanding.

The material in Chaps. 2 and 3 may appear, at first sight, to be esoteric to some. However, in order to obtain the best value from this book it is recommended that these two chapters are not only read but also reflected upon. The presented material provides the basis to the overall approach of the book.

Chapters 4 to 6 deal with material properties, the erosive capacity of water and erosion thresholds by expanding on the basic concepts introduced in Chap. 3. Chapter 4 expands on the approach to material characterization and on methods for quantifying the relative ability of earth materials to resist erosion. Quite a lot of effort is put forward in Chap. 5 to explore the character of the erosive capacity of water and to quantify it for a variety of flow scenarios. Chapter 6 defines erosion thresholds by combining the concepts for quantifying the relative magnitude of the erosive capacity of water and the relative ability of earth materials to resist erosion.

Chapters 7 and 8 provide practical approaches, using the theory and concepts developed in Chaps. 3 to 6 to quantify the extent of scour and its temporal aspects respectively. Methods and concepts for calculating the scour depth in a variety of earth materials and under a variety of flow scenarios are presented in Chap. 7. Chapter 8 provides practical approaches for calculating the rate of scour in earth materials that range from cohesive soils to intact rock. Chapter 9 provides concepts that can be used to engineer protection against scour and Chap. 10 offers case studies.

Chapter

2

Engineering Judgment

Introduction

Successful practice of civil engineering entails maintaining a delicate balance between understanding of the physical sciences and mathematics, and applying empiricism and experience. Although education in civil engineering principally focuses on rigid theoretical training, the dominant role of experience and empiricism and abilities in objective and subjective reasoning and in synthesis are of prime importance in practice. These requirements do not diminish the role of theoretical education. In fact, without a solid theoretical background it is impossible to gain and apply appropriate experience in engineering practice.

This chapter presents a view of the civil engineering decision-making process that serves as a background to this book. The objective of the book is to provide practicing engineers with tools that will assist them in the development of defensible solutions to problems in the field of scour technology. A defensible solution is one that can stand up against scrutiny, either by professional colleagues or in a court of law, and is consistent with basic principles of physics, the engineering knowledge base, and accepted practice.

The need for the development of *defensible* solutions is based on the realization that current technology does not enable engineers to develop *exact* solutions to scour problems. Experience and insight into the scour process are required to conceive solutions to problems that will concurrently ensure public safety and project economy. Mere application of available mathematical equations to calculate scour and using the results to design projects without adequate reflection do not lead to the development of defensible solutions to scour problems.

A unique problem has occurred in recent years with the availability of software that has the ability to provide the user with detailed and, apparently, realistic output. The output obtained from modern software systems, often provided in the form of three-dimensional images, creates a false impression of precision and realism. The perception of precision and realism can create a false sense of security and can lure the user into thinking that experience is of lesser importance when investigating scour with the use of modern software packages. Nothing is further from the truth.

Although the output from modern software packages appears to be accurate and realistic, they are still based on current sediment transport theory, which is known to be notoriously inaccurate. For example, the best sediment transport theories provide results that are within plus or minus 100 percent of measured data (e.g., comparisons of theories by Yang 1996). Software in the field of fluvial hydraulics cannot be more accurate than the essential theory on which it is based. Interpretation of the results obtained from software packages, based on a decision-making process shown in Fig. 2.1, is therefore of critical importance.

The value of a modern software package is that it allows the engineer to effectively and efficiently conduct sensitivity analyses, which is of immense value in the decision-making process. Software does not replace experience, but is useful for the development of information required in decision making that leads to the formulation of defensible solutions. It is only part of the decision-making process, which can now

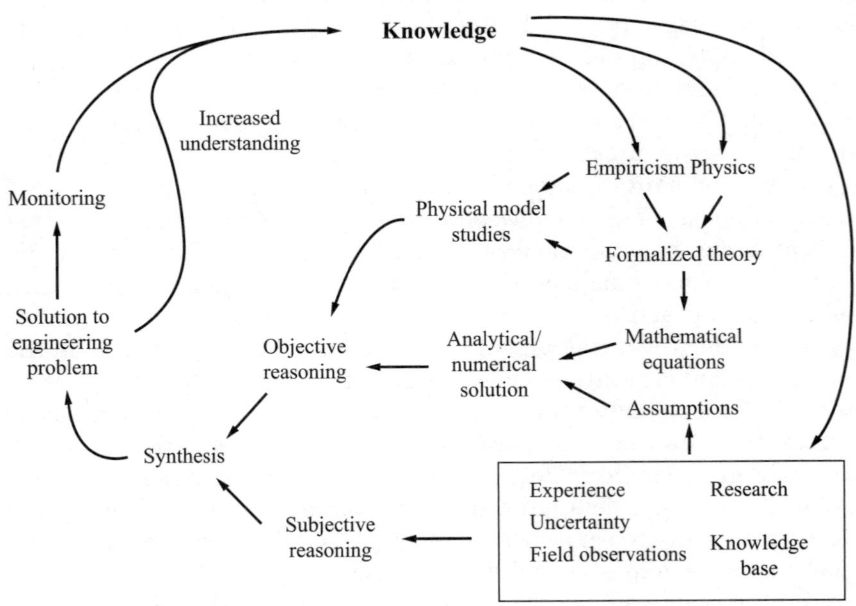

Figure 2.1 Decision-making process in civil engineering.

be executed more effectively because of the computational power available to the engineer.

Defensible Decision Making

Development of defensible solutions to engineering problems entails application of established theory, experience, appropriate assumptions, analysis and experimentation, objective and subjective reasoning, synthesis, and monitoring. This should, if applied correctly, lead to understanding of problems and development of defensible solutions that are internally consistent. A solution to a problem is internally consistent if it satisfies the basic laws of physics and leads to consistent understanding and confirmation of the perceived solution when considered from various points of view.

Civil engineering, an art based on the physical sciences, conventionally deals with the design, construction, and maintenance and operation of expensive infrastructure, such as dams, bridges, harbors, and so forth. It differs from other engineering disciplines that have the privilege to optimize and refine a product prior to mass production. For example, computer systems can be subjected to significant testing and optimization, and once perfected can be mass produced at minimum cost and appropriate reliability. Projects in civil engineering are often one of a kind, and large amounts of money are spent to design, construct, and maintain systems that are often not fully understood.

Although working with incomplete knowledge and often under great uncertainty, everything possible is done by the profession to prevent project failure while still economizing. This requires experience, detailed knowledge and insight, and decision-making capability. The issue of decision making is a vast topic and will not be covered in significant detail in what follows. However, the importance of this topic in scour technology justifies additional reading and reflection. The process of decision making in the presence of uncertainty is a topic that has been dealt with by various authors, with Vick's (2002) book on subjective probability and engineering judgment probably the most relevant to civil engineering.

Decision-making process

A view of the decision-making process implemented by civil engineers is presented in Fig. 2.1. This figure shows that knowledge in civil engineering is dynamic and continuously evolving, therefore the circular reference. The process of decision making in civil engineering does not only lead to development of solutions to a particular problem at hand, but results in continuous improvement of the individual's and the profession's knowledge base as more insight is gained through time.

Development of solutions to problems commences with an existing knowledge base. This knowledge base, which includes empirical knowledge, and understanding of physics and mathematics, combined with practical experience, either documented or personal, forms the basis of progress. As experience is gained in solving practical engineering problems, more insight is developed and the knowledge base improves.

Formalized theory developed from physical principles and empirical observation leads to the development of mathematical equations that are used, with appropriate assumptions, to develop quantitative analytical or numerical solutions to problems. In addition to using analytical and numerical techniques, the use of physical model studies in the field of hydraulic engineering is common.

Formalized theory, similitude, and empiricism are often implemented to develop physical models that can be used to study problems in hydraulics and scour. However, it should be noted that the implementation of physical models is more successful when studying fixed boundary hydraulic problems rather than when studying loose boundary problems, of which scour is part. The finding from loose–boundary hydraulic models are, at best, only an indication of what might happen in the prototype. The quantitative solutions developed by numerical or analytical studies, possibly combined with the insight obtained from physical model studies, form the basis of objective reasoning.

Objective reasoning entails using the results of mathematical solutions to problems, whether they originate from empirical equations, or analytical or numerical model studies, and the results obtained from physical model studies to reason about the possible outcome of prototype conditions. Objective reasoning uses quantitative information to develop understanding of anticipated future conditions and assess how solutions to problems that might arise can be developed.

In addition to calculating discrete scour magnitudes, e.g., the maximum scour depth that is anticipated under design conditions, it is also useful to develop graphs showing the trends in scour depth development for varying assumptions. Studying such trends does not only provide the engineer with an indication of how sensitive the results are to assumptions, but it also provides an indication as to whether the analysis procedure is justifiable, internally consistent, and defensible.

For example, if the trends in a scour analysis indicate that the extent of scour reduces as discharge increases, it provides, at first sight, a counter intuitive result. Such a trend should be investigated in more detail to determine if there are any specific reasons for this to be so. It is of course possible that the most severe scour does not necessarily occur during maximum flow conditions, depending on the problem under investigation. This determination can be made with objective reasoning.

For example, flow over a headcut in a channel can be such that lower flows, such as, say, the 2-year recurrence interval flow, can lead to more severe scour than the flow associated with a 100-year recurrence interval. Studying the flow conditions associated with these two flow events might explain why the scour under low flow conditions is more severe than that under higher flows. An explanation of this phenomenon provides the basis for a defensible solution, which, by explaining the flow conditions and reasons for the apparent anomaly, also leads to an argument that demonstrates that the result is internally consistent. The results of objective reasoning, once formulated, are subsequently used with the results of a subjective reasoning process, to devise the final solution to the problem by means of synthesis.

Even if a theory has been carefully and correctly formulated by making use of basic physics principles it is still required to make assumptions to quantify input parameters and boundary conditions. These assumptions are based on the engineer's and engineering community's existing knowledge base. In addition to engineering experience, which forms a large part of the knowledge base, the required assumptions often require probabilistic and uncertainty analysis. This is particularly relevant when considering the hydraulic loading on structures, i.e., flood magnitude and frequency of occurrence. Additionally, the knowledge base relies heavily on current and past research, which studies processes and physical relationships. The intuitive interpretation of processes, combined with experience, and probabilistic and uncertainty analysis, forms the basis for subjective reasoning, which, by its nature, is qualitative.

Once the problem has been considered from various points of view, using both objective and subjective reasoning, synthesis is used to formulate a defensible solution to the problem at hand. This entails combining the conclusions and insight developed during the processes of subjective and objective reasoning to develop a justifiable solution. The process can be quite laborious and involved when considering complicated problems, and often entails discussion between a group of engineers, raising and evaluating various concerns and points of view. It is not a well-defined process, but necessary. The combined insight of a group of engineers, sometimes aided by facilitation, leads to clarification of concerns and common decision making of what entails an appropriate solution to the problem.

After implementation of a proposed solution, the project performance is monitored and the information thus gathered either confirms or rejects the value of the knowledge that was developed during the course of the project. If the project fails, the understanding that was developed during the development of the project solution is flawed. If the project succeeds, it can be claimed that increased understanding has been accomplished. Even if a project fails, an investigation into the reasons for failure contributes to increased understanding and expansion of the profession's knowledge base.

Summary

The most important elements of the decision-making process, as far as engineering decision-making is concerned, are objective and subjective reasoning, and synthesis. The basis for objective reasoning is the quantitative results originating with analytical and numerical investigations, and from physical model studies, should these be executed.

Subjective reasoning is largely based on experience, the use of probability theory and uncertainty analysis, one's existing knowledge base, research (one's own or that of others) and, what can simply be described as, reflection. The reliance on past experience primarily provides a "gut feel" of what might or might not work, based on existing and past project performance. It is used to consider project understanding and solution development from various points of view, asking random questions and executing brain storming exercises to determine if all reasonable scenarios have been considered, and how the proposed solution might perform under each of those scenarios.

Synthesis occurs once advanced understanding of the physical processes and sufficient understanding of the performance of potential solutions have been obtained through the processes of objective and subjective reasoning. Combination of the conclusions made by means of objective and subjective reasoning leads to the development of a defensible solution to a problem that is internally consistent.

Chapter 3

Scour Processes

Introduction

This chapter deals with conceptual issues that are important for understanding the basic elements of the scour process. Scour by water can be viewed as excessive erosion, i.e., erosion that leads to removal of large masses of earth material from a particular location, such as around a bridge pier, at a bridge abutment, or from a plunge pool downstream of a dam. The basic information required for the analysis of scour includes quantification of the ability of earth material to resist the erosive capacity of water, quantification of the erosive capacity of water itself, and a threshold relationship. A threshold relationship relates the erosive capacity of water to the ability of earth material to resist it at the point of incipient motion.

Throughout this book the terms scour and erosion will be used interchangeably. For example, when the term scour resistance is used it is considered to have the same meaning as the term erosion resistance. Both refer to the ability of earth material to resist the erosive capacity of water. In the same vein, the terms erosive capacity and scour capacity of water both refer to the potential ability of flowing water to dislodge earth or engineered earth materials.

Incipient motion occurs when the erosive capacity of the water just exceeds the ability of the earth material to resist removal, and signals the beginning of the scour process. Scour will continue until a stage is reached when the erosive capacity of the water is lower than the ability of the earth material to resist it. At that stage the maximum extent of scour has been reached.

The meaning of the term *erosive capacity* of water is somewhat elusive, as will be illustrated in what follows. However, the general meaning that is attached to the term is that the chances for scour of earth material to occur increases if the erosive capacity of the water increases, and vice versa. Additionally, it is understood that the extent of scour will most probably also increase if the erosive capacity of the water increases. The extent of scour refers to how deep and how wide a scour hole might be, i.e., its spatial extent.

Quantification of the erosive capacity of water is a challenge. The general approach is to use indicator parameters that are believed to increase concurrently with the perceived magnitude of the erosive capacity of water. The indicator parameters currently used in practice include shear stress, average flow velocity, and stream power. It is shown further on that current methods to quantify these indicator parameters do not provide consistent trends when used to quantify the relative magnitude of the erosive capacity of water. This shortcoming presents engineers with a significant practical problem.

An attempt to address this problem is made by examining the essential nature of flow processes that leads to scour. Although it is common knowledge that fluctuating pressures in flowing water play a dominant role in the process of incipient motion and therefore scour, the importance of using representative indicator parameters quantifying its effect is often not acknowledged in research or in practice. The mathematical derivation of representative indicator parameters presented in this book acknowledges the role of fluctuating pressures.

Quantification of the relative ability of earth materials to resist scour demands understanding of the fundamental processes determining how earth materials resist the erosive capacity of water and how they scour. Such understanding has not been advanced significantly in the past. The reason for this is that most of the research into incipient motion has been empirical and has been conducted using non-cohesive granular earth materials, such as sand.

This shortcoming in available knowledge presents a problem because the earth materials most often dealt with in practice are not non-cohesive soils but cohesive, cemented, and vegetated soils, and rock. Additionally, the scour resistance offered by engineered earth materials, such as concrete, has received very little attention. A need therefore exists to address this shortcoming, which is one of the principal aims of this book.

This chapter introduces essential material characteristics that determine their ability, in general, to resist the erosive capacity of water. Practical methods for quantifying the ability of diverse earth and engineered earth materials such as concrete, rock, vegetated soils, and cohesive, cemented, and non-cohesive soils to resist the erosive capacity of water are presented in Chaps. 4 and 6, with applications provided in Chaps. 7 to 10.

The third piece of information required to assess scour, i.e., a threshold relationship, receives some attention in this chapter by principally focusing on its nature. It is demonstrated how incipient motion conditions differ for laminar and turbulent flow, and how material type affects it. Practical approaches for quantifying threshold conditions are dealt in Chap. 6.

Erosive Capacity of Water

Engineers intuitively know that the chances for scour to occur increases as flow conditions become more intense, i.e., when the erosive capacity of the water increases. The exact meaning of the term *erosive capacity* is somewhat elusive and engineers use indicator parameters to quantify its relative magnitude. The indicator parameters most often used in current practice are average flow velocity, shear stress, and stream power. The general expectation is that if the magnitude of an indicator parameter increases so will the erosive capacity of water (Fig. 3.1). In what follows it is shown that this is not necessarily the case when using conventional techniques to quantify the magnitudes of shear stress, average flow velocity, and stream power.

Inconsistencies of indicator parameters used in current practice

Scour equations used by practicing engineers are usually based on laboratory experiments. Say three researchers are commissioned to develop bridge pier scour equations, one using shear stress, another stream power, and the third average flow velocity as indicator parameters. Each of the researchers will very carefully measure the magnitudes of these variables in their experiments and correlate them to the observed scour depths around the bridge pier. Generally, the expectation is that scour depth will increase as each of the selected indicator parameter values increase, which indeed is the case in the experiments.

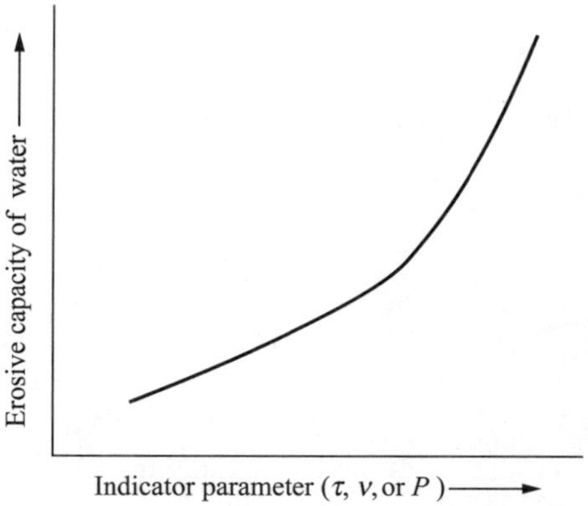

Figure 3.1 Anticipated relationship between the relative magnitude of the erosive capacity of water and indicator parameters.

A practicing engineer commissioned to estimate the scour at a bridge pier decides to use all three equations. In order to cross-check her scour depth estimates the engineer furthermore decides to conduct a sensitivity analysis by varying the assumed hydraulic roughness in the river channel. The hydraulic roughness can be represented in her calculations by making use of the absolute roughness k_s, Manning's n, or the Chezy coefficient C. If Manning's equation is used, assuming a wide channel, the following applies:

$$U = \frac{y^{2/3} s_f^{1/2}}{n} \tag{3.1}$$

$$q = Uy \tag{3.2}$$

$$\tau = \gamma y s_f \tag{3.3}$$

$$P = \tau U = \gamma y s_f U = q s_f \tag{3.4}$$

where U = average flow velocity
y = flow depth in a wide channel
s_f = energy slope
n = Manning's n
q = unit flow
τ = shear stress
γ = unit weight of water
P = stream power per unit area

Figure 3.2 shows the relationships between shear stress, average flow velocity, and stream power as functions of Manning's n. Using the equations listed earlier the engineer finds that shear stress increases, average flow velocity decreases, and stream power remains constant with

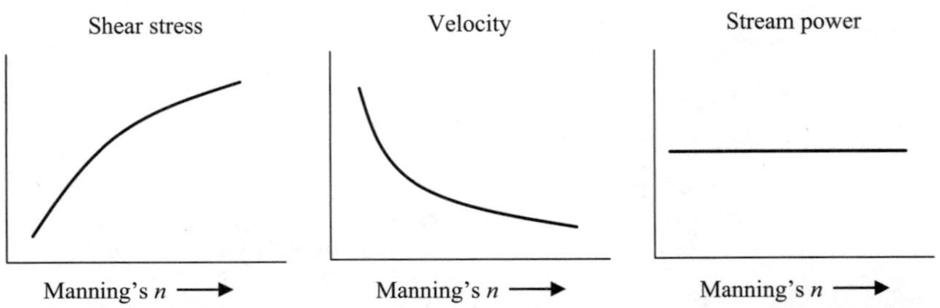

Figure 3.2 Trends in indicator parameters conventionally used to quantify the relative magnitude of the erosive capacity of water as a function of Manning's n.

increasing values of Manning's n. (The unit discharge and energy slope do not change). Increasing values of Manning's n are associated with increasing hydraulic roughness.

For increasing hydraulic roughness this analysis implies that scour will increase if the shear stress equation is used, will decrease according to the average flow velocity equation, and will remain constant for the stream power equation. These trends are inconsistent and provide the engineer with a dilemma.

The trends shown in Fig. 3.2 are also found if either the absolute roughness or the Chezy coefficient is used to calculate the change in the indicator parameters as a function of hydraulic roughness. The problem thus identified points to an inconsistency in current understanding and practice of scour technology. This problem is investigated by considering the essential elements of the scour process in this chapter and by investigating the nature of the erosive capacity of water in more detail in Chap. 5. A solution addressing this inconsistency is provided in Chap. 5.

Requirements for internal consistency

The example provided in the previous section illustrates a problem often encountered in civil engineering. Solutions to most civil engineering problems are characterized by inadequate availability of data, a need for numerous assumptions, and application of approximate methods. The reliability of solutions thus conceived can only be ensured if these methods are internally consistent. Continued application of internally consistent methods, combined with the experience gained through time by applying them, eventually result in confident development of reliable solutions to engineering problems.

An emphasis on internal consistency during the development of solution procedures for application in civil engineering is therefore of prime importance. Explanation of what is meant by internal consistency is challenging. A sincere attempt has been made in this book to develop solution procedures that are internally consistent. An explanation of what is meant by the need for internal consistency in the case of scour technology is therefore warranted.

Indicator parameters, in the case of scour, are considered to be internally consistent if they concurrently and consistently change in concert. This happens when, at the threshold of erosion, higher values of the indicator parameter representing the relative ability of earth material to resist scour is consistently associated with higher values of the indicator parameter representing the erosive capacity of water.

This in essence means that if the erosion resistance of an earth material is high, then the indicator parameter representing its relative ability to resist erosion should also be high. Concurrently, the indicator parameter representing the relative magnitude of the erosive capacity of water that will be able to lead to scour of this material should also be high.

Additionally, the indicator parameter used to quantify the *relative* magnitude of the erosive capacity of water should consistently change to the same degree as the *actual* erosive capacity of the water. And the magnitude of the indicator parameter representing the relative ability of earth or engineered earth material to resist the erosive capacity of water should change in concert with the actual ability of the earth material to resist scour. The relationship between these two parameters, representing water and material properties at the point of incipient motion, defines a consistent erosion threshold.

The selection of internally consistent indicator parameters is not necessarily an easy process. It requires careful reflection of the basic physical principles governing the behavior of a phenomena. Once the behavioral trends are understood approximations can be made and indicator parameter can be selected so as to mimic their actual behavior.

Boundary flow processes

Scour is initiated when the erosive capacity of flowing water at the boundary between the water and the solid exceeds the ability of the solid to resist removal. The interaction between flowing water and earth or engineered earth material right at the boundary determines whether it will be removed or will remain intact. It is therefore important to understand how flowing water interacts with the boundary, and how boundary conditions may affect the relative magnitude of the erosive capacity of water. The character of the erosive capacity of water is dealt with conceptually in this section. Practical methods to quantify its magnitude (either actual or relative) are presented in Chap. 5. Conceptual discussions of the ability of earth and engineered earth materials to resist the erosive capacity of water are presented in the next section.

The discussion starts by considering the development of a boundary layer over a plate, which is laminar at first, moves through a transition zone, and eventually changes into fully developed turbulent flow (Fig. 3.3). The boundary layer continually grows in the direction of flow as it moves

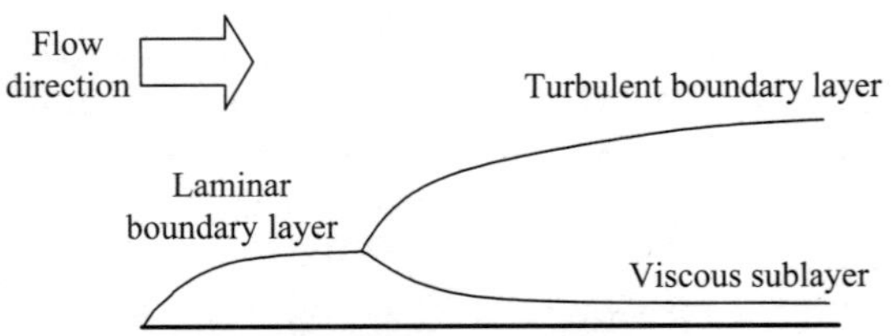

Figure 3.3 Schematic presentation of boundary layer development over a flat plate in parallel flow.

through the different stages. What is important to the engineer analyzing scour in practice is to set a scale to boundary layer thickness. In most open channel flow cases where scour is a concern the flow is fully developed and the turbulent boundary layer fills the entire depth of flow. For example, if water flows over a spillway into a channel the boundary layer is initially relatively thin, but gradually grows until it eventually occupies the total depth of flow, once fully turbulent.

The interaction between turbulent flow and the fixed boundary is facilitated by a theoretical concept known as the viscous sublayer (also known as the laminar sublayer). The viscous sublayer is a very thin layer of fluid located directly adjacent to the boundary where viscous dissipation of energy occurs. Under ideal conditions this layer is of uniform thickness and behaves in a stable manner. The presence of the layer explains why it is possible for the flow velocity to be zero right at the bed, while having a positive value a small distance from the bed. The velocity distribution over the thickness of the viscous sublayer is close to linear, from zero at the boundary to a positive value at its upper edge. (The distribution is not exactly linear, but its exact shape is not of concern in scour technology). Figure 3.4 illustrates both the velocity distribution and the location of the viscous sublayer within the turbulent boundary layer.

In rough turbulent flow the viscous sublayer is unstable. Research has shown that the viscous sublayer is hardly ever stable under normal, rough turbulent flow conditions and that it is characterized by a phenomenon known as *bursting* (Einstein and Li 1956; Kim et al. 1971; Offen and Kline 1974; Offen and Kline 1975). The bursting effects in the laminar sublayer in rough turbulent flow are caused by the introduction of instabilities into the viscous sublayer. Whether these instabilities grow to cause bursting of the laminar sublayer or whether they eventually die down, resulting in continued laminar flow, depends on the magnitude of the Reynolds number.

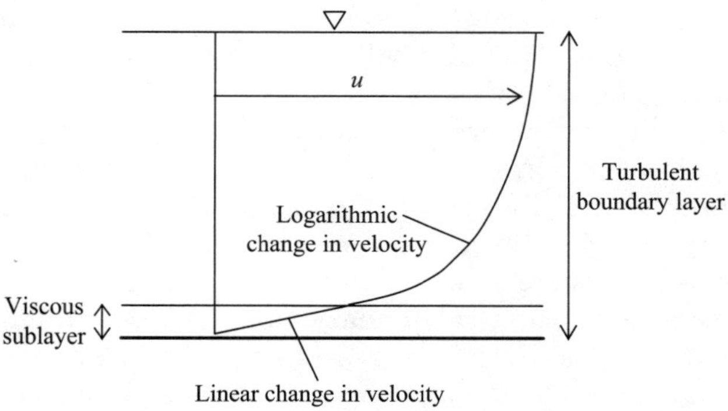

Figure 3.4 Velocity distribution in fully developed turbulent flow.

In order to assess what will happen to the viscous sublayer under laminar and rough turbulent flow conditions when subjected to the introduction of instabilities, it is good to recall that the Reynolds number is, in essence, a ratio between inertial and viscous forces. If the Reynolds number is large, as in the case of rough turbulent flow, the effects of the inertial forces are greater than the effects of the viscous forces. In the case of pure laminar flow, the effects of the viscous forces are greater than the effects of the inertial forces.

Therefore, if the Reynolds number is low, as in the case of purely laminar flow, the viscous forces will dampen out instabilities introduced into the laminar sublayer and the flow will remain laminar. In rough turbulent flow the effects of the viscous forces are inferior to the effects of the inertial forces, with the result that they can no longer dampen the effects introduced by instability into the laminar sublayer. The inertial forces therefore overcome the viscous forces and the viscous sublayer becomes unstable. In some cases the instability might be intermittent, with the viscous sublayer becoming turbulent at times and laminar at other times—a phenomenon known as *bursting*.

The process by which turbulence and fluctuating pressures are generated at the boundary due to instabilities in the viscous sublayer is schematically shown in Fig. 3.5. The flow direction in the figure, which represents conditions close to the boundary, is from left to right. Initially the flow presents itself in the form of a stable viscous sublayer (1). If the flow is disturbed, instabilities occur in the laminar sub-layer, which forms an undulating sublayer surface (2). Should the Reynolds number be low enough such that the effects of the viscous forces are greater than those of the inertial forces, the dampening effect of the viscous forces will result in the undulations subsiding. In such a case, the undulating surface of the viscous sublayer will revert back to a smooth surface, as existed prior to the disturbance [as shown by (1)].

However, if the Reynolds number increases enough the effects of the inertial forces are greater than those of the viscous forces and the instability that was introduced into the laminar sublayer will not experience dampening. In such cases the instability will grow, leading to the formation of hairpin vortices (3).

The apexes of the hairpin vortices are lifted upwards, creating a space between them and the boundary. From a continuity of mass point of view this will result in water rushing in from behind and above the vortices into the space thus created (4). This happens at great speed and gives rise to the name of this phenomenon, i.e., high-velocity sweeps.

The high-velocity sweeps impact the boundary and give rise to high-pressure zones (5). Should any sediment be present on the boundary, the impact on the boundary caused by the high-velocity sweeps can lead to

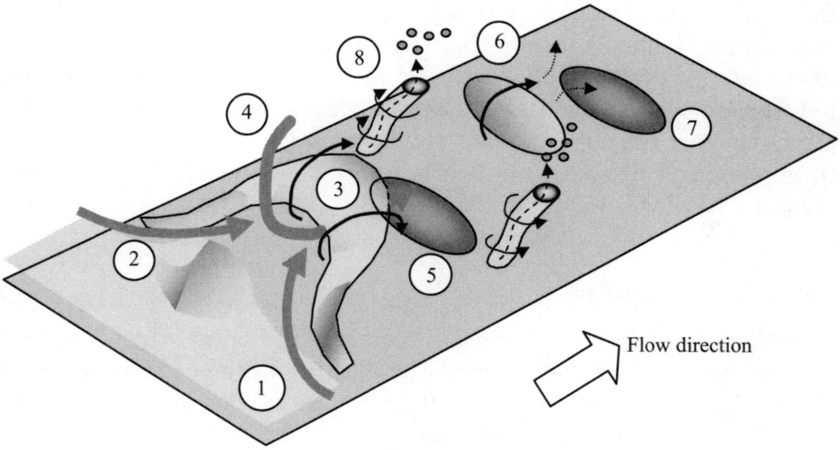

Figure 3.5 Flow processes at the boundary. (1) Laminar sub-layer. (2) Instabilities introduced into the laminar sub-layer results in an irregular surface. (3) The irregular laminar sub-layer surface leads to development of hairpin vortices with their most downstream ends lifted upwards. (4) High-velocity sweeps of water flowing into the space below the hairpin vortices appear. (5) The high-velocity sweeps interact with the bed and lead to a high-pressure zone as it collides with the bed. (6) As the hairpin vortex moves further downstream the central part of the vortex breaks loose and forms an eddy that can move either upward, parallel to or toward the boundary. (7) Eddies that move toward the boundary collide with it and lead to the formation of low-pressure zones. (8) The two remaining legs of the hairpin vortex attach temporarily to the bed where they cause low-pressure zones at the points of attachment. The small vortices are known as low-velocity streaks. The negative pressure within these vortices can suck sediment upwards and discharge it into the upper body of the flowing water.

radial outward movement of such particles. Turbulence that develops at the outer edges of the high-pressure impact zones can direct these particles upward if the action is strong enough.

As the hairpin vortex develops further, the central apex breaks loose. This results in eddies that can move either away from, parallel to, or toward the boundary (6). If an eddy moves toward the boundary and collides with the boundary it gives rise to a low-pressure zone (7) (Hofland et al. 2005).

The two remaining "legs" of a hairpin vortex form two smaller vortices (8). These vortices attach to the bed at their upstream ends, and, if parts of the viscous sublayer still exist, their downstream ends will attach to the top of that layer. The effect of these vortices is that they act like small "vacuum cleaners," sucking sediment (should it be present) into them from the bed and spewing it out into the flow above. The attachment locations of the vortices on the boundary are characterized by low pressure due to the suction within the vortices. These vortices are known as *low velocity streaks.*

Bursting of the viscous sublayer results in the creation of turbulence at the boundary; this is transferred into the rest of the flow. In addition, it also leads to pressure fluctuations on the boundary itself, varying between negative and positive pressures. The existence of pressure fluctuations on boundaries in turbulent flow plays an important role in incipient motion of earth materials and, consequently, in the scour process.

Material Characteristics

Incipient motion of earth materials occurs at the point where the erosive capacity of water just exceeds the ability of the earth material to resist its effect. The role of fluctuating pressures in determining the magnitude of the erosive capacity of water has conceptually been discussed in the previous section. This section deals with conceptual aspects of material characteristics that determine a particular type of material's ability to resist the erosive capacity of water. Practical methods to characterize and quantify erosion resistance, based on the concepts explained in this section, are presented in Chaps. 4 and 6.

Physical and chemical gels

In its most basic form any material can be characterized as one of the two material types, i.e., physical or chemical gels. This approach to characterizing earth materials is based on modern advances in physics, particularly in the field of percolation theory (PT) and network theory (see Sahimi 1994 and Rucker 2004). PT deals with phase transitions, while network theory explains spatial relationships.

A network consists of a lattice of points in space (also known as sites) that are either occupied or not occupied, and, when occupied, are either connected or not connected to other occupied points. If the lattice under consideration represents an earth material, the percentage of sites that are occupied determines its porosity. The nature of the connections between occupied sites determines the network's behavior.

In order to develop understanding of the concept it is helpful to consider the images in Fig. 3.6. The open circles in Fig. 3.6*a* illustrate a lattice of potential sites that can be occupied by physical elements. In the case under consideration, i.e., an earth material, the sites can potentially be occupied by minerals making up materials such as a rock mass, cohesive, non-cohesive, or cemented soil.

If some of the sites in Fig. 3.6*a* are occupied by non-cohesive soil minerals that are connected to each other by mere touching, the network is known as a *physical gel* (Fig. 3.6*b*). It is possible that not all of the sites will be occupied, i.e., some of the sites may remain empty. The porosity

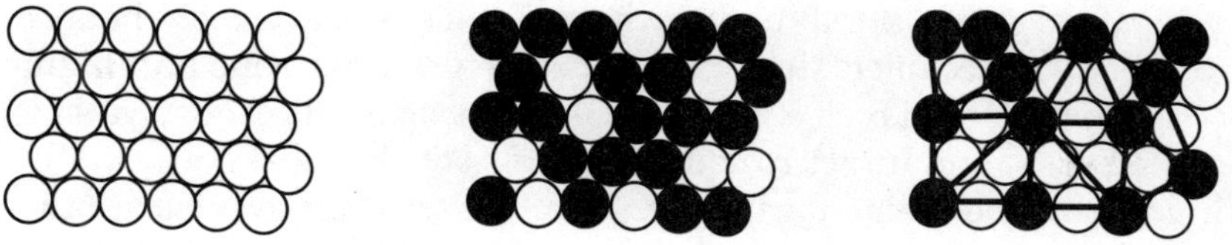

(a) Sites in a network lattice that may or may not be *occupied*, and may or may not be *connected*

(b) *Physical gel* occupied sites are connected when they touch each other

(c) *Chemical gel* occupied sites are connected with fixed bonds, which could be cohesive or welded bonds

○ Site not occupied

● Occupied site

— Fixed bond

Figure 3.6 (*a*) Lattice network with potential sites that can be occupied. (*b*) Physical gel showing occupied sites. (*c*) Chemical gel showing occupied sites and fixed bonds.

of the assembly is determined by the number of sites that are occupied as a function of the total number of available sites.

Large porosities are associated with low material bulk density, and vice versa. An assembly of particles with high-porosities will be weaker than one with low porosities. It is therefore reasonable to anticipate the existence of a critical porosity that is required to allow the assembly of occupied sites to perform a certain function, such as resisting the erosive capacity of water. If the porosity decreases below the critical porosity it is most probably reasonable to expect that the erosion resistance of the physical gel would increase.

Figure 3.6*c* illustrates a *chemical gel*. As in the case of a physical gel, the porosity is determined by a function relating the number of occupied sites and the total number of available sites. However, the nature of the connections between the sites differs. Unlike the case of physical gels where the connections merely exist because the occupied sites touch each other, the sites in chemical gels are connected with *fixed bonds*. These are chemical bonds, such as cohesion or cementation.

For example, in a chemical gel like clay the bonds between individual clay platelets are due to cohesion formed by electrical charges on the edges and flat sides of the platelets. In the case of a cemented soil it can be due to the presence of lime ($CaCO_3$) that results in chemical bonding between individual soil particles. An engineered earth material like concrete is held together by chemical bonds between individual aggregate particles provided by hydrated cement.

It is possible that a chemical gel with a high-porosity can be stronger than a physical gel with a lower porosity. The higher potential strength of a chemical gel is attributed to the presence of the fixed, chemical bonds between occupied sites. Non-cohesive granular soils and highly fractured, weathered rock are examples of physical gels. Cohesive soils, cemented soils, and intact rock are examples of chemical gels.

In general, the elements in a physical gel will roll over each other when subjected to an external force; provided that they touch at least three other occupied sites (consider this from a three-dimensional point of view). If a chemical gel is subjected to an external force the fixed bonds between elements will prevent them from rolling over each other. These materials are subject to bending, brittle fracture, or fatigue failure when impacted by an external force. The nature of the response depends on the characteristics of the fixed bonds and on the nature of the loading.

Having explained the basic characteristics of physical and chemical gels, what remains is to assess how their differing characteristics influence response to the erosive capacity of water. In this regard it is important to recall that the erosive capacity of turbulent flow is characterized by fluctuating pressures that interact with the surface layer of an earth material.

The properties of the surface layer of a material (i.e., the interface between the water and solid), and the properties of subsequent surface layers that might be exposed as each layer is removed during the scour process, determines the erodibility of a material. In this regard, a difference exists between the characteristics of surface layers of physical and chemical gels. A physical gel consists of a number of individual elements while a chemical gel principally consists of a solid that might contain imperfections. The following two subsections describe the basic character of the response of physical and chemical gels to the erosive capacity of water.

Physical gel response to scour

The surface layer of a physical gel consists of discrete elements that are connected to one another by touching. The erosion resistance of the surface layer of a physical gel is therefore determined by the submerged weight of the individual elements and the friction forces that can develop between them. The contribution made by the friction forces will depend on the gradation of the physical gel (i.e., whether it is well graded or uniformly graded) and the degree of compaction.

For example, compare the potential erosion resistance between two physical gels. Say, the one consists of non-cohesive, coarse sand with uniform gradation and the other gel consists of well graded, non-cohesive sand. It is reasonable to expect that the friction forces that will develop between the elements of the latter will be higher than those that will develop in the former. The well graded, non-cohesive sand would be characterized by higher erosion resistance (Fig. 3.7). This is principally due to more particles touching each other, thus generating larger net internal friction and interlocking force.

Other factors that come into play include the degree of saturation of the soil and whether it is compact or loosely assembled. Consider two non-cohesive soils, both completely saturated with water with the one loosely packed (Fig. 3.8*a*) and the other compacted (Fig. 3.8*c*). Applying a shear stress to the loosely packed array of particles will result in them falling into the spaces between the particles below them. This leads to the particles occupying the space that was previously taken up by the water (Fig. 3.8*b*). The expulsion of the interstitial water leads to a reduction in the effective stress between the particles, and therefore a reduction in the resistance offered by the soil to the applied shear stress.

When the compact material is subjected to a shear stress (Fig. 3.8*c*), it will dilate when the individual particles move onto the ones below them (Fig. 3.8*d*). Dilation results in the material sucking in water from its surrounds. This leads to an increase of the effective stress between the particles, and increased resistance to the shear stress applied to the compacted soil.

However, it should be noted by the reader that although shear stress is the principal activator leading to incipient motion in laminar flow, it is not the case in turbulent flow. The principal force in turbulent flow leading to incipient motion originates from fluctuating turbulent pressures. These fluctuating forces result in the particles "popping" out of

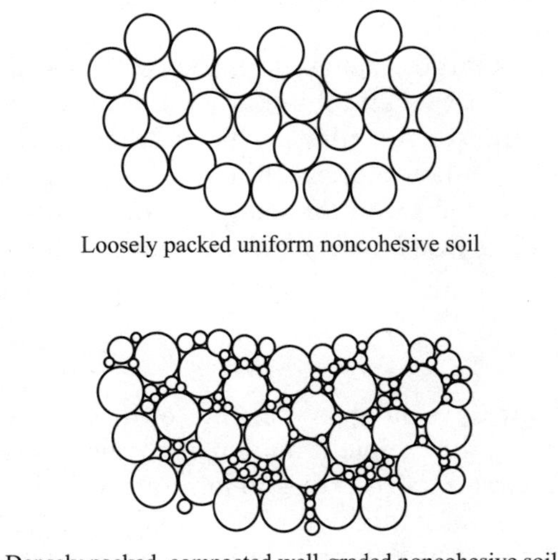

Figure 3.7 Loosely packed uniform non-cohesive soil has lower resistance against erosion than well graded, densely packed, compacted non-cohesive soil.

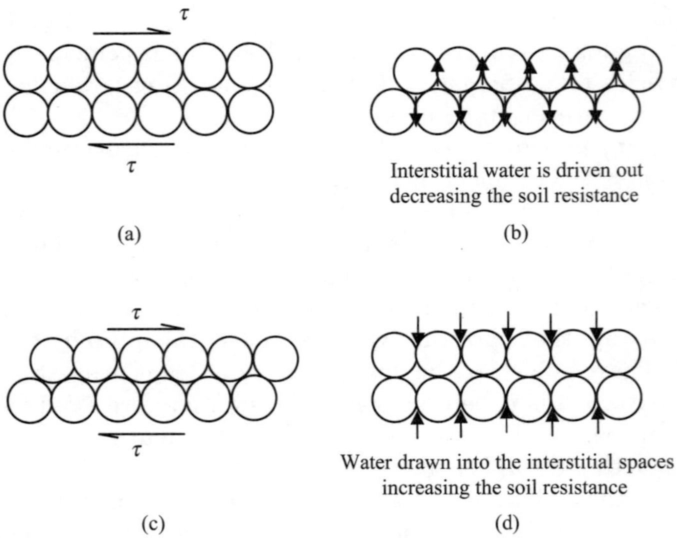

Figure 3.8 (*a*) Loose, non-cohesive, saturated soil subject to shear stress. (*b*) The loose soil compresses due to the action of the shear, resulting in water driven out of the interstitial spaces and a subsequent decrease in resistance to the shear stress applied in (*a*). (*c*) Compacted, non-cohesive, saturated soil subject to shear stress. (*d*) The compacted soil dilates due to the shear, sucking water into the interstitial spaces and increasing the resistance offered by the soil to the shear stress in (*c*).

the matrix in a vertical direction. The mechanism explained in Fig. 3.8 is relevant to incipient motion under laminar flow conditions and not under turbulent flow conditions.

Incipient motion of physical gels largely depends on the relationship between the surface layers of the material and the character of the flowing water, i.e., whether it is laminar or turbulent. This is particularly true in the case of non-cohesive granular material (like sand). Incipient motion characteristics of non-cohesive granular material and jointed rock masses, when the latter behave like physical gels, are discussed in what follows.

Non-cohesive soils

Flow conditions play a role in how physical gels respond to the erosive capacity of water. When considering the incipient motion of physical gels it is important to distinguish between its response to laminar and turbulent flow respectively. In the case of laminar flow the fluid interacts with an assembly of particles, while in the case of turbulent flow pressure fluctuations interact with individual particles. This affects the magnitude of the erosive capacity of flowing water required to cause incipient motion.

Laminar flow. Laminar flow is characterized by an absence of pressure fluctuations. The dominance of the effects of viscous forces results in the flowing water attempting to drag the *assembly* of particles at the surface along with it. The erosive capacity of the water required to cause incipient motion of surface elements in a physical gel will, in the case of laminar flow, be higher than in the case of turbulent flow.

Figure 3.9 illustrates a layer of non-cohesive soil particles resting on another layer with water under laminar flow conditions flowing over it. One of the particles, with forces acting on it, is shown in the lower portion of the figure. The force F represents the action of the laminar flow, and the force F_R the resistance offered by the soil particle. The force W_g is the submerged weight of the particle, and the angle ϕ represents the angle of friction between this particle and the ones adjacent to and below it. The shear stress can be found by dividing the forces F and F_R by the projected horizontal surface area of a particle A, i.e., $\tau_o = \frac{F}{A}$ and $\tau_R = \frac{F_R}{A}$. When the bed shear stress τ_o is just large enough to initiate movement of the assembly of particles it is known as the critical shear

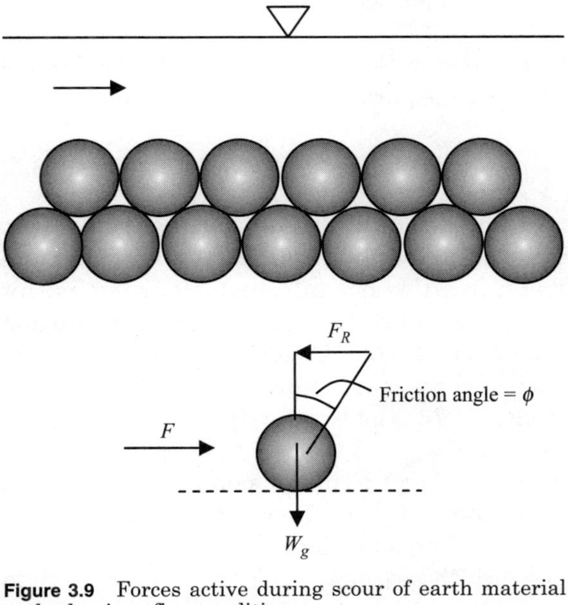

Figure 3.9 Forces active during scour of earth material under laminar flow conditions.

stress, represented by the symbol τ_c. Incipient motion is imminent when $\tau_c = \tau_o = \tau_R$.

The resisting force F_R is a function of the submerged weight of the soil particle and the angle of friction, which is expressed as follows:

$$F_R = W_g \tan\phi = \frac{\pi d^3}{6}(\rho_s - \rho) g \tan\phi \tag{3.5}$$

where d = diameter of the soil particle
ρ_s = mass density of the soil
g = acceleration due to gravity
ϕ = angle of friction of the soil

The critical shear stress τ_c that should be exceeded for incipient motion to occur can be expressed as

$$\tau_c = \frac{4F_R}{\pi d^2} = \frac{2}{3} gd(\rho_s - \rho)\tan\phi \tag{3.6}$$

It is often useful to express the critical shear stress in dimensionless form, as proposed by Shields (1936), i.e.

$$\theta = \frac{\tau_c}{(\rho_s - \rho)gd} = \frac{2}{3}\tan\phi \tag{3.7}$$

θ is known as the Shields parameter. If the angle of friction of the soil $\phi = 30°$ then the value of the Shields parameter becomes

$$\theta = 0.4 \tag{3.8}$$

The value of θ calculated above is quite significant as it has been found experimentally to equal the upper limit of the value of the dimensionless critical shear stress, i.e., the critical Shields parameter, for laminar flow conditions. Shields himself never measured this exact value but extrapolated his data. His extrapolation implied a higher critical dimensionless shear stress for laminar flow conditions. However, subsequent research by Mantz (1973) and White (1970) indicates a tendency to a maximum value of 0.4.

The value of $\theta = 0.4$ indicates that the dimensionless shear stress required for incipient motion under laminar flow conditions is greater than that required for turbulent flow conditions. This is due to the fact that laminar flow interacts with the assembly of particles rather than with individual particles, as is the case in turbulent flow.

The fact that the dimensionless shear stress required for the incipient motion of particles under laminar flow conditions is greater than that required under turbulent flow conditions can be seen in Fig. 3.10, which is known as the Shields diagram. This diagram relates the particle Reynolds number $\frac{u_* d}{\nu}$ and the Shields parameter θ and can be used to identify conditions that will lead to the incipient motion of non-cohesive soil particles for both laminar and turbulent flow. This diagram will be discussed in more detail in Chap. 6. For now it suffices to state that erosion will occur if data points relating dimensionless shear stress and the particle Reynolds number are located above the threshold line, and that no erosion will occur when such points are located below the threshold line. The relationship between dimensionless shear stress and the particle Reynolds number represented by the threshold line identify conditions of incipient motion.

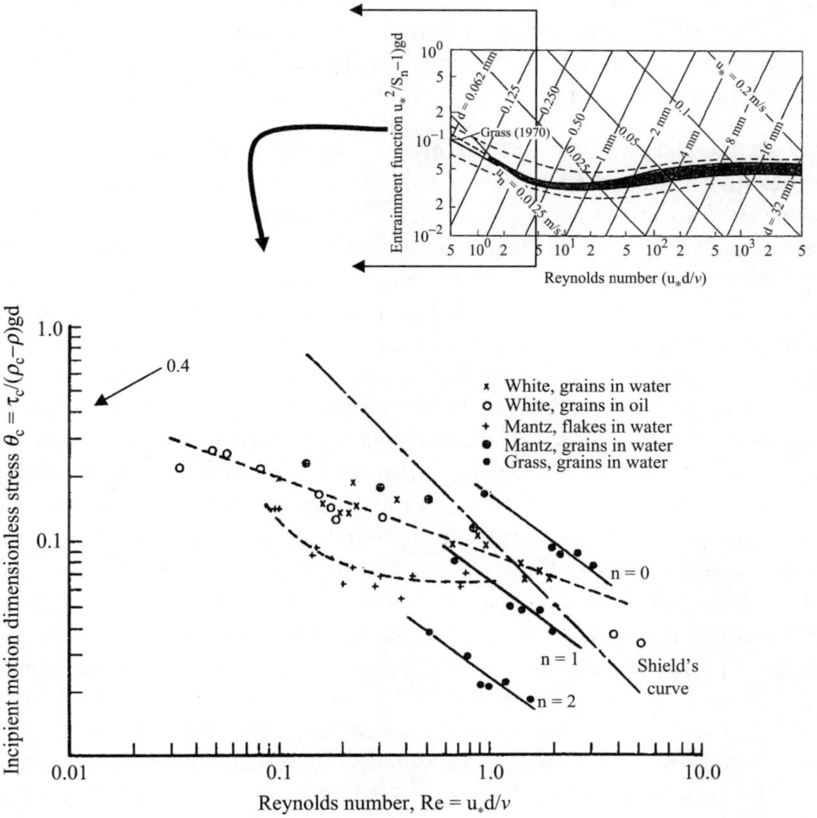

Figure 3.10 Shields diagram (top) for incipient motion, and detail for laminar flow according to Mantz (1973) (below).

It is useful to note that the particle Reynolds number relates the particle diameter to the wall layer thickness δ (see Chap. 5), i.e.,

$$\frac{d}{\delta} = \frac{du_*}{\nu}$$

As the particle Reynolds number increases, the flow at the boundary changes from laminar to turbulent conditions. It has been found experimentally by Colebrook and White (1937) that individual sediment particles start to shed eddies, i.e., turbulence starts to play a role, when

$$\frac{u_* d}{\nu} > 5 \tag{3.9}$$

This is roughly equal to the thickness of the viscous sublayer (see Chap. 5) and confirms that when the particle diameter is roughly equal to the viscous sublayer thickness, turbulence starts to develop. This cutoff value occurs roughly at the lowest point on the Shields diagram. Incipient flow conditions to the left of the value of five are characteristic of laminar flow and those on the right are characteristic of turbulent flow.

Turbulent flow. In the case of turbulent flow the magnitude of the erosive capacity of water that leads to the incipient motion of a physical gel is lower than in the case of laminar flow. The pressure fluctuations that develop during the course of turbulent flow processes interact with individual elements in a physical gel, as opposed to an assembly of elements as is the case with laminar flow. Less erosive capacity is required to move individual elements (turbulent flow) than that which is required to move an assembly of elements (laminar flow).

The forces acting on a sediment particle under turbulent flow conditions are its self-weight, friction, and hydraulic forces, characterized as fluctuating drag and lift forces. The lift forces on a particle originate from two sources. The first source consists of a pressure difference that develops due to steady flow of water over a sediment particle and the second source is due to pressure fluctuations that develop in turbulent flow. The net lift force therefore consists of a steady part and a fluctuating part.

When water flows over a matrix of sediment particles in a steady, time-invariant manner it leads to the development of a pressure differential over the particles. The reason for this is that the water between the interstices within the matrix is stationary, while the water flowing over the particles moves. Based on the Bernoulli principle it follows that the pressure below the particle will be higher than the pressure on top of the particle. The magnitude of the pressure difference depends on the flow velocity.

Application of the Bernoulli principle in this case essentially reduces to relating kinetic and pressure energy. If the kinetic energy increases the pressure energy will decrease and vice versa. The pressure in the water flowing over the particles will therefore be lower than the pressure of the stationary water underneath the particles. This leads to a pressure differential over the particles, in an upward direction.

In addition to the pressure difference induced by the steady portion of the flowing water, additional fluctuating pressures are induced by vortices that develop in the turbulent flow. Booij and Hofland (2004) investigated the causes of incipient motion in non-cohesive gravel and published a map of turbulent flow, showing how an eddy in the flow moves downward toward the bed and interacts with a stone (Fig. 3.11). This mapping concurs with the explanation of eddies interacting with the bed provided previously (Fig. 3.5).

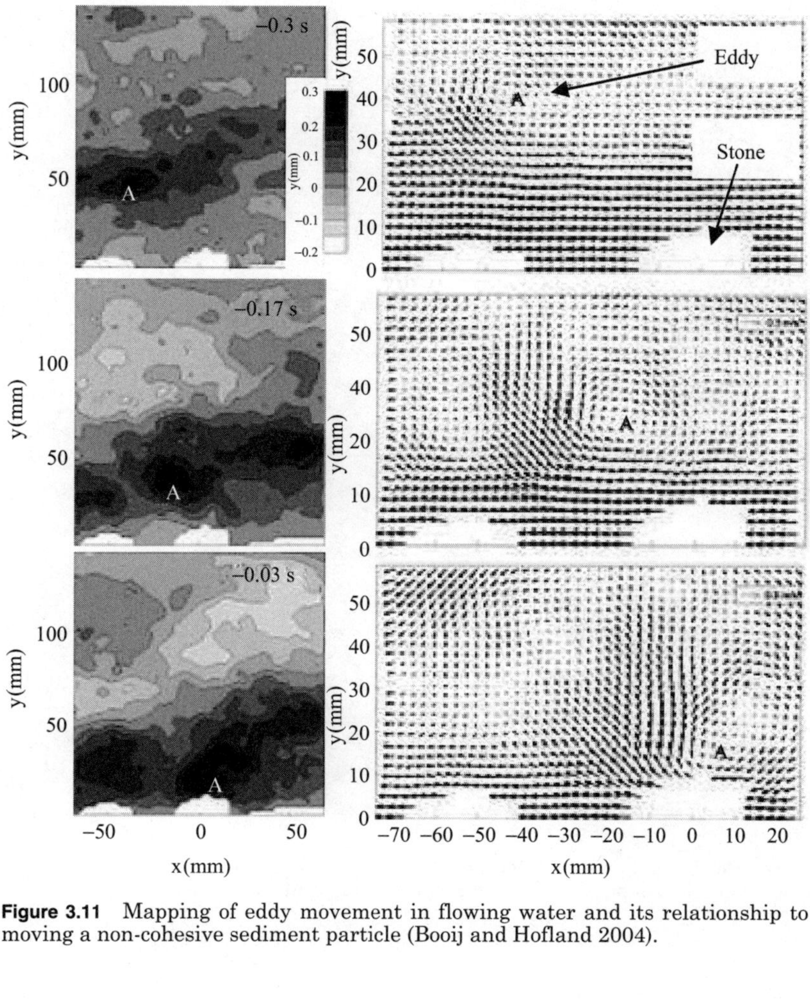

Figure 3.11 Mapping of eddy movement in flowing water and its relationship to moving a non-cohesive sediment particle (Booij and Hofland 2004).

These random vortices introduce pressure differentials onto the bed. Booij and Hofland (2004) explained the relationship between boundary pressure and vortex flow close to the boundary by making use of experimental investigations and potential flow theory. One of their results is presented in Fig. 3.12, which shows the potential flow lines of the vortex in the upper portion of the figure, the pressure variation on the boundary in the middle image and the effects of the pressure variation on a sediment particle in the lower image.

The interaction between the vortex and the boundary introduces a reduction in pressure on the boundary surface. The middle image in Fig. 3.12 shows that the pressure reaches a minimum underneath the eddy when

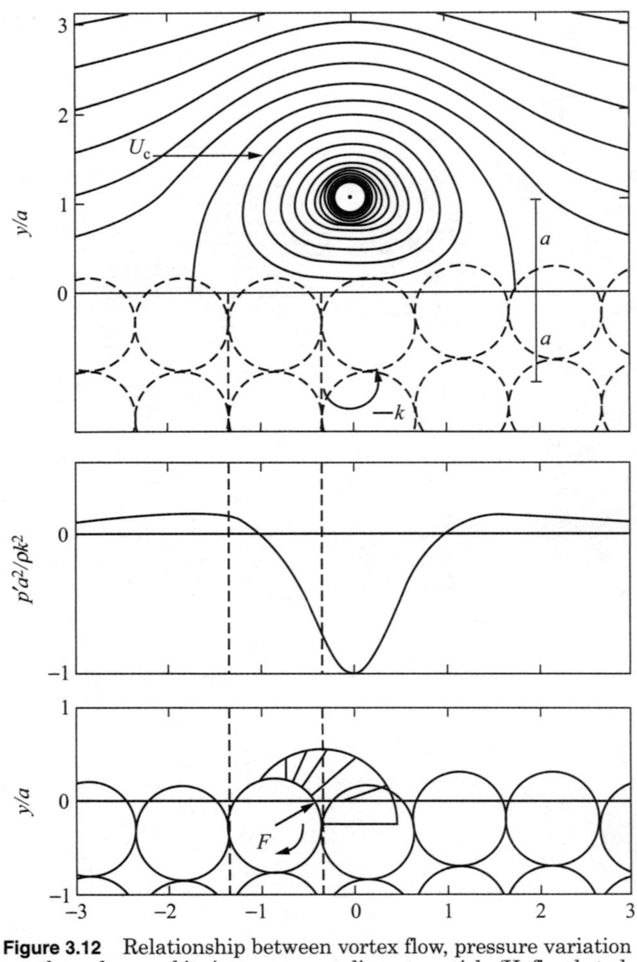

Figure 3.12 Relationship between vortex flow, pressure variation on a boundary and its impact on a sediment particle (Hofland et al. 2004).

it interacts with the boundary, which indicates development of varying low pressure on the top surface of the stone. This causes a differential distribution of pressure over the stone, resulting in a net force that tends to rotate the particle in the direction shown by the arrow in the bottom image of Fig. 3.12. This rolling tendency can lead to removal of the stone from the matrix.

Particles located immediately below the vortex, which experiences a complete reduction of pressure over their top surface (not a differential between high and low pressure as discussed in the previous section) can result in the particle popping vertically out of the matrix. The pressure below the particle is higher than the pressure induced on its top surface by the vortex.

Croad (1981) illustrates the role of fluctuating pressures by calculating the incipient flow conditions for turbulent flow and comparing it with the Shields diagram. He did this by making use of Hinze's (1975) finding that the fluctuating portion of pressures on the bed in turbulent flow could, on average, be correlated as

$$p' = 3\tau_t \tag{3.10}$$

This means that the root mean square of the fluctuating pressures p' is approximately equal to three times the turbulent boundary shear stress τ_t. It is important to note that the pressure fluctuation is proportional to the turbulent boundary shear stress and not the drag on the boundary (i.e., the wall shear stress τ_w). The importance of this observation is further elaborated on in Chap. 5.

Emmerling (1973) further found that the positive and negative pressure peaks can be up to $6p'$, which means that pressure peaks up to 18 times the turbulent boundary shear stress can be reached at the boundary, i.e.,

$$p_{\max} = 18\tau_t \tag{3.11}$$

If one now considers a sand grain with diameter d that is acted upon by an upwards pressure equal to $p_{\max}$, the total uplift force active on the particle can be expressed as

$$F_L = \frac{\pi d^2}{4}(18\tau_t) \tag{3.12}$$

At incipient motion this force is balanced by the submerged weight of the particle, which is calculated as

$$W_g = \frac{\pi d^3}{6}(\gamma_s - \gamma) \tag{3.13}$$

Therefore, right at incipient motion the lift force equals the submerged weight of the particle, i.e.,

$$\frac{\pi d^2}{4}(18\tau_t) = \frac{\pi d^3}{6}(\rho_s - \rho)g$$

from which follows

$$\frac{\tau_t}{(\rho_s - \rho)gd} = \frac{4}{6 \cdot 18} = 0.037 \tag{3.14}$$

The ratio on the right-hand side is known as the Shields parameter θ. Inspection of the Shields diagram (Fig. 3.13) indicates that the minimum value of the shields parameter for incipient motion occurs roughly at a particle Reynolds number of about 10, equaling approximately 0.037.

This force balance will obviously only be valid when the negative pressure fully encapsulates the particle. The conditions when this is true can be estimated by making use of measurements of the dimensions of fluctuating pressures on a boundary under turbulent flow conditions. Emmerling (1973), Kim et al. (1971) and Willmarth and Lu (1972) made such measurements and found that the size of pressure spots in rough turbulent flow ranges between:

$$20\frac{\nu}{u_*} \le \xi_1 \le 40\frac{\nu}{u_*} \tag{3.15}$$

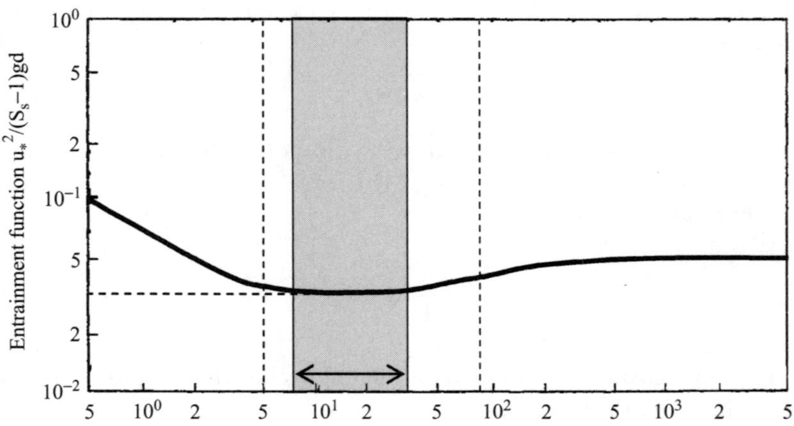

Figure 3.13 Shields diagram, illustrating range where fluctuating pressures can completely encapsulate non-cohesive soil particles.

and

$$7\frac{\nu}{u_*} \leq \xi_3 \leq 30\frac{\nu}{u_*} \tag{3.16}$$

where ξ_1 denotes dimension in the direction of flow and ξ_3 denotes dimension transverse to the direction of flow. This means that the particle diameters that could be fully affected by negative pressure fluctuations range between

$$7\frac{\nu}{u_*} \leq d \leq 30\frac{\nu}{u_*} \tag{3.17}$$

which can be rewritten in terms of the particle Reynolds number, showing that

$$7 \leq \frac{u_* d}{\nu} \leq 30$$

where $\mathrm{Re}_* = \frac{u_* d}{\nu}$ is the particle Reynolds number.

The latter represents the variable on the ordinate of the Shields diagram. Careful consideration of the Shield diagram shows that the minimum value of the Shields parameter θ lies between values of Re_* ranging between 7 and 30 (Fig. 3.13). As the values of the Reynolds particle number increases beyond the value of 30 the Shields parameter increases, indicating that higher erosive capacity is required for incipient motion beyond this point. This will happen because the pressures do not completely encapsulate the particles anymore.

In order to derive additional equations that mathematically describe incipient conditions in turbulent flow, consider the non-cohesive sediment particle arrangement in Fig. 3.14*a*, which is subject to *fluctuating* shear (F) and lift forces (F_L). These forces are resisted by the submerged weight of the particle (W_g) and the friction forces between the particle and those surrounding it (F_1, F_2, and F_3). [In a three-dimensional arrangement other friction forces surround the particle (not shown in the figure.)]

The shear and lift forces act on the particle with varying frequencies, as induced by the turbulence of the water flowing over it. These forces induce pulses onto a particle and if the magnitude of an upward pulse over a pulse period Δt is large enough to overcome the resisting forces, the particle will be ejected from the matrix. In order to determine whether this will happen, it is necessary to integrate over all the forces acting on the particle and determine whether the particle will be ejected during the pulse time period Δt.

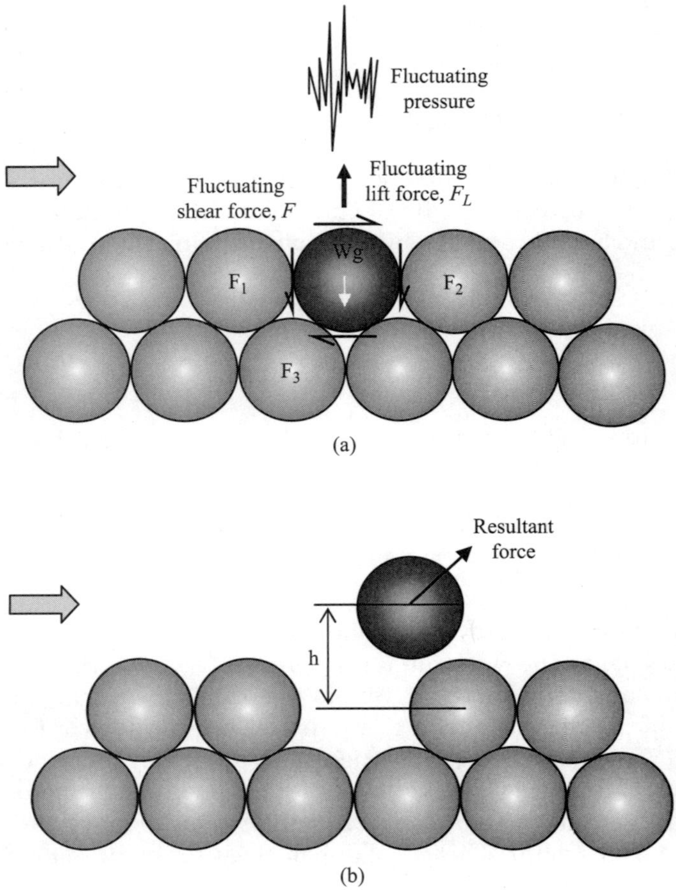

Figure 3.14 Initiation of motion of a non-cohesive sediment particle subject to the action of fluctuating pressures in turbulent flow.

It is furthermore important to note that the effect of time comes into play when considering incipient motion of earth material under turbulent flow conditions. The reason for this is that incipient motion under such flow is caused by *fluctuating* pressures and it is therefore required to consider the effect of impulses that occur over short time periods.

If the vertical distance h through which a sediment particle can be lifted by a pressure impulse is large enough the sediment particle will be mobilized (Fig. 3.14*b*), This will definitely occur when $h \geq d$, where d is the particle diameter. If $h << d$ the particle remains intact, possibly vibrating.

By integrating over all the forces acting on the particle, following an approach similar to Bollaert (2002) who considered removal of rock blocks by fluctuating turbulent pressures, it is possible to quantify the impulse on the particle that can occur over a short time period Δt (the duration of an impulse on the particle),

$$\int_0^{\Delta t} (F_L - W_g - F_1 - F_2 - F_3)dt = F_{\Delta t} = m \cdot V_{\Delta t} \tag{3.18}$$

where $F_{\Delta t}$ = net impulse on the particle during the period Δt
m = mass of the sediment particle
$V_{\Delta t}$ = average velocity achieved by the particle over the period Δt

The height through which the particle can be elevated in the pulse period Δt can be determined once the velocity $V_{\Delta t}$ is known. In order to accomplish this one can first write the kinetic energy of the particle as

$$E_{\Delta t} = \frac{1}{2} m V_{\Delta t}^2 \tag{3.19}$$

where $E_{\Delta t}$ is the kinetic energy imparted to the particle over a short pulse period Δt. By setting the kinetic energy obtained during the short pulse period equal to the potential energy attained during that same period,

$$\frac{1}{2} m V_{\Delta t}^2 = mgh \tag{3.20}$$

from which follows that,

$$h = \frac{V_{\Delta t}^2}{2g} \tag{3.21}$$

The important point in this derivation is that the height through which the sediment particle is elevated comes from the kinetic energy imparted to the particle by an impulse during the *short pulse period* Δt. This means that the *rate* at which the energy is imparted to the particle is relevant in incipient motion resulting from turbulent flow.

Recalling that energy is defined as change in work, it is possible to express the rate by which the energy is imparted to the particle as

$$\frac{E}{\Delta t} = \frac{\left(\int_0^{\Delta t} (F_L - W_g - F_1 - F_2)dt \right) \cdot h}{\Delta t} = F_{\Delta t} \cdot \frac{h}{\Delta t} = F_{\Delta t} \cdot V_{\Delta t} \tag{3.22}$$

It can therefore be concluded that the height through which a sediment particle can be elevated is a function of the rate by which energy is imparted to the particle, i.e., it is a function of the power of water:

$$\frac{E}{\Delta t} = F_{\Delta t} V_{\Delta t} = \text{power} \tag{3.23}$$

Having recognized that pressure fluctuations play a dominant role in turbulent flow it is concurrently true that the *rate of energy transfer* is relevant to incipient motion of sediment. The derivation that is presented in this section shows that stream power is a suitable indicator parameter for quantifying the relative magnitude of the erosive capacity of *turbulent flow*.

Jointed rock

Following the material characterization approach presented in this chapter it can be concluded that rock strata can be classified as either physical or chemical gels. The characterization depends on the difference between characteristic dimensions of turbulent flow and rock, i.e., it is a function of scale. In cases where the rock strata has a high strength but small block size relative to the characteristic dimension of flow, it is appropriate to characterize the rock as a physical gel. When the rock is strong and consists of large blocks relative to the characteristic dimension of turbulent flow it is more appropriate to characterize it as a chemical gel (see next section).

When conditions are such that the rock acts like a physical gel, and the fluctuating forces introduced by turbulent flow are large enough, individual blocks of rock will be removed from the rock formation in a manner similar to non-cohesive soil particles. Once the net upward impulses over blocks of rock introduced by the turbulent flowing water are greater than the resistance offered by the block of rock it will be ejected from its matrix as a solid unit.

Alternatively, if conditions are such that the rock is characterized as a chemical gel, scour involves up to three processes. These processes are brittle fracture, fatigue failure, and dynamic impulsion (Bollaert 2002). If brittle fracture or fatigue failure occurs the rock is broken up into smaller pieces. Once these pieces of rock are small enough for intact removal by the fluctuating pressures, the rock mass elements remaining after the actions of brittle fracture or fatigue failure behave like a physical gel. The brittle fracture and fatigue failure processes are discussed in the next section.

Consider the rock formation depicted in Fig. 3.15 that shows a block of rock in its matrix, subjected to impinging, pulsating, turbulent forces

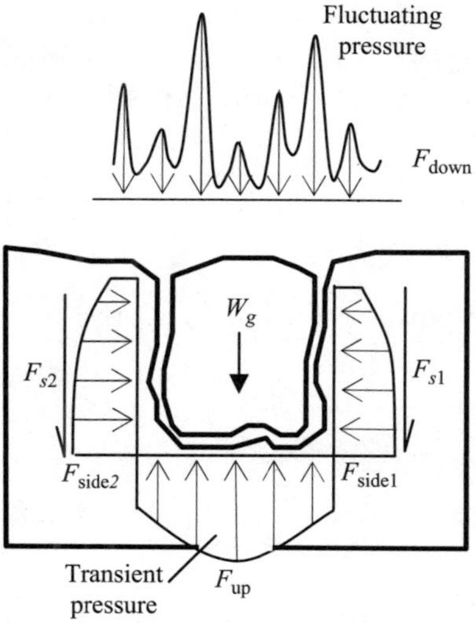

Figure 3.15 Dynamic impulsion (removal of blocks of intact rock), showing the force balance on the rock subject to, say, an impinging jet.

introduced by, say, a plunging jet. The pressure fluctuations introduced at the surface by the plunging jet impacts the pressures within the joint itself. If the transient flow within the joint leads to the development of pressures within the joint that exceed the water pressure overlying the rock, as well as the submerged weight of the rock block and the friction forces on its sides, it will be ejected.

In order to develop a relationship that can be used to calculate whether a block of rock can be removed from its matrix by pulsating forces it is necessary to set up a force balance representing the pulsating forces and integrate over the pulse period, Δt (Bollaert 2002),

$$\int_0^{\Delta t} (F_{up} - F_{down} - W_g - F_{s1} - F_{s2})\,dt = F_{\Delta t} = mV_{\Delta t} \tag{3.24}$$

where F_{up} = total upward impulse caused by the transient pressure in the joint

F_{down} = total downward impulse caused by the fluctuating pressures on top of the rock block

W_g = submerged weight of the block of rock

F_{s1} and F_{s2} = instantaneous shear forces generated on the sides of the block of rock during the pulse period Δt

$F_{\Delta t}$ = net impulse

m = mass of the rock block

$V_{\Delta t}$ = average velocity attained by the mass of rock during the time period Δt

Following the same procedure as outlined for non-cohesive granular soil one can express the height through which a block of rock will be lifted by the pulses imposed on it during this short pulse period Δt as follows:

$$h = \frac{V_{\Delta t}^2}{2g} \tag{3.25}$$

and it can be shown that the amount of power required to lift the rock block through this distance is

$$\frac{E}{\Delta t} = \frac{\left(\int_0^{\Delta t} (F_{\text{up}} - F_{\text{down}} - W_g - F_{s1} - F_{s2})dt\right) \cdot h}{\Delta t} = F_{\Delta t} \cdot \frac{h}{\Delta t} = F_{\Delta t} \cdot V_{\Delta t} \tag{3.26}$$

As in the case of granular non-cohesive soils it is concluded that stream power ($F_{\Delta t}V_{\Delta t}$) can be used to quantify the relative magnitude of the erosive capacity of water leading to incipient motion. Practical methods that implement stream power for determining the erodibility of rock are presented in Chaps. 5 to 7.

Additionally, other methods for identifying incipient conditions that can lead to scour using estimates of the magnitude of turbulent pressure fluctuations originating from plunging jets are presented in Chaps. 6 to 8.

Vegetated earth material

The principal reason for increased erosion resistance offered by vegetated soils is that the roots of the plant bind soil particles together in an assembly, i.e., it forms larger "pseudo" particles. The success that can be accomplished when using vegetation to protect soils against erosion largely depends on the root architecture and growth habit of the plant's roots. Fibrous roots with a growth habit that results in closely spaced fine roots binding soil particles together provide larger resistance against erosion than tap roots that are widely spaced and do not have the ability to bind the soil together. The additional mass strength that the roots offer to the soil is not that significant, nor is the increase in shear strength. The principal increase in erosion resistance is offered by the increase in the effective "particle size" of the soil due to the presence of fibrous roots that binds the individual soil particles. In this regard vegetated earth materials can be viewed as physical gels.

This concept is illustrated in the photograph shown in Fig. 3.16, which was taken on a floodplain one day after a flood inundated it. The soil on

Figure 3.16 Vegetated soil illustrating the effect of roots binding the soil to increase the erosion resistance of the substrate.

the floodplain consists of silty fine sand that would easily erode if not protected. The roots of the grass on the floodplain are characterized as fine fibrous roots that bind the soil. The photo in Fig. 3.16 illustrates how soil that was not bound by the roots was eroded around the plant, while the soil that was bound by the grass roots remained intact.

If the plant material is not characterized as having fibrous roots the effect of the plants will not be the same. This can be seen by viewing the erosion that occurred underneath a plant with a root system consisting primarily of tap roots, with very little evidence of fibrous roots (Fig. 3.17).

In addition to the effects of fibrous plant roots protecting the soil against erosion, it is also known that plant foliage can play an important role. This can be seen in Fig. 3.18 that shows a stand of vegetation in a channel prior to a flood, which is bent down during the course of a flood (bottom picture). Once the flood passed, the vegetation recovered again and stood tall.

The effect of foliage is that it essentially lifts the erosive capacity of the water from the bottom of the channel. Instead of being in contact with the soil in the channel bed the change in velocity distribution, due to the presence of the foliage, elevates the effect of the erosive capacity of the water to the top of the bent vegetation. Carollo, Ferro, and Termini (2002) conducted a study to determine the effect of vegetation on the vertical velocity distribution in a channel. They found that the

Figure 3.17 Nonfibrous roots do not protect soil against erosion.

velocity distribution changed quite dramatically when compared to that in a channel without vegetation (Fig. 3.19).

The change in the velocity distribution affects the vertical distribution of the erosive capacity of the water. The vertical distribution of stream power can be determined by differentiating Eq. (3.4) with respect to flow depth y, i.e.,

$$\frac{dP}{dy} = \tau \frac{dU}{dy}$$

The difference in vertical stream power distribution for the two flow scenarios, i.e., flow with and without vegetation is presented in Fig. 3.20. From this graph it is seen that the maximum concentration of the erosive capacity of the water is located above the soil in the case of a vegetated channel bed, while the maximum value for a nonvegetated bed is located right at the bed.

Figure 3.18 The effect of plant foliage in protecting a channel bed against erosion.

Chemical gel response to scour

Scour of chemical gels such as rock or cohesive soils, like clay, is largely dependent on their surface properties and how they interact with the fluctuating pressures of turbulent flow. However, the fixed bonds of chemical gels result in surface layer properties that differ from those of physical gels. Whereas the surface layer of a physical gel consists of discrete

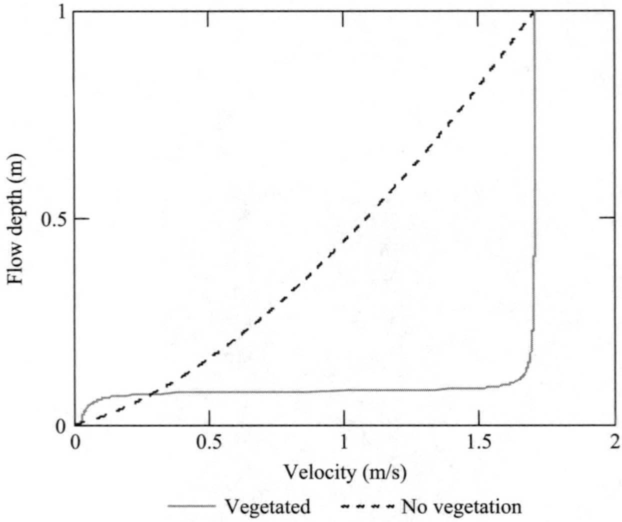

Figure 3.19 Vertical velocity distributions in a channel with and without vegetation. Based on research by Carollo, Ferro, and Termini (2002).

elements, the exposed surface of a chemical gel consists of a continuous mass that usually contains some imperfections. Additionally, it is possible that the surface properties of chemical gels can change when they come into contact with fluids, as is the case with cohesive soils such as clay. In what follows the failure mechanism of a brittle material, like

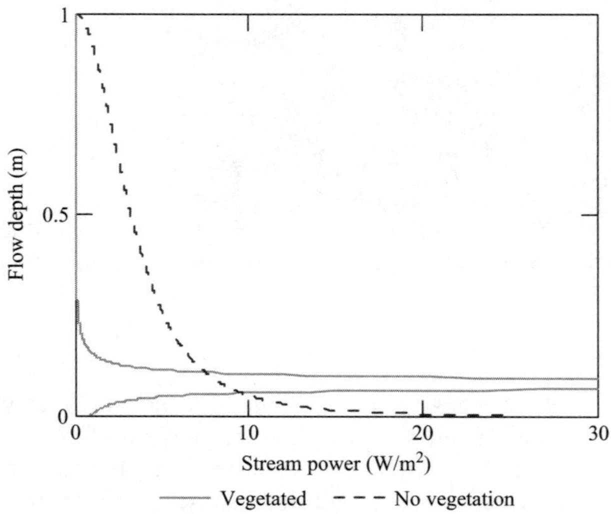

Figure 3.20 Estimated distribution of the erosive capacity of water in channel beds with and without vegetation.

rock, is first discussed. This is followed by a discussion of the erosion of cohesive soils that experience changes in their properties when in contact with fluids.

Intact rock

Intact rock, an example of a chemical gel, is used to explain how chemical gels fail. Prior to discussing rock scour processes it is prudent to first describe typical rock properties. Rock formations are usually characterized by macro-discontinuities known as joints, faults, or shear zones. These discontinuities lead to the development of rock blocks within a matrix that can range from very large to small. The rock blocks themselves often contain additional imperfections known as fissures or micro-fissures. Fissures are essentially close-ended cracks within a solid rock mass, i.e., they are not continuous throughout the whole rock mass. They are often characterized as having openings at the surface, extending some distance into the rock where they end as close-ended cracks. The distinction between micro-fissures and fissures is one of scale. Micro-fissures are identified with a microscope, whereas fissures can be identified with the naked eye.

Individual rock blocks delineated by macro discontinuities like joints, shears, and faults can be characterized as chemical gels. (This is not necessarily always the case, as is discussed further on.) Such blocks, when behaving like a chemical gel, can be broken up into smaller pieces by the erosive capacity of water if the conditions are right. In order for this to happen the presence of fluctuating pressures generated by turbulent flow and the presence of imperfections in the rock surface (fissures and micro-fissures) are required.

The strength of close-ended fissures in chemical gels is characterized by their fracture toughness. This is a strength measure that determines if a chemical gel (such as intact rock) is strong enough to resist stress intensities that can develop within close-ended fissures. If the stress intensity developed at the tip of a fissure is greater than the fracture toughness of the chemical gel, it will fail in brittle fracture (Fig. 3.21), i.e., in an explosive manner. Figure 3.22 shows rock that failed in brittle fracture after being subjected to a plunging jet at Santa Luzia Dam in Portugal. The signs of brittle fracture can be seen on the surface of the rock facing the group of men at the bottom, with one man standing at the top. The surface of the rock is rough and irregular, characteristic of brittle fracture.

The theory describing fracture toughness and stress intensity development in fissures, and practical methods for quantifying the same, are presented in Chaps. 5 to 8. For purposes of distinguishing between failure characteristics of physical and chemical gels the failure criterion used here merely states that failure will occur if the stress intensity developing in a fissure exceeds the fracture toughness of a chemical gel.

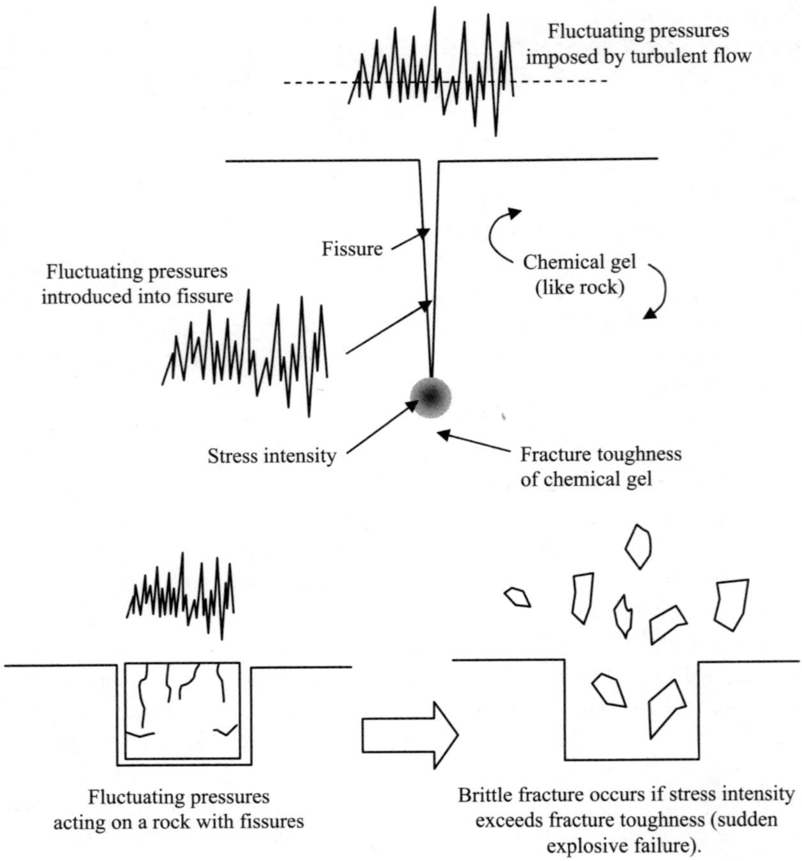

Figure 3.21 A chemical gel with a close-ended fissure impacted by fluctuating pressures. If the stress intensity developed at the end of the fissure is greater than the fracture toughness of the chemical gel it will fail in brittle fracture.

In cases where the stress intensity that develops in a close-ended fissure does not exceed its fracture toughness the chemical gel will not fail immediately, i.e., it will not fail in brittle fracture. However, the repeated application of the fluctuating stress intensity developed due to continuous application of fluctuating pressures over long enough periods of time leads to the development of slip-surfaces within the atomic structure of the chemical gel at the tips of close-ended fissures. If the fluctuating pressures are applied long enough the development of these slip-surfaces eventually leads to cracks that continuously and progressively increase in length until they eventually split the block of rock. This is known as fatigue failure (see Fig. 3.23).

Figure 3.22 Santa Luzia Dam, Portugal. Scour of rock downstream of the spillway due to brittle fracture. Note the irregular surface on the rock, which is characteristic of brittle fracture (Photo: Dr. Jose de Melo, Portugal).

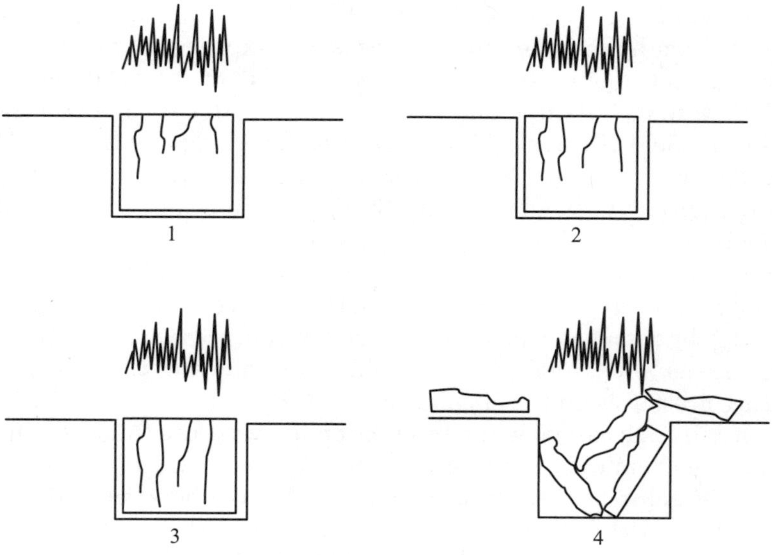

Figure 3.23 Subcritical failure of a block of rock (chemical gel) subject to fluctuating pressures. Also known as fatigue failure.

The processes of brittle fracture and fatigue failure in chemical gels form an important component of the scour process. If the extent of a chemical gel at commencement of the application of fluctuating pressures is such that it cannot be dislodged en mass, brittle fracture or fatigue failure can lead to the formation of smaller elements that are subsequently removed more easily. In essence what happens is that the character of the chemical gel can change to that of a physical gel during the scour process. If brittle fracture or fatigue failure leads to the formation of gel elements that are small enough relative to the scale of the fluctuating pressures to allow removal of individual intact units, the material's character has changed to that of a physical gel.

For example, individual rock blocks in a formation can be too large for removal by the erosive capacity of the water. When the application of fluctuating pressures leads to brittle fracture or fatigue failure the rock blocks are broken up into smaller pieces that are more easily removed as individual units.

Cohesive soils

It is useful to distinguish between chemical gels that experience change in their properties when in contact with water and those that do not. An example of a chemical gel that experiences change in strength when in contact with water is clay. Most rock are examples of a chemical gels that do not experience such changes.

Distinguishing between the failure mechanisms in chemical gels that experience a change in their properties when exposed to water and those that do not is important. Full knowledge is not yet available on how cohesive materials like clay scour. Research that has been conducted in this field of investigation indicates that clay potentially fails like a brittle material, but that its fracture toughness is dependent on the interaction between the eroding fluid and the cohesive material itself (see Raudkivi et al. 1998).

It appears as if the scour process in clay commences with "washing off" of individual clay elements that lie on the surface. This is then usually followed by removal of clods of material by a process often referred to in practice as "plucking." The term "plucking" indicates the effects of fluctuating pressures on the clay fabric.

Conceptually one can view the interaction between the fluid causing scour and the chemical gel making up the clay as shown in Fig. 3.24. The chemical gel is held together by chemical forces between individual elements. As the fluid leading to scour flows over the gel it can also penetrate the spaces between the elements, affecting the nature and strength of the chemical bonds. In some cases the strength of the chemical bonds can increase and in other cases it can decrease.

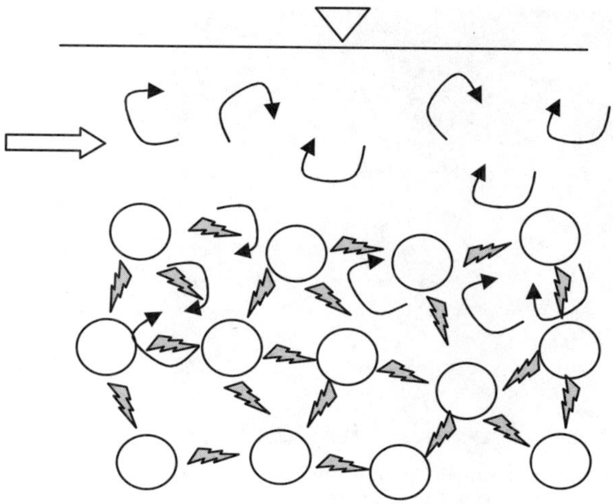

Figure 3.24 Cohesive soil, a chemical gel. This sketch conceptually illustrates chemical bonds between individual soil elements, with water subject to turbulent pressure fluctuations flowing over it and also entering the chemical gel matrix.

The clay itself may either be intact, with no imperfections (which is unlikely), or it can contain imperfections. The imperfections can exist within the clay with no expression to the surface (i.e., fissures contained within the clay), or it can exist as close-ended fissures with openings to the surface. As fluctuating water pressures interact with the imperfection it can lead to either brittle fracture of fatigue failure. Brittle fracture will occur when the stress intensities at the tips of the imperfections are greater than the fracture toughness of the clay. Fatigue failure will occur if the stress intensities are lower than the fracture toughness of the clay, but are applied in a fluctuating manner long enough for it to eventually fail in fatigue. Figure 3.25 shows the character of a clay surface that has been subjected to scour. The rough surface resulted from the effects of "plucking."

The fracture toughness of the clay is dependent on the strength of the chemical bonds between individual clay elements. When water comes into contact with clays it can lead to changes in the chemistry of the material, which can be characterized by either strengthening or weakening of the material, depending on the chemistry of the water. For example, Croad (1981) found that the rate of erosion of all clays investigated by him decreased if the salt content of the water increased. He also found that the rate of erosion was dependent on temperature. If the temperature increases from a low value it is found that the rate of erosion first increases until it reaches a maximum. It then decreases until

Figure 3.25 Scour of clay downstream of a model pier. Evidence of "plucking" is seen in the foreground, characterized by the irregular surface. (Photo: Dr. Jean-Louis Briaud, Texas A&M)

it reaches a minimum, whereafter it increases again. The relationship first has an upside-down U shape, and then a U shape.

As the fracture toughness of clay is affected by the chemistry of the water that gets into touch with it, the potential for scour and the rate by which clay may scour is dependent on the chemical properties of the clay and the water. Croad (1981) showed that the rate of scour of clay can be explained as a chemical process, which is discussed in more detail in Chaps. 4 and 8.

Summary

This chapter introduces basic concepts that are important in scour technology, with more detail provided in other chapters. One of the most important aspects that is dealt with includes the apparent inconsistency of conventional indicator parameters used to quantify the relative magnitude of the erosive capacity of water. It is shown that calculated values of shear stress, flow velocity, and stream power provide inconsistent trends when considered as a function of hydraulic roughness. When using calculated values of these parameters as indicators of the relative magnitude of the erosive capacity of water, their trends lead to contradictory conclusions.

When considering shear stress as a function of hydraulic roughness it implies that the erosive capacity of water increases if the hydraulic roughness of the boundary increases. When considering average flow velocity as a function of hydraulic roughness it implies that the erosive capacity of water decreases if the hydraulic roughness increases. The trend of stream power as a function hydraulic roughness is constant, implying no change in erosive capacity. These contradictions highlight the need to select indicator parameters for quantifying the erosive capacity of water that are internally consistent, easy to calculate, and reasonably represent the dominant processes characteristic of flow conditions that lead to the incipient motion of earth materials.

The discussion of boundary processes indicates that fluctuating pressures form the dominant process at the boundary that leads to the incipient motion of earth and engineered earth materials under turbulent flow conditions. It is shown that the lack of fluctuating pressures in laminar flow justifies the use of shear stress as an indicator parameter of the erosive capacity of water. The presence and important role of pressure fluctuations in turbulent flow requires recognition that impulse forces and the rate of imparting energy to the boundary becomes important. It is shown that stream power is an indicator of the relative magnitude of pressure fluctuations, and that it is justifiable to use it as an indicator parameter to quantify the relative magnitude of the erosive capacity of water. Additional evidence is also provided in Chap. 5. The apparent inconsistency in the trends of common indicator parameters of the erosive capacity of water is addressed in Chap. 5.

This chapter also distinguishes between two essential material types, namely physical and chemical gels. The difference between the two types of gels lies principally in the bonds between individual elements making up these materials. The bonds in physical gels exist merely because individual elements touch each other, and the forces thus created within the gel are characterized by friction. A typical example of a physical gel is non-cohesive sand. The bonds in chemical gels, on the other hand, are characterized as "fixed" bonds, which are principally characterized as chemical connections. Cohesive soils and intact rock are typical examples of chemical gels.

Distinguishing between physical and chemical gels creates additional insight into scour processes, particularly when viewed in terms of the ways that these two gel types fail. In the case of physical gels (e.g., noncohesive sands) incipient motion in turbulent flow is characterized by the removal of individual elements. The removal process is therefore relatively simple, as it reflects the interaction between fluctuating pressures and individual soil elements.

In the case of chemical gels (e.g., cohesive soils and intact rock) incipient motion is characterized by two potential failure mechanisms. These are brittle fracture and fatigue failure. Brittle fracture occurs when the stresses created by turbulent fluctuating water pressures lead to stress intensities in close-ended imperfections in the chemical gel (e.g., close-ended fissures) that exceed the fracture toughness of the gel. When this happens the failure occurs in an explosive manner. If the stress intensities resulting from fluctuating pressures do not exceed the fracture toughness of the chemical gel, continued application of the fluctuating pressures can lead to slow growth of cracks in the material, and eventual failure by fatigue.

It is also pointed out that the fracture toughness of certain chemical gels, like clay, is affected by the chemistry of the water with which it comes into contact. This characteristic allows one to view the scour process in clays as a chemical process, which is explained in more detail in Chaps. 4 and 8.

Classification of materials as either physical or chemical gels is dependent on scale. If the scale of individual elements of the material is much larger than the scale of the fluctuating pressures, the material can most probably be characterized as a chemical gel. When the scale of individual elements of the material is equal to or smaller than the scale of the fluctuating pressures, the material can most probably be classified as a physical gel. It is possible that a chemical gel can change into a physical gel. This happens when the chemical gel fails in brittle fracture or fatigue failure and the scale of the resulting elements are equal to or smaller than the scale of the pressure fluctuations.

Chapter

4

Material and Fluid Properties

Introduction

This chapter presents the properties of water and earth materials relevant to scour assessment. The relevant properties of water are presented first, followed by a discussion of relevant earth materials properties. The discussion of earth materials is divided into two groups—physical and chemical gels.

In keeping with the general approach in this book, an effort is made to explain material behavior in quite a lot of detail. The objective of these explanations is to provide the reader with the necessary insight that will assist in objective and subjective reasoning when implementing the decision-making process outlined in Chap. 2. Additional methods for estimating material characteristics are provided in Chap. 8. The background required to assess the value and applicability of test results discussed in Chap. 8 is contained in this chapter.

Water

The most important water properties relevant to scour are presented in what follows:

The mass density of water, ρ is 1000 kg/m^3 on average.
The unit weight of water, $\gamma = \rho g$ is 9820 N/m^3 on average.
The dynamic viscosity of water μ is related to the kinematic viscosity ν by the relationship

$$\nu = \frac{\mu}{\rho}$$

The dynamic viscosity of water can be estimated with the equation

$$\mu = \frac{1.79 \times 10^{-3}}{1 + 33.69 \times 10^{-3} T' + 2.21 \times 10^{-4} T'^2} \frac{\text{kg}}{\text{m} \cdot \text{s}} \tag{4.1}$$

where T' is the temperature in degrees Centigrade. Kinematic viscosity of water is also sensitive to temperature and its value can be estimated with the equation

$$\nu = \frac{40 \times 10^{-6}}{20 + T'} \frac{\text{m}^2}{\text{s}} \tag{4.2}$$

The air concentration α_i in water is calculated as

$$\alpha_i = \frac{\beta}{1 + \beta} \tag{4.3}$$

and the free air content in water as

$$\beta = \frac{q_a}{q} \tag{4.4}$$

where β = free air content
q = unit flow of water
q_a = unit flow of air

Alternatively, the free air content can also be calculated as

$$\beta = \frac{V_a}{V_w} \tag{4.5}$$

where V_a is the volume of air and V_w is the volume of water.

The pressure wave celerity in a mixed fluid, i.e., the speed by which a pressure wave can move through it, determines how quickly massive materials like rock fails under the impact of the erosive capacity of water. The need for representative equations that can be used to calculate the pressure wave celerity in mixed fluids, e.g., a mixture of air and water, is therefore obvious. Pressure wave celerity is also known as the speed of sound.

In scour technology the pressure wave celerity becomes important when considering the impact of pressures and pressure fluctuations in scour of rock. The relationship between a fissure or joint aperture, which is relatively small, and the thickness of the rock itself, which is relatively large and massive, is therefore such that the elasticity of the rock does not affect the pressure wave celerity within a fissure or joint significantly. We can therefore resort to using equations that directly represent the magnitude of pressure wave celerity in a mixed medium, without allowing for the effects of the surrounding material, e.g., the rock.

In general, the pressure wave celerity, i.e., the speed of sound, through a material is calculated as

$$c_{\text{mix}} = \sqrt{\frac{K_e}{\rho_e}} \tag{4.6}$$

where K_e is the effective bulk modulus of the mixed fluid and ρ_e is the effective density of the mixed fluid. The effective bulk modulus of the mixed fluid is calculated as (Kafesaki, et al. 2000)

$$K_e = \frac{1}{\frac{\beta}{K_a} + \frac{(1-\beta)}{K_l}} \tag{4.7}$$

where K_a is the bulk modulus of air and K_l is the bulk modulus of the liquid. Bulk modulus represents the "springiness" of a material. In the case of air, the bulk modulus is set equal to the pressure p_a of the air. The bulk modulus of water is

$$K_l = 1\text{GPa} = 1 \cdot 10^9\,\text{Pa}$$

and the effective mass density of the mixed fluid is calculated as (Kafesaki, et al. 2000)

$$\rho_e = \rho \frac{[\beta(\rho_a - \rho) + 2\rho_a + \rho\ ]}{[2\beta(\rho - \rho_a) + 2\rho_a + \rho]} \tag{4.8}$$

where ρ_a is the density of air. Air is a compressible fluid and its density depends on its pressure and temperature, which can be estimated by making use of the Universal Gas Law. For compressible fluids, i.e., air, its density can be calculated as,

$$\rho_a = \frac{p_a M}{\gamma' R T} \tag{4.9}$$

where p_a = air pressure
M = molar mass of air
γ' = compressibility factor
R = universal gas constant
T = absolute temperature in Kelvin

The molar mass of air, a mixture of nitrogen and oxygen, is approximately,

$$M = 0.029 \frac{\text{kg}}{\text{mole}}$$

and the Universal Gas Constant

$$R = 8.315 \frac{\text{J}}{\text{K} \cdot \text{mole}}$$

A compressibility factor $\gamma' = 1.4$ can be used for air. The effective density of the mixture can therefore be calculated as

$$\rho_e = \rho \frac{\left[\frac{p_a M}{\gamma' RT}(\beta + 2) + \rho(1 - \beta)\right]}{\left[\frac{2p_a M}{\gamma' RT}(1 - \beta) + \rho(2\beta + 1)\right]} \tag{4.10}$$

and the effective bulk modulus as

$$K_e = \frac{1}{\frac{\beta}{p_a} + \frac{(1-\beta)}{K_l}} \tag{4.11}$$

The absolute temperature in Kelvin is calculated as

$$T = 273 + T' \tag{4.12}$$

where T' is the temperature in degrees Centigrade.

Using Eqs. (4.6) and (4.10) to (4.12) it is possible to calculate the value of the pressure wave celerity in air as 343 m/s at 20°C and atmospheric pressure. The pressure wave celerity in pure water is 1000 m/s, at the same temperature. Equations (4.6) to (4.9) are only valid for air content values ranging between 0 and 0.50 and for a unique value of 1.0 (Krokhin et al., 2003).

Figure 4.1 is a plot of Eqs. (4.6) to (4.9), which shows that the pressure wave celerity in water containing free air changes rapidly as a

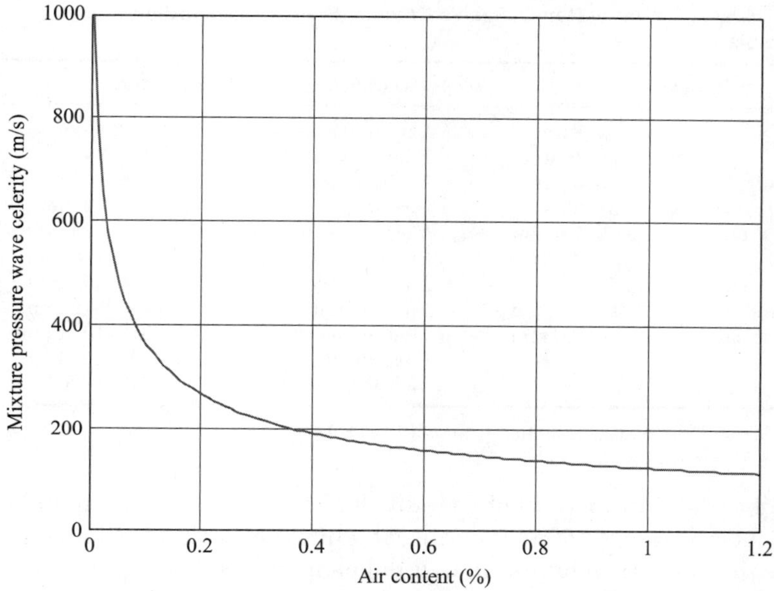

Figure 4.1 Change in pressure wave celerity in an air-water mixture as a function of air content.

function of air content. The pressure wave celerity in water changes from 1000 m/s, when it contains no air, to about 100 m/s when it contains only 1 percent of free air by volume.

Other parameter values

The value of the Avogadro number is

$$L = 6.022 \times 10^{23} \frac{1}{\text{mol}} \tag{4.13}$$

Physical Gels

Physical gels consist of elements in occupied spaces that are connected to each other by touching (see Chap. 3). The two types of physical gels of interest in scour technology are non-cohesive granular material and jointed rock.

Non-cohesive granular material

Non-cohesive granular materials can be classified as shown in Table 4.1.

TABLE 4.1 Classification of Non-Cohesive Granular Sediment According to Particle Size

Silt and sand		Gravel, cobbles, and boulders	
Very fine silt	4–8 μm	Very fine gravel	2–4 mm
Fine silt	8–16 μm	Fine gravel	4–8 mm
Medium silt	16–31 μm	Medium gravel	8–16 mm
Coarse silt	31–62 μm	Coarse gravel	16–32 mm
Very fine sand	62–125 μm	Very coarse gravel	32–64 mm
Fine sand	125– 250 μm	Small cobbles	64–128 mm
Medium sand	250– 500 μm	Large cobbles	128–256 mm
Coarse sand	0.5–1.0 mm	Small boulders	256–512 mm
Very coarse sand	1.0–2.0 mm	Medium boulders	512–1024 mm
		Large boulders	1024–2048 mm
		Very large boulders	2048–4096 mm

SOURCE: From British Standards Institution, BS 1377, 1975.

Unit weight. The unit weight of an individual non-cohesive granular element is the unit weight of its mineral content. A value that is commonly used for the specific weight of individual elements is 2650 kg/m^3.

Specific gravity. Specific gravity is the unit weight of a soil element divided by the unit weight of water. The unit weight of water is usually assumed to be 1000 kg/m^3. The specific gravity of a soil element is therefore 2.65 (assuming a unit weight for the soil of 2650 kg/m^3).

Jointed rock

When assessing scour of rock it is important to consider the properties of the rock mass as a whole. This requires consideration of not only the mass strength of the rock, but also the impact of discontinuities on its scour resistance. Discontinuities, consisting of micro-fissures and fissures, and macro-discontinuities like joints, bedding planes, foliations and faults play important roles in defining a rock mass' engineering properties and behavior.

The *mass strength of rock* determines whether it is possible for the erosive capacity of the water to break large, solid pieces of rock into smaller pieces. If this is possible, the smaller broken pieces of rock will be more susceptible to removal by the flowing water. The mass strength of rock is commonly expressed in terms of its unconfined compressive strength (UCS).

Breakup of larger blocks of rock requires the presence of *close-ended fissures and micro-fissures* impacted by fluctuating water pressures. The fluctuating water pressures acting within the fissures can lead to brittle fracture or fatigue failure of the rock. Fissures are planar cracks that are common in hard rock having experienced internal deformation.

They usually occur as crystal boundary cracks, but can also extend through crystals in hard rock. The difference between micro-fissures and fissures is one of scale. Fissures can be observed in hand samples, while identification of micro-fissures requires the use of a microscope.

Macro-discontinuities like joints, bedding planes, foliations and faults play important roles in defining a rock mass' scour resistance. *Joints* are discontinuities that originate from brittle fracture of the rock and are characterized as features that have not experienced displacement of the rock on opposite sides of the discontinuity. *Bedding-planes* occur in sedimentary deposits, characterizing the sequential deposition of deposited sedimentary material. *Foliation* is a planar arrangement of structural or textural features in any rock type, but particularly that resulting from the alignment of constituent mineral grains of a metamorphic rock. Foliation is exhibited most prominently by sheety minerals, such as mica or chlorite. *Faults* (also sometimes referred to as *fractures*), on the other hand, are characterized as discontinuities where the rock on the opposite sides of the discontinuity have experienced displacement.

The principal cause of *jointing* in both stratified and igneous rocks is crustal movement, although the specific origin of the movement may not always be apparent. Expansion and contraction from the intrusion of hot igneous rocks leads to jointing, as does crystallization and contraction of consolidating sediment. Joints are present in nearly all surface rocks and extend in various directions, generally more toward the vertical than to the horizontal.

Sedimentary rocks are usually characterized by two sets of joints at right angles to one another, each extending down perpendicular to the bedding. One joint set often extends in the direction of the dip and the other in the direction of strike. In igneous rocks, jointing is generally quite irregular. However, in granite, two vertical sets forming right angles to one another on the top surface and another set of cross-joints that are approximately horizontal occur frequently.

Sills and dikes, resulting from the intrusion of molten rock, usually form columnar jointing when it cools. Three sets of joints perpendicular to the cooling surfaces usually intersect each other roughly at angles of about 120°. The dimensions of the polygonal columnar features range in size from about 70 mm to 6 m, depending on its cooling rate. Joints may have smooth, clean surfaces, or they may be irregular or scarred by slickensides.

Fractures (or *faults*) in rock occur when tectonic compression or tension forces displace rock on the opposite sides of the fracture or fault. They can occur either along a distinct plane, known as a fault plane, or as movement along a number of subparallel surfaces. The latter leads to a fault zone of fractured rock. Faults can be relatively short or can be hundreds of kilometers long (e.g., the San Andreas Fault in California). Their widths vary, ranging from several centimeters to hundreds of meters.

Bedding planes can either be intact or separated. When separated they are usually interpreted as a joint set. For example, sedimentary rock normally has two joint sets perpendicular to the bedding plane. If the bedding planes are separated, the interpretation for purposes of scour assessment is that the formation is characterized by three joint sets.

Foliation often occurs parallel to bedding planes, but it may not necessarily be related to any other structural rock features. For example, schist exhibits strong foliation with partings along well-defined planes of medium-grained mica or hornblende. Gneiss on the other hand, which is characteristically rich in feldspar and quartz, tends to be coarse-grained, stronger, and less distinctly foliated. Gneiss does not split along its planes like schist.

Other features related to discontinuities affecting the scour resistance of rock are aperture spacing of micro- and macro-discontinuities, dip and dip direction of discontinuity planes, and the shape of rock blocks.

Aperture spacing, which is the distance between the opposite faces of joint, fracture, and fault planes, plays a role in determining the ease by which rock blocks can be removed. Large aperture spacing will make rock removal easier than when the joints are tight. The filling in an aperture also affects its ability to resist scour. Apertures can be filled with earth material such as rock flour or clay, known as *gouge*, or can be free of any material. If it is filled with clay gouge, it will lead to easier removal of the rock than when the gouge consists of, say, rock flour.

The *roughness* of joint and fracture (fault) planes determine, jointly with the type of gouge, the shear resistance offered by these surfaces. For example, if the plane is smooth and contains a clay gouge its shear resistance will be lower than that of a rough, tight joint with no gouge.

The *dip and dip direction* of rock (its orientation) relative to flow direction also impacts the ease of removal by water. If rock is dipped in the direction of flow it is easier to remove than when it is dipped against the direction of flow, or if the dip is at right angles to the flow.

The impact of the *shape of a rock block* on its erosion resistance depends on whether it is equi-sided or elongated. Equi-sided blocks of rock will be easier to remove than elongated shapes. Additionally, the shape of a rock and its location within the rock matrix could lead to it acting as a *key block*. The orientation of the faces of a key block within its matrix is such that they provide additional resistance to scour, but, once removed, lead to general failure of the surrounding rock (from there the name key block).

In summary, the principal rock characteristics playing a role in resisting the erosive capacity of water are mass strength, block size and shape, and discontinuity characteristics, which include the number of joint sets, joint set spacing and orientation, aperture spacing, joint alteration (gouge), and joint roughness. If the rock is massive, with few discontinuities, it is most probably appropriate to classify it as a chemical gel. In such a case scour of the rock will occur if it fails in either brittle fracture or fatigue failure. When rock is characterized by the presence of frequent discontinuities, it is possible that the rock can be classified as a physical gel. In such a case, the dominant scour process will be characterized as removal of intact blocks of rock.

Chemical Gels

Chemical gels, as pointed out previously (Chap. 3), consist of elements in occupied spaces that are connected to each other by fixed bonds. These bonds are chemical in nature and in order to break up a chemical gel it is necessary to break these bonds. A conceptual way to look at the bonds is to view them as small elastic bands holding the elements together. The stiffness of these "elastic bands" depends on the strength of the fixed chemical bonds. In the case of fresh, hard rock, the strength of the fixed chemical bonds between individual crystals is very high, and the stiffness of the "elastic bands" can be viewed as very high. In the case of a clay, the fixed chemical bonds between individual clay platelets are not as strong as in rock and the stiffness of the "elastic bands" can be viewed as more pliable, i.e., not as stiff.

In order to understand how the fixed chemical bonds between individual elements fail one can draw a graph of the energy required to break a bond as a function of displacement. By using the analogy of an elastic band the change in energy required to stretch and eventually break the band will follow a curve as a function of displacement as shown in Fig. 4.2. At first the band is merely held taut (a) and the energy required to hold it in that shape has a certain value illustrated by the horizontal portion of the energy curve. Then, as one starts to apply more force to the ends of the elastic band it starts to stretch and the relationship between the energy required to stretch the band further and further and the distance through which it is stretched takes on a positive upward curve (b). The maximum amount of energy is reached just prior to the elastic band snapping (c). Once the band snaps this energy is released in the form of heat and sound, and the energy level remaining in the elastic band drops to (d). Viewed in the context of the behavior of a chemical gel, the elements held together by this particular fixed bond are now separated.

The shape of the energy versus displacement curve shown in Fig. 4.2 is typical of the energy relationships in chemical reactions. The amount of energy to set a chemical reaction in motion (known as the activation

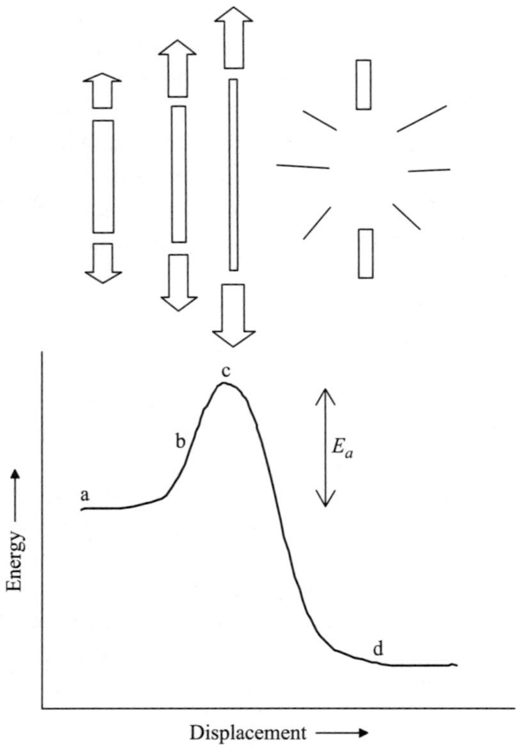

Figure 4.2 Energy as a function of displacement required for stretching and breaking an elastic band.

energy E_a) is represented by the vertical distance between points (a) and (c). Using this relationship one can now develop models of the failure mechanisms of chemical gels. For example, consider a piece of rock with a close-ended fissure that extends to the surface of the rock. If the activation energy that is applied to the rock (e.g., pressure building up in the crack due to the presence of turbulent pressure fluctuations) is large enough, it is conceivable to imagine that all the chemical bonds that keep the mineral elements in the rock together in the region within the rock below the end of the fissure can all snap at once. If this happens, the rock will suddenly burst apart in an explosive manner. Such a failure mode is known as brittle fracture.

One can also imagine another failure mode in such a brittle material. Say, the energy imparted to the chemical bonds is only slightly larger than the activation energy required to break the bonds. In such a case it is possible that some of the bonds might be broken while others still remain intact. For example, the stresses at the tip of the close-ended crack could reasonably be expected to be larger than the stresses deeper into the rock below the close-ended crack. This is likely to result in the bonds breaking in a sequential manner, and not all at once. As the crack grows, the edge of the close-ended fissure moves deeper into the rock, causing progressively higher stress intensities at different locations within the rock as the tip of the growing crack moves deeper and deeper into the rock. The sequential failure of chemical bonds as the crack grows deeper and deeper into the rock results in sequential failure of the rock. This failure mode is known as fatigue failure or sub-critical failure.

In order to gain deeper understanding of the failure mechanisms characteristic of chemical gels we will first consider the erosion of cohesive earth materials, such as clay, and then move on to scour of rock. The theory presented to explain the failure mechanisms characteristic of cohesive earth materials might, at first glance, seem somewhat esoteric. However, in order to explain the apparent anomalies sometimes observed in erosion of cohesive earth materials, it is necessary to understand these failure mechanisms.

For example, Croad (1981) found that the erosion rate of clays sometimes has a concave and other times a convex shape when plotted as a function of shear stress (Fig. 4.3). It is shown further on that these shapes can be explained in terms of the variables that affect the erosion resistance of clays.

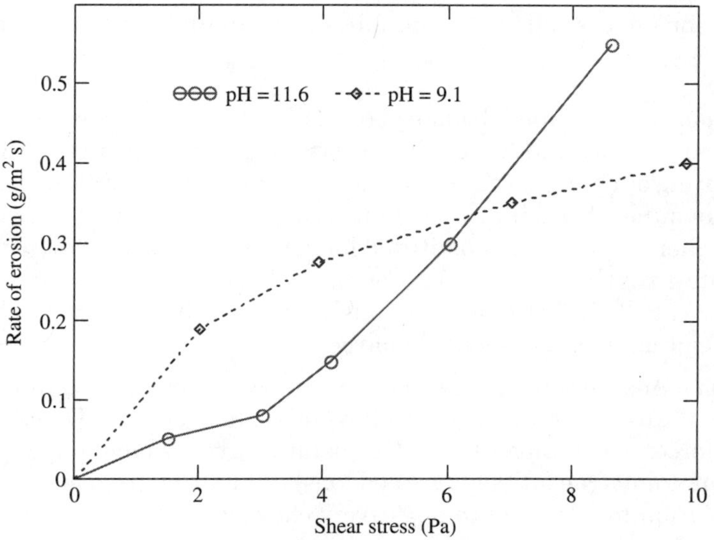

Figure 4.3 Convex and concave shapes of erosion rate expressed as a function of shear stress for Na-Bentonite clay with varying pH of the interstitial and ambient water (Croad 1981).

The explanation of the essential nature of erosion of cohesive materials that follows reveals a very complex process. The rate of erosion of cohesive earth materials is a function of the activation energy and the number of bonds per unit area between clay particles. The chemical nature of the bonds, characterized by the activation energy, is a function of temperature, pH, and salinity. Accurate prediction of the erodibility of cohesive earth materials is no simple matter. For example, it has been found in some cases that the effects of pH and salinity are greater than that of shear stress (see, e.g., Tan 1983 and further on in this chapter). Prediction of the anticipated salinity, pH, and temperature anticipated to occur during an erosion event in practice can be challenging.

Erosion of cohesive soils

The rate of erosion of cohesive soils can be viewed in the context of a chemical reaction. It therefore seems reasonable to use concepts for predicting the rate of chemical reactions to develop methods for calculating the rate of erosion of cohesive materials. This approach has been taken by Croad (1981), based on an idea by Raudkivi and Hutchison (1974).

Prior to developing the erosion rate expression, it is useful to reflect on the essential nature of clay particles and their chemical bonds. Clay particles are very small and flat, and usually have opposite charges on their faces and edges. The differing charges on clay particles in electrolyte solutions can result from the following (Raudkivi 1976; Grim 1968):

- *Isomorphous substitution* Isomorphous substitution is the replacement of one atom by another of similar size in a crystal lattice without disrupting or changing the crystal structure of the mineral. In the case of clay minerals such substitution occurs when cations (positive ions) of higher valency are substituted for cations with lower valency, i.e., the latter replaces the former: For example, Mg^{2+} replaces Al^{3+} or Al^{3+} replaces Si^{4+}. In both cases a deficiency of positive charge will develop, leading to a net negative charge.
- *Adsorption* Adsorption occurs when ions from an electrolyte solution adhere to a surface and cause it to have a surface charge. Usually chemical forces are stronger than electrostatic forces, which makes it possible for anions (negatively charged ions) originating from an electrolyte solution to adsorb onto negatively charged surfaces, increasing its negative charge. It is also possible for anions to adsorb onto positively charged surfaces. If enough anions adsorb to such surfaces it is possible to change their sign from positive to negative.

- *Ion-exchange* Ion-exchange occurs when ions from an electrolyte solution are swapped with those from within or on the surface of a clay crystal lattice.
- *Ionization of broken bond surfaces* When solid surfaces break it usually results in ionization. The ion charge depends on the pH of the electrolyte solution surrounding the particles. Two possible reactions can take place, depending on the pH of the solution:

$$M]-OH+H_2O \rightarrow M]-O^-+H_3O^+ \tag{4.14}$$

Or

$$M]-OH+H_3O^+ \rightarrow M]-OH_2^+ +H_2O \tag{4.15}$$

Reaction (4.14) results in a negative charge to the metal species M], while reaction (4.15), which is a reaction in acidic conditions, results in a positive charge to the surface of the metal species. It is possible that the surface charge can become zero for certain pH values. When this occurs, i.e., the pH of the electrolyte solution results in no charge on the surface, the condition is called the isoelectric point (i.e.p.).

Another important characteristic of the surfaces of clay particles is the formation of what is known as a "double-layer." This layer does not only form on clay particles but on any surface in contact with an electrolyte. Within an electrolyte (the liquid phase) the ions move around in a random fashion following a random walk known as Brownian motion, while the solid phase (e.g., the basal surface of a clay particle) can be charged. At the interface between the charged solid phase and the liquid phase a rearrangement of ions occurs to balance the attractive forces by diffusion. The exact detail of how the double-layer forms and its arrangement is not discussed here, save to say that its thickness can change as a function of the nature of the electrolyte solution.

For example, it is known that the double layer is suppressed in the presence of salt, i.e., it becomes thinner. Similarly, in some cases the double layer is also suppressed in the presence of an electrolyte with a low pH. The impact of this thinning of the double layer can be significant as it can affect the strength of the clay. The reason for this behavior is explained below.

The attraction between clay particles is dependent on electrostatic forces and van der Waal's forces. The van der Waal's force is an attractive potential energy force that develops between two particles that are close enough together to activate the attractive forces between constituent atoms of the respective particles. These forces can be very large relative to the electrostatic forces if the particles are very close together.

This therefore implies that the influence of the van der Waal's forces will increase if two clay particles move closer to each other.

Such an increase in attraction can occur when the thickness of the double layer decreases. Say, for example, that the charges on the basal surfaces of clay particles are negative and those on the edges are positive. The arrangement of particles will in such a case be characterized by a so-called cardhouse structure, i.e., edges of particles will be attached to the basal surfaces of others.

If the thickness of the double layer on both the basal surfaces and the edges of the particles reduce due to the presence of salt in the electrolyte or a lowering of the pH the basal surfaces and edges will move closer to each other. If they move close enough the van der Waal's forces will start to dominate and make the clay much stronger.

Consider the case where the pH of the solution is such that the i.e.p. is reached. The basal and edge surfaces of the clay particles will have no charge. In such a case it is possible that the clay particles may arrange themselves in edge to edge and basal surface to basal surface arrangements. When close enough the van der Waal's forces will determine the attraction between particles. Such an arrangement of particles under i.e.p. conditions can result in an increase in the strength of the clay.

The brief discussion above illustrates the sensitivity of clay behavior to the chemistry of the electrolyte solution (the liquid phase) surrounding it. It also emphasizes the important role of the double-layer on the surfaces of the clay particles, and the respective roles that electrostatic and van der Waal's forces play. Changes in salt concentration or pH of the electrolyte solution result in changes in the thickness of the double-layer. When this occurs the van der Waal's forces can either dominate or become less important. The van der Waal's forces, once activated, are much greater than the electrostatic forces and can result in an increase in the strength of the clay.

However, it is important not to be deceived by the simplified explanation provided above. The behavior of clays, as illustrated in this chapter, is very complex. The strength of the material depends on a number of factors, including the arrangement of clay platelets, the salt concentration of the electrolyte surrounding the platelets, its temperature, and its pH. It is always required to conduct tests on site-specific clay samples to determine their characteristics.

When discussing the chemical concepts that will be used to formulate expressions describing the erosion rate of cohesive materials it is useful to recall that chemistry is somewhat of an empirical science. The discussion therefore starts off in a conceptual manner, and uses established and proven concepts in chemistry further on to derive the relationship between the erosion rate of cohesive earth materials and other variables. Practical approaches to apply the theory presented in what follows can be found in Chap. 8.

Croad (1981) used a conceptual relationship, borrowed from chemistry, to formulate the erosion of earth materials as follows:

$$S + F \rightarrow \text{Erosion products} \tag{4.16}$$

where S and F are the soil and fluid modules, respectively.

The rate at which chemical reactions occur is a function of the concentration of the respective substances that are used to cause the reaction. In the case of the formation of erosion products, the rate of erosion e can therefore be expressed as

$$e = k[S][F] \tag{4.17}$$

where k is the rate constant and $[S]$ and $[F]$ are the concentration of the soil and fluid modules, respectively. Therefore, what one now needs to do is to determine ways to quantify the rate constant and the concentrations of the soil and fluid modules in a practical manner that becomes useful in the prediction of the rate of erosion.

Referring back to the discussion of the fluctuating boundary processes when turbulent water flows over an earth material (Chap. 3), a useful way to deal with this problem is to express the concentration of the fluid module in terms of the average distribution of fluctuating pressure spots on the boundary. This can be estimated by making use of research that was performed to determine the average size and spacing of the fluctuating pressure spots on a surface subject to turbulent flow.

Consider the conceptual distribution of fluctuating pressures in Fig. 4.4. The shape and size of the spots of fluctuating pressure varies, as does

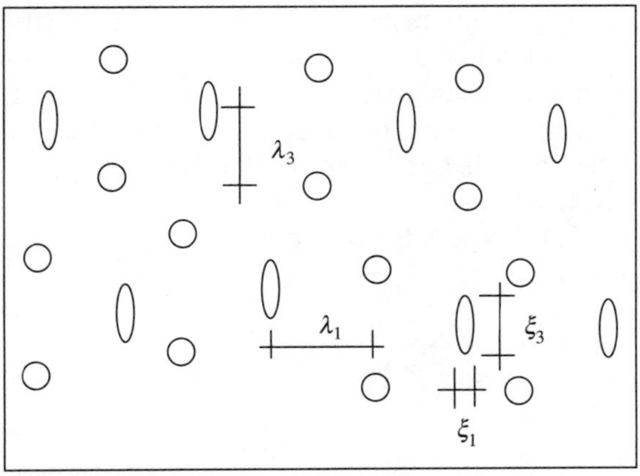

Figure 4.4 Distribution of fluctuating pressures on a boundary subject to turbulent flow.

the spacing between the pressure spots. The area of a pressure spot is expressed as $\xi_1 \cdot \xi_3$, where ξ_1 is the dimension in the direction of flow and ξ_3 is the dimension transverse to the direction of flow. The spacing between pressure spots is represented by λ_1 in the direction of flow and by λ_3 transverse to the direction of flow.

It is therefore possible to calculate the average concentration of pressure spots on the boundary as follows:

$$[F] = \frac{\xi_1 \cdot \xi_3}{\lambda_1 \cdot \lambda_3} \tag{4.18}$$

What now remains is to obtain quantitative estimates of the average size of pressure spots on a boundary and the average spacing between pressure spots. Research by Emmerling (1973); Kim et al., (1971); and Willmarth and Lu (1972) indicate the following variation in dimensions of the size of pressure spots in rough turbulent flow:

$$20\frac{\nu}{u_*} \leq \xi_1 \leq 40\frac{\nu}{u_*} \tag{4.19}$$

and

$$7\frac{\nu}{u_*} \leq \xi_3 \leq 30\frac{\nu}{u_*} \tag{4.20}$$

These expressions are quite useful because they provide the dimensions of the pressure spots as a function of the wall layer thickness, i.e., $\delta = \nu/u_*$ (see Chap. 5). The thickness of the viscous sublayer right at the boundary is about 5δ, whereas the thickness of the near-bed region (the combination of the viscous sub-layer and the buffer layer) is about 70δ (see Chap. 5). The average length of the pressure spots in the direction of flow is equal to roughly about one-half the thickness of the near-bed region. In the transverse direction it is about one-tenth to about one-half of the of the near-bed region thickness.

The distances between pressure spots for rough turbulent flow have been estimated by Hinze (1975):

$$440\frac{\nu}{u_*} \leq \lambda_1 \leq 640\frac{\nu}{u_*} \quad \text{and} \quad 72\frac{\nu}{u_*} \leq \lambda_3 \leq 200\frac{\nu}{u_*} \tag{4.21}$$

This indicates that the spacing between pressure spots in the direction of flow is roughly about 6 to 9 times the thickness of the near-bed region, while the spacing in the transverse direction is roughly 1 to 3 times the thickness of the near-bed region.

These dimensions allow calculation of the average concentration of pressure spots on the boundary of rough turbulent flow, i.e.,

$$0.001 \leq [F] \leq 0.038 \tag{4.22}$$

Croad (1981) assumed an average value of $[F] = 0.01$ in his work.

The soil module is expressed in terms of mass per unit area. In order to derive a relationship representing the concentration of the soil module it is necessary to first understand how clay erodes, which can be determined by observation. It has been reported by many researchers that clay subject to rough turbulent flow fails in the form of clods of material that are "plucked" from the boundary (e.g., Briaud et al., 2001; Croad 1981; Tan 1983).

Figure 4.5 shows a positive pressure pulse first impacting the boundary, causing an increase in the pore pressures within the soil. As the positive pressure pulse is replaced by a negative pressure fluctuation the high pore pressure within the soil does not reduce immediately and causes an uplift pressure within the soil, enhanced by the effects of the negative pressure pulse moving over it. Repeated fluctuating pressure reversal can break some of the fixed bonds between clay elements, eventually leading to fatigue failure of the clay. It commences with a small rupture within the clay that grows and, once complete, leads to the removal of a clod of clay from the boundary.

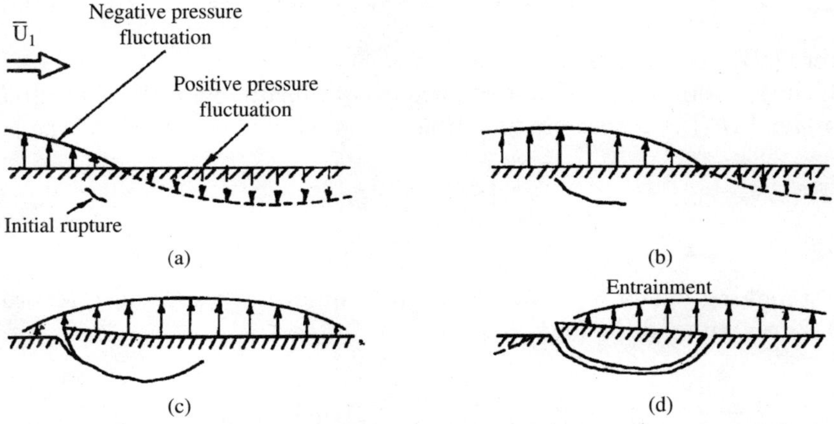

Figure 4.5 Removal of a clod of clay by the action of pressure fluctuation (Croad 1981).

The erosion process in soft clays is not as clear to observe as the edges on the boundary, remaining after the clods have been removed, are more rounded. However, the process, in essence, remains the same. Therefore, if the void ratio of the soil (i.e., the ratio between the volume of voids and the volume of solids of the soil) is represented by V', then the mass of a clod of soil with average dimensions of a_1 by a_2 by a_3 can be expressed as

$$m_s = \frac{\rho_s(a_1 a_2 a_3)}{(1+V')} \tag{4.23}$$

The concentration of the soil module S along the surface boundary, i.e., the mass per unit area of the soil, can be expressed as

$$[S] = \frac{\rho_s(a_1 a_2 a_3)}{(1+V')}\frac{1}{(a_1 a_2)} = \frac{\rho_s a_3}{(1+V')} = \frac{\rho_s a}{(1+V')} \tag{4.24}$$

where ρ_s is the mass density of the soil and a is the average thickness of the soil clod.

Under the action of negative boundary pressure fluctuation a soil module can acquire sufficient energy to surmount the energy barrier that bonds the clay elements along the lower surface of the clod to the rest of the clay body. Once the activation energy has been exceeded, the combined fluid and soil modules are said to be an activated complex. Once activated, the activated complex might become entrained into the main body of the turbulent flow. The formulation expressing the transition state is expressed as follows:

$$[S] + [F] \rightarrow [SF^*] \rightarrow \text{erosion products} \tag{4.25}$$

where $[SF^*]$ is the activated complex.

If the reactants, i.e., $[S]$ and $[F]$ are in equilibrium with the activated complex $[SF^*]$ it is possible to define an equilibrium constant K^*, i.e.,

$$\frac{[SF^*]}{[S][F]} = K^* \tag{4.26}$$

By making use of a standard result is quantum mechanics, Moore (1972) rearranged Eq. (4.26) as

$$[SF^*] = [S][F]\frac{Z'_*}{Z'_S Z'_F}\exp\left(-\frac{E_a}{RT}\right) = [S][F]P^*\exp\left(-\frac{E_a}{RT}\right) \tag{4.27}$$

where P^* = partition coefficient ratio
Z' = partition functions per unit volume with the subscripts referring to the different species
E_a = activation energy
R = universal gas constants
T = absolute temperature (measured in Kelvin)

A different way to express the rate of erosion of clay is

$$e = \vartheta^*[SF^*] \tag{4.28}$$

where ϑ^* is the frequency of passage of activated complexes over the energy barrier (i.e., in excess of the activation energy), which can also be written as

$$\vartheta^* = \frac{1}{T_B} \tag{4.29}$$

where T_B is the period of turbulent bursts.

When relating the burst period to the outer layer dimensions, it is found that (Kim et al., 1971; Croad 1981)

$$\frac{\bar{u} \cdot \bar{T}_B}{\delta_b} = 5 \tag{4.30}$$

where δ_b = boundary layer thickness (which is equal to the depth of flow in fully developed turbulent open channel flow)
$\bar{u}$ = average flow velocity
$\bar{T}_B$ = average turbulent burst period

The constant value of Eq. (4.30) varies between 3 and 7, with an average value of 5.

Using Eqs. (4.27), (4.28), and Eq. (4.17), it can be shown that

$$k = \vartheta^* P^* \exp\left(\frac{-E_a}{RT}\right) \tag{4.31}$$

The frequency of passage of activated complexes is related to the frequency of turbulent bursts, as characterized by the formation and destruction of hairpin vortices associated with the instabilities of the laminar sublayer (see Chap. 3).

In order to relate the forces associated with the fixed bonds to the rate of erosion, consider the characteristics of the activation energy required

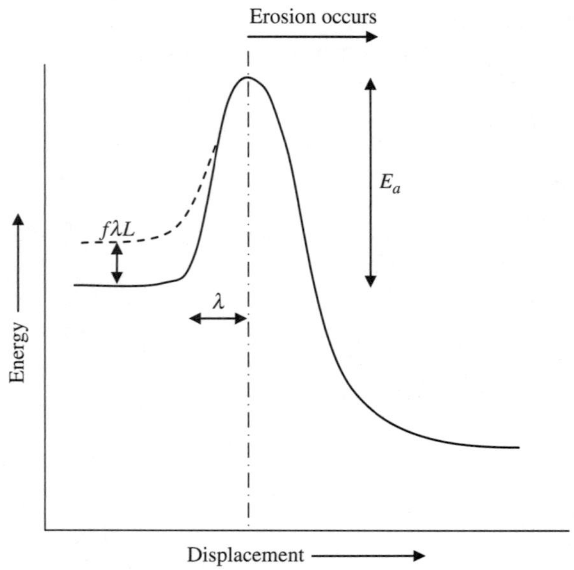

Figure 4.6 Energy profile for erosion of chemical gels (such as cohesive soils).

to break the bonds. Consider the relationship between energy and displacement characteristic of chemical reactions and, as such, the erosion of chemical gels (Fig. 4.6). As the fluctuating pressures move across a specific area on the boundary the negative pressures attempts to move the particles through a distance λ, known as the displacement distance. Once the displacement is large enough the fixed chemical bonds are broken and the individual clay elements are released. When this occurs, active erosion of the chemical gel occurs.

The energy imparted to the cohesive soil reduces the total magnitude of the activation energy every time a pressure fluctuation moves over the boundary. Recalling that energy is equal to the change in work, one can quantify the reduction in activation energy when a force f moves an element through a distance λ as $f\lambda$. The energy imparted to the particles per mole of elements can therefore be expressed by multiplying the force per element with the Avogadro number, i.e.,

$$E = f\lambda L \tag{4.32}$$

When accounting for the effect of negative fluctuating pressures, the rate constant k can be rewritten as

$$k = \vartheta^* P^* \exp\left(\frac{-E_a}{RT} + \frac{f\lambda L}{RT}\right)$$

Using Eqs. (4.17), (4.24), and (4.22) an equation for the rate of erosion of cohesive soils can be written as

$$e = k[S][F] = \vartheta^* P^* \exp\left(\frac{-E_a}{RT} + \frac{f\lambda L}{RT}\right) \cdot \frac{\rho_s a}{(1+V')}(0.01) \tag{4.33}$$

(In the above equation the assumption is made that $F = 0.01$, as proposed by Croad (1981). It could also be assumed to vary between 0.001 and 0.038).

Should one assume that the clod of clay removed from the boundary roughly represents a half sphere, the value of a is equal to about one-half the average diameter of the surface expression of the clod of clay (Fig. 4.7). It is also reasonable to expect that the expression of the failure cracks on the surface should at least follow the boundaries between the positive and negative pressures. Obviously the pressure fluctuations do not occur in the same place every time, but moves randomly around over the boundary.

Croad (1981) observed crack boundaries enclosed by areas with diameters ranging between $50\nu/u_* = 50\delta$ and 70δ for flows with shear velocities varying between $u_* = 0.032$ and 0.100 m/s. This dimension will vary, depending on the turbulence structure and intensity close to the boundary. For the time being, assume that the value of 70δ is representative. This means that the average depth of a typical soil module is approximately

$$a = 35\frac{\nu}{u_*} = 35\delta \tag{4.34}$$

The soil module concentration can therefore be expressed as

$$[S] = \frac{35\rho_s}{(1+V')} \cdot \frac{\nu}{u_*} \tag{4.35}$$

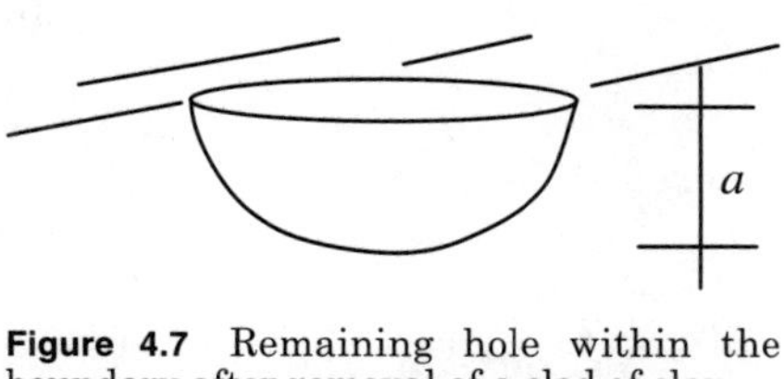

Figure 4.7 Remaining hole within the boundary after removal of a clod of clay.

Inserting this value for [S] into Eq. (4.33) one finds

$$\dot{e} = k[S][F] = \vartheta^* \frac{0.35\rho_s}{(1+V')} \cdot \frac{\nu}{u_*} \cdot P \exp\left(\frac{-E_a}{RT} + \frac{f\lambda L}{RT}\right) \tag{4.36}$$

Additionally it is also possible to derive an expression for the force per bond f. If the negative pressure acting on a clod of cohesive soil is p, then the force per bond can be expressed as

$$f = \frac{p(2a)^2\pi/4}{n_B(2a)^2\pi/2} = \frac{p}{2n_B} \tag{4.37}$$

where n_B is the number of bonds per unit area. [$\pi(2a)^2/2$ = area along the surface of the half spherical shape of the clod; $(2a)^2\pi/4$ = area over which the negative pressure p acts.]

Hinze (1975) found that the average fluctuating pressure can be expressed as a function of the turbulent boundary shear stress τ_t, i.e.,

$$p' = 3\tau_t \tag{4.38}$$

where p' is the fluctuating pressure, also known as the root mean square (rms) value of the fluctuating pressure.

The fact that the fluctuating pressure can be written as a function of the turbulent boundary shear stress *does not mean that the erosion results because of a shear process*. It is merely a convenient way to provide an idea of the relative magnitude of the average fluctuating pressure in rough turbulent flow. It is also known that the fluctuating pressure can be as high as $6p'$ (Emmerling 1973). It is therefore appropriate to express the fluctuating pressure as

$$p' = k'\tau_t \tag{4.39}$$

where $3 \le k' \le 18$ is a magnification coefficient. The value most often used to express extreme pressure fluctuation in open channel flow is $k' = 18$.

The force per bond is expressed as

$$f = \frac{p'}{2n_b} = \frac{k'\tau_t}{2n_b} \tag{4.40}$$

and the rate of erosion equation can therefore be written as

$$e = k[S][F] = \vartheta^* \frac{0.35\rho_s}{(1+V')} \cdot \frac{\nu}{u_*} \cdot P^* \exp\left(\frac{-E_a}{RT} + \frac{k'\lambda L\tau_t}{2n_B RT}\right) \tag{4.41}$$

In Chap. 5 it is shown that

$$\frac{\tau_t}{\rho u_*^2} \cdot \frac{\bar{u}}{u_*} = 7.853 \tag{4.42}$$

from which follows

$$\tau_t = 7.853\left(\tau \cdot \frac{u_*}{\bar{u}}\right) \tag{4.43}$$

where $\bar{u}$ = average flow velocity

τ = tractive shear stress; also sometimes referred to as the wall shear stress (= ρgys in the case of open channel flow)

y = flow depth in a wide open channel

s = energy slope

The erosion rate equation can therefore be expressed in terms of the tractive shear stress, i.e.,

$$e = k[S][F] = \vartheta^* \frac{0.35\rho_s}{(1+V')} \cdot \frac{\nu}{u_*} \cdot P^* \exp\left(\frac{-E_a}{RT} + \frac{3.9265k'\tau\lambda L u_*}{n_B RT\bar{u}}\right) \tag{4.44}$$

It can also be written in terms of the applied stream power at the boundary. Using Eq. (4.42) it can be shown that the applied stream power at the boundary p_{applied} is expressed as (also see Chap. 5)

$$P_{\text{applied}} = \tau_t \bar{u} = 7.853 \frac{\tau^{3/2}}{\sqrt{\rho}} \tag{4.45}$$

Equation (4.41) can therefore be written in terms of the applied stream power as

$$e = k[S][F] = \vartheta^* \frac{0.35\rho_s}{(1+V')} \cdot \frac{\nu}{u_*} \cdot P^* \exp\left(\frac{-E_a}{RT} + \frac{k'\lambda L}{2n_B RT\bar{u}}(P_{\text{applied}})\right) \tag{4.46}$$

Characteristics of clay erosion. Equations (4.41), (4.44), and (4.46) relate the variables that affect the erosion of clays. This section illustrates some of the relationships between erosion rate, temperature, pH, and the erosive capacity of water. These relationships were prepared by rewriting Eq. (4.41) in the form

$$e = \gamma \cdot \left(\frac{\nu}{u_*} \frac{\bar{u}}{\delta_b} \right) \cdot \exp\left(\frac{-E_a}{RT} + \frac{\beta \tau_t}{T} \right) \tag{4.47}$$

where $\gamma = 0.35\rho_s P^*/5(1 + V')$ (Recall that $\vartheta^* = 1/T_b$ and $\bar{T}_b = 5\delta_b/\bar{u}$).

The turbulent boundary shear stress τ_t for the Couette Flow Device (see Chap. 8), which is used to measure the rate of erosion of clay, was calibrated by Croad (1981), who expressed it as a function of the average flow velocity of the fluid in the device, i.e.,

$$\tau_t = \begin{cases} 0.413 \cdot \rho \bar{u}^2 \cdot \left(\dfrac{\bar{u}\delta_b}{\nu} \right)^{-1} & \text{if} \quad \dfrac{\bar{u}\delta_b}{\nu} < 2203 \\ 4 \cdot 10^{-6} \cdot \rho \bar{u}^2 \cdot \left(\dfrac{\bar{u}\delta_b}{\nu} \right)^{0.5} & \text{if} \quad 2203 < \dfrac{\bar{u}\delta_b}{\nu} < 19953 \\ 0.01102 \cdot \rho \bar{u}^2 \cdot \left(\dfrac{\bar{u}\delta_b}{\nu} \right)^{-0.3} & \text{if} \quad 19953 < \dfrac{\bar{u}\delta_b}{\nu} < 5 \cdot 10^4 \\ 4.29 \cdot 10^{-4} \cdot \rho \bar{u}^2 & \text{if} \quad \dfrac{\bar{u}\delta_b}{\nu} > 5 \cdot 10^4 \end{cases} \tag{4.48}$$

The term $\bar{u}\delta_b/\nu$ is an expression for the Reynolds number in the device, where δ_b = distance between the sample and the outer boundary of the device. The four categories of Reynolds number given above represent the following:

$$\frac{\bar{u}\delta_b}{\nu} < 2203 \text{ laminar flow}$$

$$2203 < \frac{\bar{u}\delta_b}{\nu} < 19953 \text{ critical flow}$$

$$19953 < \frac{\bar{u}\delta_b}{\nu} < 5 \times 10^4 \text{ transitional}$$

$$\frac{\bar{u}\delta_b}{\nu} > 5 \times 10^4 \text{ rough turbulent}$$

The shear velocity is expressed as

$$u_* = \sqrt{\frac{\tau_t}{\rho}}$$

Figure 4.8, developed from Eqs. (4.47) and (4.48) shows a relationship between flow velocity, absolute temperature, and erosion rate of clay in a Couette flow device. It illustrates that the temperature can have a more significant impact on the rate of erosion than the erosive capacity of the water. This happens because the chemical bonds between the clay particles are affected by temperature.

However, the surface shown in Fig. 4.8 is not a generic relationship. Varying the values of the variables in the erosion rate equation, the three-dimensional relationship in Fig. 4.8 changes. For example, Fig. 4.9 shows the relationship between erosion rate, temperature, and flow velocity for a different set of parameters. The surface has a completely different form, and if a relationship between erosion rate and shear stress is drawn for a temperature of 273 K it is found to be concave (Fig. 4.10).

Figure 4.11 shows the erosion rate relationships for yet another set of variables. It shows that the relationship between erosion rate and the erosive capacity of the water is convex for this particular case.

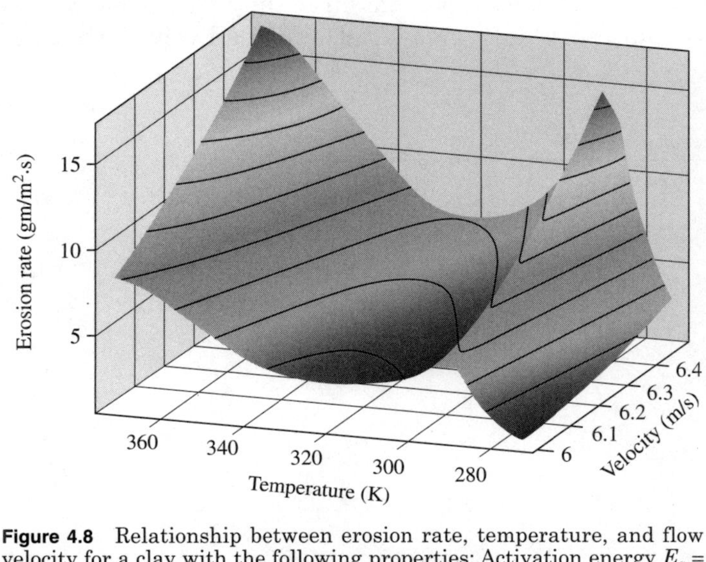

Figure 4.8 Relationship between erosion rate, temperature, and flow velocity for a clay with the following properties: Activation energy E_a = 50 kJ/mol; δ_b = 20 mm; β = 100 K/Pa; $\gamma = 2 \times 10^8$ g/m^2.

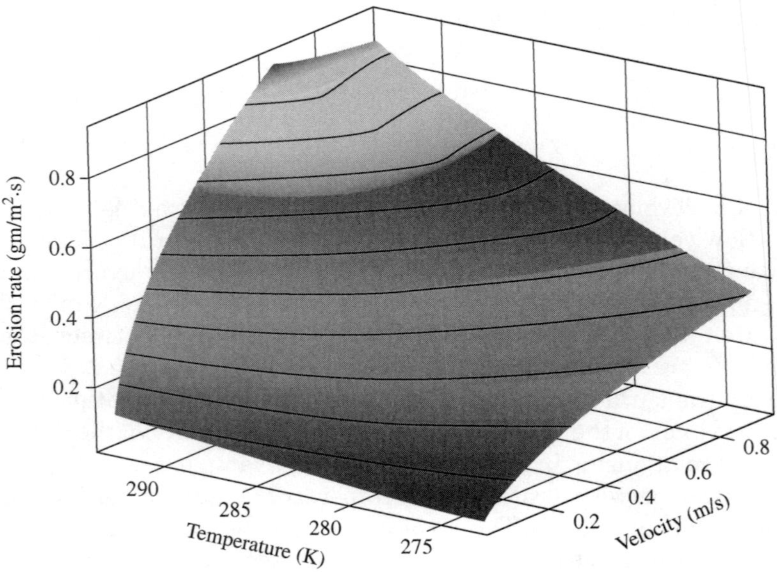

Figure 4.9 Relationship between erosion rate, temperature, and flow velocity for activation energy $E_a = 50$ kJ/mol; $\beta = 60$ K/Pa; $\delta_b = 20$ mm; $\gamma = 0.05 \times 10^8$ g/m^2.

Figure 4.12 shows the convex relationship between shear stress and erosion rate for a temperature of 273 K.

Figures 4.10 and 4.12 have shapes similar to the concave and convex shapes of the laboratory data measured by Croad (1981) that are

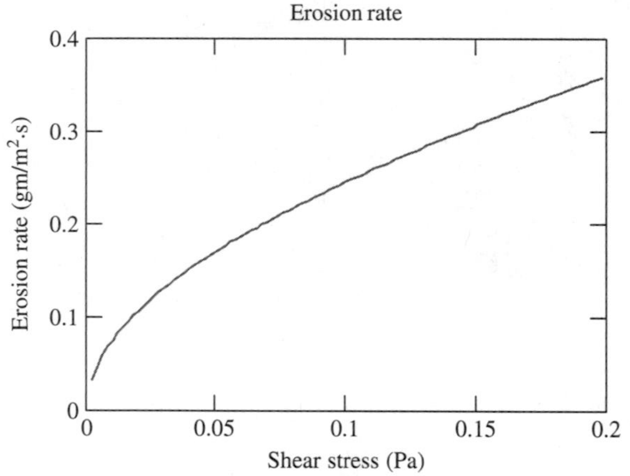

Figure 4.10 Concave relationship between erosion rate for a temperature of 273 K and for activation energy $E_a = 50$ kJ/mol; $\beta = 60$ K/Pa; $\delta_b = 20$ mm; $\gamma = 0.05 \times 10^8$ g/m^2.

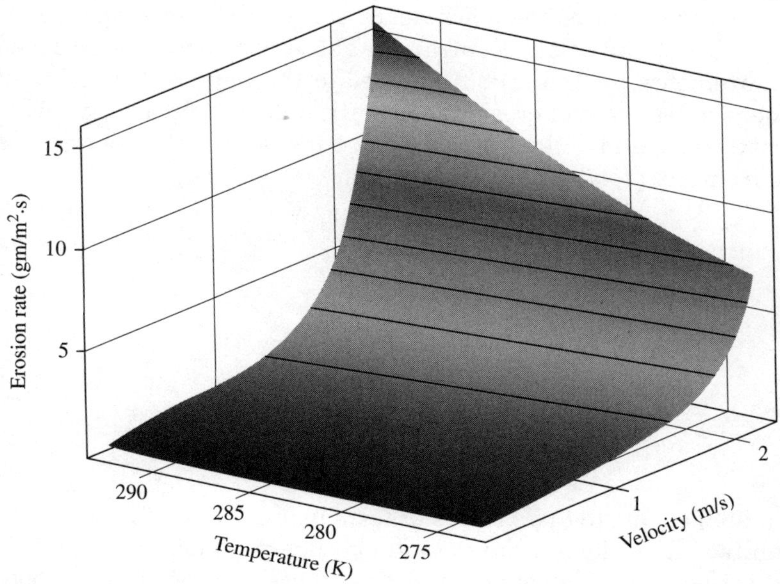

Figure 4.11 Concave relationship between turbulent shear stress and erosion rate activation energy $E_a = 30$ kJ/mol; $\beta = 500$ K/Pa; $\delta_b = 20$ mm; $\gamma = 0.05 \times 10^8$ g/m^2.

shown in Fig. 4.3. Although these curves were not fitted to the data, it does illustrate that the shape of the relationship between erosion rate and shear stress is a function of the inherent material properties of clays.

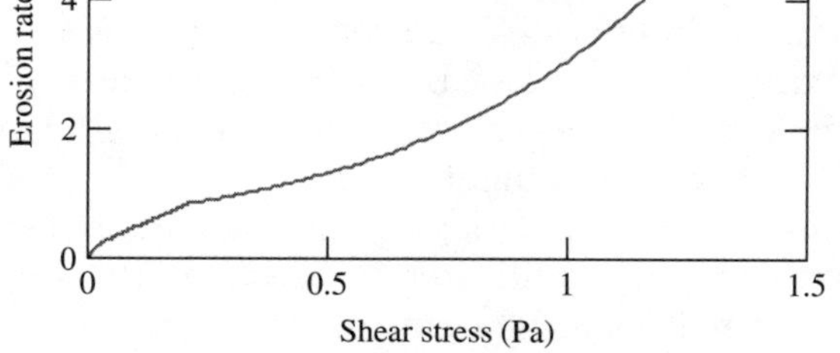

Figure 4.12 Convex relationship between erosion rate and shear stress for a temperature of 273 K and activation energy $E_a = 30$ kJ/mol; $\beta = 500$ K/Pa; $\delta_b = 20$ mm; $\gamma = 0.05 \times 10^8$ g/m^2.

Figure 4.13 illustrates the dependence of the rate of erosion of clays to changes in pH and salt concentration. These relationships are based on measurements by Tan (1983) and relate the natural logarithm of the dimensionless rate of erosion to the pH and the salt concentration of the interstitial and ambient water. The dimensionless erosion rate is derived from Eq. (4.47) and is expressed as

$$E = \frac{e}{A_1} \tag{4.49}$$

where

$$A_1 = \frac{0.35\rho_s}{5(1+V')} \frac{\nu}{u_*} \frac{\bar{u}}{\delta_b} \tag{4.50}$$

Tan (1983) conducted two series of tests using three different clays, a bentonite and two kaolin clays (kaolin koclay and kaolin ball clay). In the first series of tests, he maintained a salt concentration of zero in both the pore- and ambient water, while varying the pH between 3 and 11 using HCl to control the latter. In the other series of tests he maintained the pH at a value of 7.0, while varying the salt concentration of the pore- and ambient water between 10^{-1} and 10^{-5} mole of sodium chloride per liter of water. In all cases, he varied the shear stress in a Couette flow device between very small values (close to zero) to a maximum value of 8 Pa.

Figure 4.13 indicates that the rate of erosion of clay can be more sensitive to the variation of pH than it is to the variation in shear stress. It is also seen that the erosion commences almost immediately as soon as shear stress is applied to the clay. This implies that the clay might not have a critical shear stress below which erosion does not occur. In all the cases tested, it also appears as if the relationship between the rate of erosion and pH is concave in nature. The value of the pH where the maximum rate of erosion occurs differs for the three clays tested. In the case of the bentonite clay the maximum rate of erosion appears to be associated with a pH of approximately 5. In the case of the kaolin koclay it occurs at a pH of approximately 9, while in the case of the kaolin ball clay it appears to occur at a pH of approximately 7.

Inspection of the response of erosion of clay to varying salt concentration of the pore- and ambient water indicates a convex relationship. It is also concluded that the sensitivity of the rate of erosion to changes in salt concentration is greater than what it is to variation of shear stress. In the case of the bentonite clay, the minimum rate of erosion occurs at a salt concentration of about 10^{-3} mole of sodium chloride per

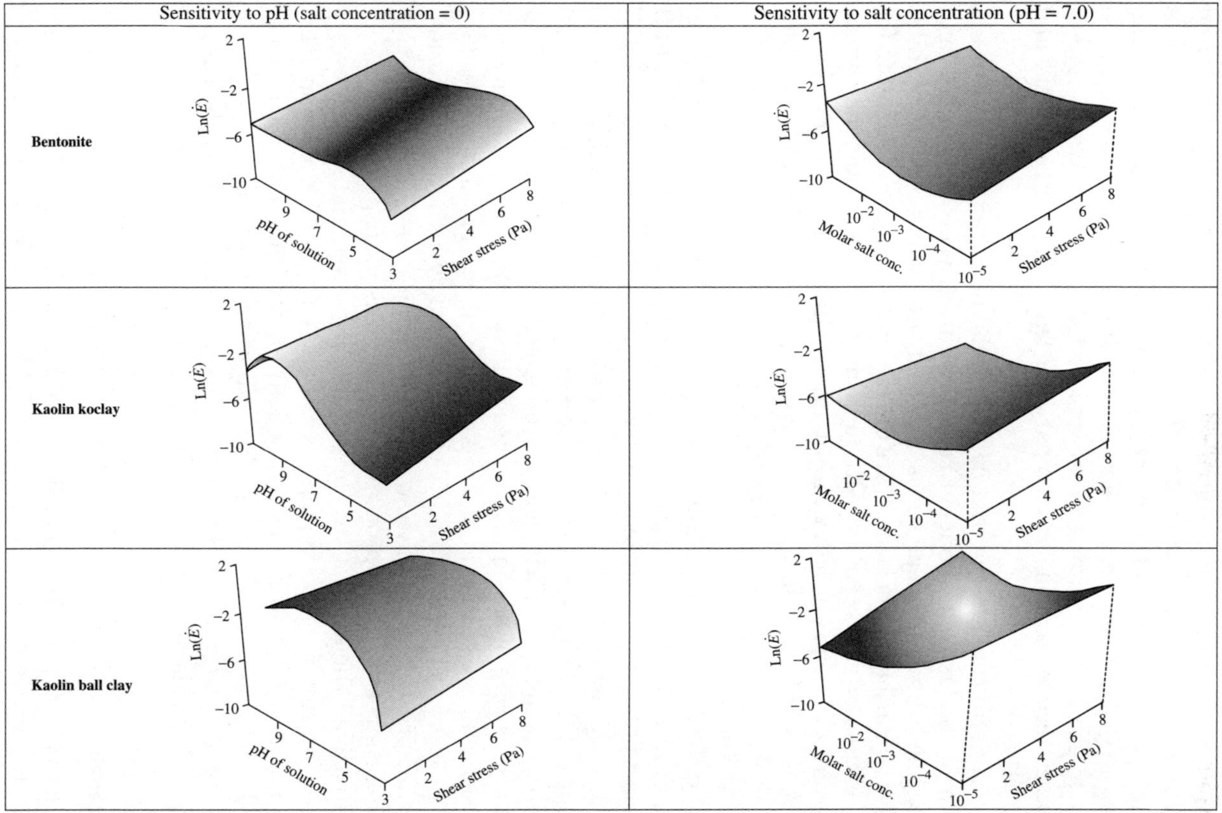

Figure 4.13 Relationships between dimensionless erosion rate, pH, salinity, and shear stress for three clays (based on data by Tan 1983).

liter of water, while the maximum erosion rate occurs at the maximum salt concentration. In the cases of the kaolin clays the maximum rates of erosion occur at the minimum salt concentrations.

Tan (1983) investigated the structure of the clays for varying pH and salt concentration and concluded that changes in these parameters affected the relative positioning of individual clay platelets. When the lower rates of erosion were observed, the clay platelets appeared to be closer to each other, leading to increases in the Van der Waal's forces. This increased attraction between particles lead to higher resistance and lower rates of erosion.

The general conclusions that can be made indicate that clays would most probably exhibit a concave relationship between erosion rate and pH for fixed values of shear stress, and that they would probably exhibit a convex relationship between rate of erosion and salt concentration for fixed values of shear stress. Although no general conclusions can be made as to what pH or salt concentration will result in the maximum rate of erosion for all clays it can be concluded that these two variables most probably have greater impact on erosion rates of clay than what shear stress has.

The differences between the graphs indicate that site specific testing is required when considering erosion characteristics of clay, and that it is important to have knowledge of the pH and salt concentration anticipated at the project under consideration. It is also concluded that the concept of a threshold shear stress or stream power possibly does not exist for clays. An appropriate measure of erosion resistance for clays is most probably its rate of erosion. This means that clays that are highly erosion resistant are characterized by very low rates of erosion, while clays that are less erosion resistant are characterized by high rates of erosion.

From a practical point of view, in projects where the concept of threshold shear stress or stream power is useful, it is most probably reasonable to assign the shear stress or stream power associated with a very low erosion rate for a particular clay as its "threshold" condition. This assignment will obviously be dependent on the judgment of the individual conducting the tests if such an approach is selected.

The reader is reminded that practical approaches to apply the theory presented in this section can be found in Chap. 8.

Intact rock

Fracture mechanics. In general, materials can fail in either a plastic mode or a brittle mode. When a material fails in plastic mode it is predominantly characterized by yielding. Failure in a brittle mode is subject to fracture. Failure of an intact rock mass during the scour process is principally characterized by brittle fracture or fatigue failure. Fatigue failure occurs when a cyclic loading is applied long enough to result in a slow growth of cracks that eventually leads to fracturing of the material.

The first successful analysis of a fracture problem was executed by Griffith in the 1920s when he investigated the propagation of brittle cracks in glass. He used an energy approach based on the concept that a crack will propagate if it results in the lowering of the total energy of the system. Application of the energy-based approach encountered some difficulties in practice and the general approach changed to the use of stress intensity, mainly because of the work of Irwin in the 1940s and 1950s. He was able to show that his stress intensity approach was similar to the energy approach, with the added advantage that it was more oriented toward practice. The approach to brittle fracture and fatigue failure used in this book is based on the stress intensity approach and the reader who wishes to obtain more in-depth information is referred to Ewalds and Wanhill (1989) for an introduction to the topic.

The theory of fracture mechanics is not repeated here. The approach that is followed is that stress intensity is viewed as an indicator parameter that provides an indication of the relative magnitude of the forces present at the tip of a crack. (This is similar to the use of shear stress or steam power as indicator parameters of the relative magnitude of the erosive capacity of water).

From the strength of materials point of view, a concept of fracture toughness is used as an indicator parameter of the relative ability of the material to resist the stress intensity at the tip of a crack. Simple, approximate methods for quantifying both the magnitude of the stress intensity at the tip of a crack and the magnitude of the fracture toughness of the material are presented. It is important for the reader to recognize that the state of the art when using brittle fracture or fatigue failure concepts to investigate the potential for rock scour is not well advanced at this stage. Although application of the concepts has been shown to lead to apparently realistic results, it requires making substantial and significant assumptions. Enough data are often not available to fully quantify the fracture toughness of the material, nor the stress intensity that can develop in rock due to the presence of fluctuating turbulent pressures.

Nevertheless, the approaches that are presented here to quantify fracture toughness and stress intensities have been used successfully in practice (see, e.g., Annandale and Bollaert 2002). The method has been applied to case studies and also compares favorably with Annandale's (1995) erodibility index method (Bollaert 2002).

Brittle fracture of rock occurs when the stress intensity K in a close-ended fissure exceeds the fracture toughness K_I of the rock, i.e.,

$$K \geq K_I \tag{4.51}$$

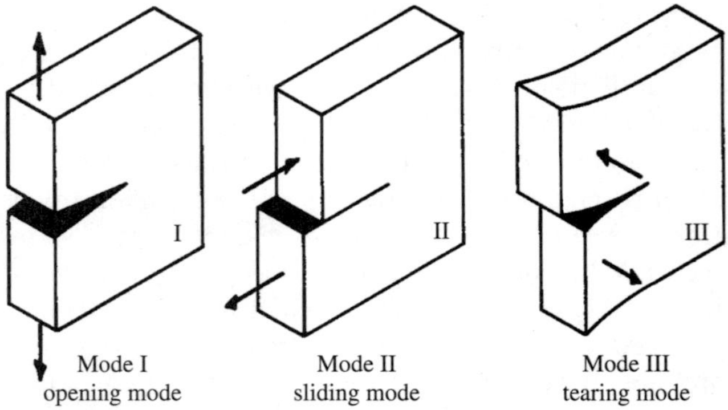

Figure 4.14 Loading modes in fracture mechanics.

Stress intensity *K*. Calculation of the magnitude of stress intensity within a fissure could have been covered either in the chapter dealing with hydraulics (Chap. 5) or in this chapter. The reason for this is that stress intensity is not only a function of the pressure in the fissure, but also depends on the fissure geometry, which is a material property.

Three loading modes are possible when considering fracture: an opening mode, sliding mode, and a tearing mode (Fig. 4.14). Development of an approach to calculate the stress intensity in rock scour assumes that the rock will fail in pure tension (i.e., mode I), and that the rock is characterized by a homogeneous, linear elastic, isotropic, and impermeable medium. The assumption of loading mode I is reasonable because the turbulence in the flowing water introduces fluctuating pressures in the close-ended fissures in the rock, forcing them to fail in the opening mode. The assumptions regarding the material are most probably also reasonable when the stress zone at the tip of the crack is small relative to the crack dimension and the overall dimension of a block of rock.

By making these assumptions the following equation for calculating stress intensity has been developed (Atkinson 1987),

$$K = \sigma_{\text{water}} \cdot \sqrt{\pi a} f \tag{4.52}$$

where K = stress intensity (MPa $\sqrt{\text{m}}$)
a = crack length (m)
σ_{water} = water pressure in the close ended crack (MPa)
f = a function that accounts for the geometry of the rock block and its crack extension, the loading conditions and the edge effects

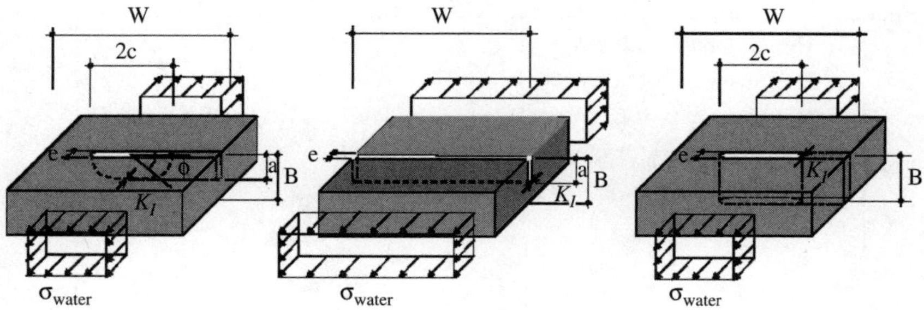

Figure 4.15 Close-ended fissure-representative geometries, from left to right: Semi-elliptical, single edge, and center-cracked fissures (Bollaert 2002).

The values of f can be estimated for three fissure conditions that could occur in rock (Fig. 4.15), as follows:

- Semi-elliptical fissure

$$f\left(\frac{a}{B},\frac{a}{c},\phi\right) = C\cdot\frac{\left(\sin^2\phi+\frac{a^2}{c^2}\cdot\cos^2\phi\right)^{1/4}}{\frac{3\cdot\pi}{8}+\frac{\pi}{8}\cdot\frac{a^2}{c^2}} \tag{4.53}$$

- Single-edge fissure

$$f\left(\frac{a}{B}\right) = \left[1.12-0.231\left(\frac{a}{B}\right)+10.55\left(\frac{a}{B}\right)^2-21.72\left(\frac{a}{B}\right)^3+30.39\left(\frac{a}{B}\right)^4\right] \tag{4.54}$$

- Center-cracked fissure

$$f\left(\frac{c}{W}\right) = 1+0.256\left(\frac{c}{W}\right)-1.152\left(\frac{c}{W}\right)^2+12.2\left(\frac{c}{W}\right)^3 \tag{4.55}$$

The edge conditions of fissures in rock are rarely known when conducting scour analysis, and one way to approach quantification of the value of f is to select values that are deemed prudent and representative. In order to assist in such selection Bollaert (2002) plotted potential values of f as a function of a/B or c/W, where B is the thickness of the rock, W is the rock width, and c is the half-width of the fissure on

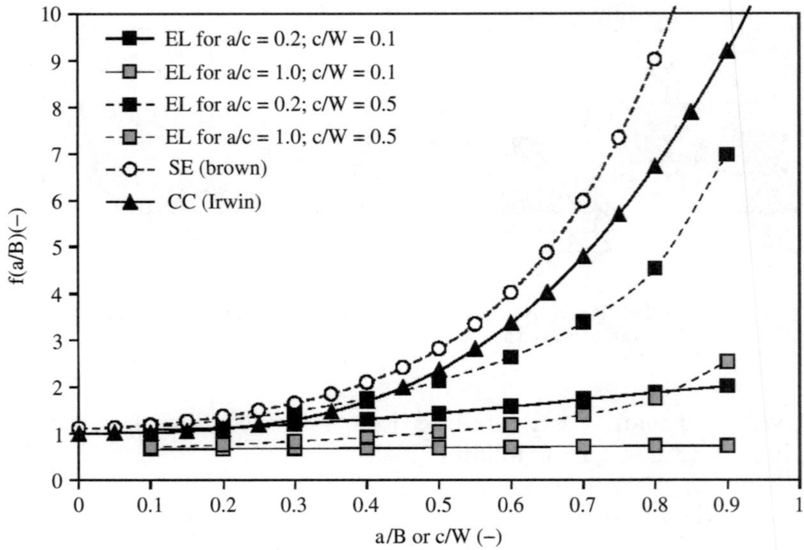

Figure 4.16 Comparative values of the factor f for elliptical (EL), single edge (SE), and center-cracked (CC) fissures (Bollaert 2002).

the surface of the rock. From this graph (Fig. 4.16) it can be seen that reasonable values of *f* most probably range between 1 and 3; and conservative values would hardly be much higher than 10.

In order to fully quantify the magnitude of the stress intensity factor it is necessary to estimate the magnitude of the stress caused by water pressure inside a fissure. By assuming that the distribution of instantaneous dynamic pressure within a fissure is sinusoidal, and if the pressure at the entrance to a fissure is represented by the symbol p_0, and that the maximum pressure at the end of the fissure is represented by the symbol $p_{\max}$ (Fig. 4.17), the variation of pressure in a fissure can be expressed as

$$p(x) = p_0 + (p_{\max} - p_0) \cdot \sin\left(\frac{\pi}{2} \cdot \frac{x}{a}\right) \tag{4.56}$$

where a is the length of the close-ended fissure (m) and x is the variable distance along the fissure, from the opening to the close-ended side (m).

By setting $x = a$, i.e., the length L, it is possible to calculate the average pressure in the fissure (Bollaert 2002),

$$p_{\text{average}} = 0.36p_0 + 0.64p_{\max} \tag{4.57}$$

For calculating the stress intensity factor, set the value of $\sigma_{\text{water}} = \sigma_{\text{average}}$ in Eq. (4.52).

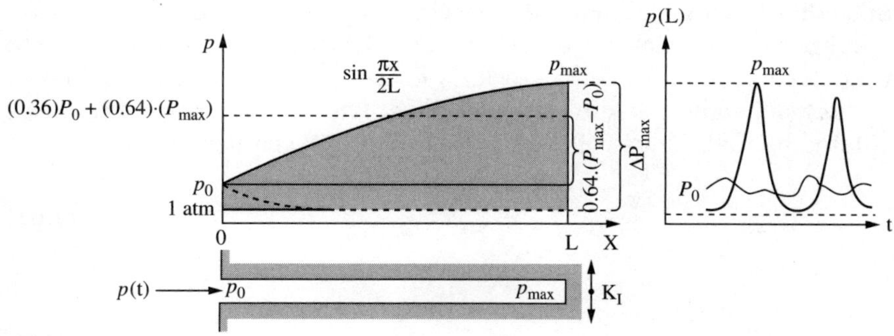

Figure 4.17 Pressure distribution of a sinusoidal pressure wave in a close-ended fissure (Bollaert 2002).

Fracture Toughness K_I. Atkinson (1987) published the results of fracture toughness tests on rock performed by a number of researchers. Estimates of the fracture toughness of rock can be made by referring to these tables, or by using regression functions developed by Bollaert (2002) that are based on these data. Either of the following two equations can be used (Bollaert 2002):

$$K_{I,\ insitu,T} = (0.105 \text{ to } 0.132) \cdot T + (0.054\sigma_i) + 0.5276 \qquad (4.58)$$

$$K_{I,\ insitu,UCS} = (0.008 \text{ to } 0.010) \cdot UCS + (0.054\sigma_i) + 0.42 \qquad (4.59)$$

where T = tensile strength of the rock (MPa)
UCS = unconfined compressive strength of the rock (MPa)
σ_i = confining stresses in the rock (MPa)

The confining stress in the rock is often assumed to be zero. This is most probably a reasonable assumption because scour of the rock occurs once it is exposed to the surface. When this happens, the confining stresses in the rock are most probably relaxed, and approximately equal to zero.

Brittle fracture. Brittle fracture of the rock will occur when the stress intensity in the close-ended rock fissure is equal to or greater than the fracture toughness of the rock, i.e.,

$$K \geq K_I \qquad (4.60)$$

When the rock breaks up in brittle fracture, it does so in an explosive manner. In cases when the stress intensity is less than the fracture toughness of the rock the rock could potentially fail in fatigue, also known as sub-critical failure.

Sub-critical failure. When the stress intensity is lower than the fracture toughness of the rock, repeated cyclic application of stress could lead to time-dependent failure of the rock, by fatigue. The failure mode is known as time-dependent or sub-critical failure. The time to failure can be calculated by making use of the Paris et al., (1961) equation, i.e.,

$$\frac{da}{dN} = C \cdot (\Delta K)^m \tag{4.61}$$

where a = crack length
N = number of cycles
C and m = rock material parameters that can be determined by experiment
ΔK = the difference between the maximum and minimum stress intensity factors at the tip of the crack

The maximum stress intensity is associated with the value of K determined with the value of p_{average} using Eqs. (4.52) and (4.57). The minimum value of the stress intensity factor is assumed to equal zero.

The values of C and m depend on rock type and quality. Typical values would be on the order of $C = 10^{-8}$ to 10^{-10} and $m = 8$ to 10. Test values for a number of rock types are presented in Table 4.2.

TABLE 4.2 Values of *C* and *m* for Various Rock Types

Type of rock	Fatigue exponent m	Coefficient C
Arkansas novaculite	8.5	1.0E-8
Mojave quartzite	10.2 to 12.9	3.0E-10
Ruhr sandstone	2.7 to 3.7	2.0E-6 to 1.0E-6
Tennessee sandstone	4.8	4.0E-7
Solenhofen limestone	8.8 to 9.5	1.1E-8
Carrara marble	5.1	2.5E-7
Falerans micrite	8.8	1.1E-8
St-Pons marble	8.8 to 9.9	1.1E-8 to 4.0E-9
Tennessee marble	3.1	2.0E-6
Merrivale granite	13.6 to 23.1	1.5E-10 to 4.0E-14
Westerley granite	11.8 to 11.9	8.0E-10
Yugawara andesite	8.8	1.1E-8
Black gabbro	9.9 to 12.2	4.0E-9 to 5.0E-10
Kinosaki basalt	11.2	1.2E-9
Ralston basalt	8.2	1.8E-8
Whin Sill dolerite	9.9	4.0E-9

SOURCE: From Atkinson 1987.

Empirical Characterization of Physical and Chemical Gels

The erodibility index is a geomechanical index that is used to quantify the relative ability of earth and engineered earth materials to resist the erosive capacity of water. This section presents the index and provides guidance on its use. After introducing the erodibility index its composition is explained in terms of physical and chemical gel concepts presented in Chap. 3.

Application of the erodibility index method is based on an erosion threshold that relates the relative magnitude of the erosive capacity of water and the relative ability of earth and engineered earth materials to resist scour. The correlation between stream power (P), representing the relative magnitude of the erosive capacity of water (see Chap. 5), and a mathematical function [$f(K)$] that represents an earth material's relative ability to resist erosion can, at the erosion threshold, be expressed by the relationship

$$P = f(K) \tag{4.62}$$

If $P > f(K)$, the erosion threshold is exceeded, and the earth material is expected to erode. Conversely, if $P < f(K)$, the erosion threshold is not exceeded, and the earth material is expected not to erode.

Annandale (1995) established a relationship between stream power and the erodibility index by analyzing published and field data for a wide variety of earth material types and flow conditions and found the relationship shown in Fig. 4.18. Two data types are plotted on the graph,

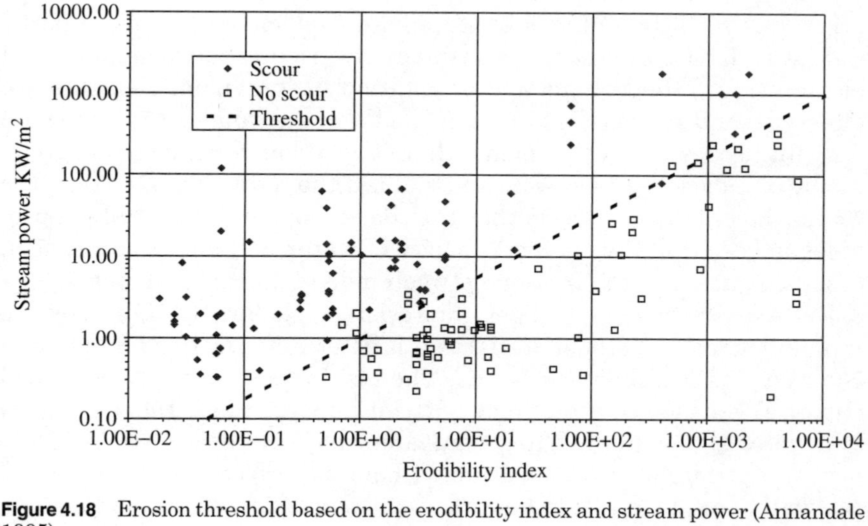

Figure 4.18 Erosion threshold based on the erodibility index and stream power (Annandale 1995).

events where scour occurred and events where scour did not occur. The dotted line is the approximate location of the erosion threshold, which indicates the separation between events that scoured and those that did not. This threshold relationship is discussed in more detail in Chap. 6 and its use demonstrated in Chaps. 7, 9, and 10.

$f(K)$, i.e., the erodibility index, is defined as

$$K = M_s \cdot K_b \cdot K_d \cdot J_s \tag{4.63}$$

where M_s = mass strength number
K_b = block size number
K_d = discontinuity bond shear strength number
J_s = relative ground structure number

Equation (4.63) was originally developed by Kirsten (1982) to characterize the excavatability of earth materials. It has also been found that this index provides a good indicator of the relative ability of earth materials to resist the erosive capacity of water (Annandale 1995).

Insight into the selection of parameters that are used to quantify the relative magnitude of the erosion resistance offered by earth and engineered earth materials can be found by referring to concepts of chemical and physical gels (Chap. 3). It is useful to recall that one of the principal differences between physical and chemical gels is how the occupied sites in their respective lattice networks are connected. In the case of physical gels, the connections exist merely because occupied sites touch each other. In the case of chemical gels, the connections are characterized as "fixed bonds," which are usually chemical bonding.

The resistance to the erosive capacity of water offered by physical gels originates with the self-weight of individual elements (sediment particles, gravel, blocks of rock, and so forth), the friction between individual elements, and their shape and orientation. If individual elements are elongated and stacked on top of one another at an angle, it will be more difficult to remove such elements than it would be to remove round elements or cubes (see Figs. 4.27 and 4.28). If the individual elements are round the only added value that the connections in a physical gel provides in resisting the erosive capacity of water is the friction between occupied sites, i.e., the friction between individual earth elements.

In the case of a chemical gel, the resistance is largely offered by the inherent strength of the material that is determined by the strength and character of the chemical bonds. The erosion resistance of a typical chemical gel such as intact rock without imperfections (the ideal case) is usually higher than that of equivalent physical gels. If water flows over the surface of a rock stratum without imperfections, there is little the water can do to scour the rock.

However, the existence of rock without discontinuities or imperfections is rare. In the best case, massive rock will contain close-ended fissures that are open to the surface of the rock. If water flows over such rock the pressure fluctuations characteristic of the turbulent flowing water will introduce fluctuating pressures into these fissures. When the fluctuating pressures become very large, it is possible that the development of stress intensities could exceed the fracture toughness of the rock, causing brittle fracture. Once brittle fracture occurs, the chemical gel is converted to a physical gel if the individual elements resulting from brittle fracture are small enough.

When developing a geomechanical index to quantify the relative ability of earth materials to resist the erosive capacity of water, such as the erodibility index, it is necessary to acknowledge that earth and engineered earth materials can be characterized as chemical or physical gels, or a combination of the two. By quantifying the erodibility index as the product of the four parameters indicated in Eq. (4.63) recognition is given to the roles of physical and chemical gel characteristics in determining the relative ability of materials to resist erosion.

The relative influence of chemical gel characteristics is represented by the mass strength number, which is directly related to the unconfined compressive strength (UCS) of the material. The UCS is representative of the chemical bonding properties of the material, as already shown.

The physical gel characteristics of the material are accounted for by parameters representing the material's block or particle size, the friction between such elements, and their orientation. These are respectively quantified by the block size number, the discontinuity bond shear strength number, and the relative ground structure number.

Mass strength number (M_s)

Rock. As far as empirical characterization of earth materials is concerned, the discussion on brittle fracture and fatigue failure presented earlier on showed that the use of UCS as a relevant indicator parameter of the relative magnitude of the mass strength of rock is appropriate. Table 4.3 contains the values of M_s for *rock.* These are related to field identification and the *UCS* of the rock, expressed in MPa. The latter can be quantified by making use of the procedures described in ASTM D-2938 (Standard Test Method for Unconfined Compressive Strength of Rock Core Specimens).

The values of M_s for rock can also be quantified by making use of the following equations:

$$M_s = C_r \cdot (0.78) \cdot (UCS)^{1.05} \qquad \text{when } UCS \leq 10 \text{ Mpa} \tag{4.64}$$

TABLE 4.3 Mass Strength Number for Rock (M_s)

Hardness	Identification in profile	Unconfined compressive strength (MPa)	Mass strength number (M_s)
Very soft rock	Material crumbles under firm (moderate) blows with sharp end of geological pick and can be peeled off with a knife; is too hard to cut tri-axial sample by hand.	Less than 1.7 1.7–3.3	0.87 1.86
Soft rock	Can just be scraped and peeled with a knife; indentations 1 mm to 3 mm show in the specimen with firm (moderate) blows of the pick point.	3.3–6.6 6.6–13.2	3.95 8.39
Hard rock	Cannot be scraped or peeled with a knife; hand-held specimen can be broken with hammer end of geological pick with a single firm (moderate) blow.	13.2–26.4	17.70
Very hard rock	Hand-held specimen breaks with hammer end of pick under more than one blow.	26.4–53.0 53.00–106.0	35.0 70.0
Extremely hard rock	Specimen requires many blows with geological pick to break through intact material.	Larger than 212.0	280.0

SOURCE: From Kirsten 1982.

and

$$M_s = C_r \cdot (UCS) \qquad \text{when } UCS > 10 \text{ MPa} \tag{4.65}$$

where C_r is a coefficient of relative density, defined as

$$C_r = \frac{g \cdot \rho_r}{27 \times 10^3} \tag{4.66}$$

where ρ_r = mass density of the rock in kg/m^3
$g = 9.82$ m/s^2, the acceleration due to gravity
27.10^3 N/m^3 = reference unit weight of rock

Weathering can impact values assigned to M_s. Exposed rock is subject to weathering during the lifetime of a project, an aspect that should be considered in analysis and design. Rock weakens as it weathers with a concomitant decrease in the value of M_s. Assignment of appropriate values of M_s to account for weathering is a matter of professional experience and judgment. It can be accomplished by either testing the strength of samples of weathered rock similar to that under consideration, or by estimating the strength reduction that could be expected, and assigning appropriate M_s values purely based on engineering judgment.

Cohesive soils. Practical experience over the last 15 years has shown that quantification of the shear strength of a cohesive material, a chemical gel, provides a reasonable indication of its inherent mass strength for the purpose of erosion assessment. Vane shear strength and field descriptions of cohesive soils can be used to quantify the value of M_s for *cohesive* soils with the aid of Table 4.4. The vane shear-strength is determined in accordance with ASTM D-2573 (Standard Test Method for Field Vane Shear Test in Cohesive Soil) or ASTM D-4648 (Standard Test Method for Laboratory Miniature Vane Shear Test for Saturated Fine-grained Clayey Soil). Estimates of the undrained shear strength of the cohesive material can also be used to estimate shear strength if vane shear-strength data are unavailable. Such estimates can be made with information obtained from the UCS test using ASTM D-2166 (Standard Test Method for Unconfined Compressive Strength for Cohesive Soil).

Non-cohesive soils. A simple test for quantifying the relative magnitude of the mass strength of non-cohesive soil can be performed by means of

TABLE 4.4 Mass Strength Number for Cohesive Soil (M_s)

Consistency	Identification in profile	Vane shear strength (kPa)	Mass strength number (M_s)
Very soft	Pick head can easily be pushed in up to the shaft of handle. Easily molded by fingers.	0–80	0.02
Soft	Easily penetrated by thumb; sharp end of pick can be pushed in 30–40 mm; molded by fingers with some pressure.	80–140	0.04
Firm	Indented by thumb with effort; sharp end of pick can be pushed in up to 10 mm; very difficult to mold with fingers. Can just be penetrated with an ordinary hand spade.	140–210	0.09
Stiff	Penetrated by thumbnail; slight indentation produced by pushing pick point into soil; cannot be molded by fingers. Requires hand pick for excavation.	210–350	0.19
Very stiff	Indented by thumbnail with difficulty; slight indentation produced by blow of pick point. Requires power tools for excavation.	350–750	0.41

NOTE: Cohesive materials of which the vane shear strength exceeds 750 kPa to be taken as rock—see Table 5.1.
SOURCE: From Kirsten 1982.

TABLE 4.5 Mass Strength Number for Non-cohesive Granular Soils (M_s)

Consistency	Identification in profile	SPT blow count	Mass strength number (M_s)
Very loose	Crumbles very easily when scraped with geological pick	0–4	0.02
Loose	Small resistance to penetration by sharp end of geological pick	4–10	0.04
Medium dense	Considerable resistance to penetration by sharp end of geological pick	10–30	0.09
Dense	Very high resistance to penetration of sharp end of geological pick—requires many blows of pick for excavation	30–50	0.19
Very dense	High resistance to repeated blows of geological pick—requires power tools for excavation	50–80	0.41

NOTE: Granular materials in which the SPT blow count exceeds 80 to be taken as rock—see Table 4.1.

SOURCE: From Kirsten 1982.

the standard penetration test (SPT). The values of M_s for *non-cohesive* granular soils in Table 4.5 are correlated with field profile identification tests and SPT blow counts. The latter is determined in accordance with ASTM D-1586 (Standard Test Method for Penetration Test and Split Barrel Sampling of Soils). Increases in the value of SPT blow counts correspond to increases in the value of M_s. When the SPT blow count exceeds 80, the non-cohesive granular material is considered to be equivalent to rock, requiring application of Table 4.3 to quantify the value of M_s. Field identification tests referred to in these tables are in accordance with Korhonen, et al., (1971), Jennings, et al., (1973), and the Geological Society of London (1977).

Block or particle size number (K_b)

The value of K_b is determined in different ways for rock and for granular soil. In the case of rock, it is a function of rock joint spacing and the number of joint sets, whereas it is a function of particle size in the case of non-cohesive granular soil. The value of K_b is set equal to one in the case of fine-grained, homogeneous cohesive granular soil.

Rock. Joint spacing and the number of joint sets within a rock mass determines the value of K_b for rock. Joint spacing is estimated from borehole data by means of the rock quality designation (*RQD*) and the number of joint sets is represented by the joint set number (J_n). *RQD* is a standard parameter in drill core logging and is determined as the ratio between the sum of the lengths of pieces of rock that are longer than 0.1 m and the total core run length (usually 1.5 m), expressed as a percent (Deere and Deere 1988). *RQD* values range between 5 and 100. A *RQD* of 5 represents very poor quality rock, and a *RQD* of 100 represents very good quality rock. For example, if a core contains four pieces of rock longer than 0.1 m, with lengths of 0.11, 0.15, 0.2, and 0.18 m then the cumulative length of rock longer than 0.1 m is 0.64 m and the *RQD* is 0.64 m/1.5 m × 100 = 43.

Schematic presentations explaining the joint set concept are shown in Fig. 4.19 and in the photographs in Figs. 4.20 and 4.21. The values of the J_n are found in Table 4.6. J_n is a function of the number of joint sets, ranging from rock with no or few joints (essentially intact rock), to

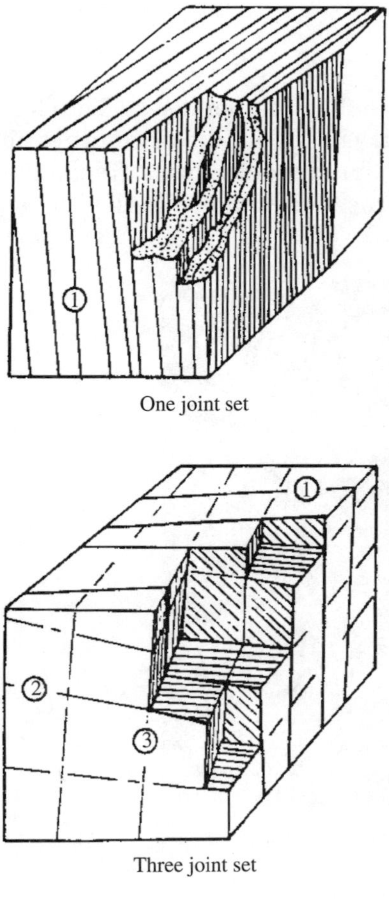

Figure 4.19 Schematic presentation illustrating the concept of joint sets.

Figure 4.20 A rock formation with one joint set.

rock formations consisting of one to more than four joint sets. The classification accounts for rock that displays random discontinuities in addition to regular joint sets. Random joint discontinuities are discontinuities that do not form regular patterns. For example, rock with two

Figure 4.21 A rock formation with three joint sets. Two orthogonal sets and one in the plane of the paper.

TABLE 4.6 Joint Set Number (J_n)

Number of joint sets	Join set number (J_n)
Intact, no, or few joints/fissures	1.00
One joint/fissure set	1.22
One joint/fissure set plus random	1.50
Two joint/fissure sets	1.83
Two joint/fissure sets plus random	2.24
Three joint/fissure sets	2.73
Three joint/fissure sets plus random	3.34
Four joint/fissure sets	4.09
Multiple joint/fissure sets	5.00

SOURCE: From Kirsten 1982.

joint sets and random discontinuities is classified as having two joint sets plus random (see Table 4.6). Having determined the values of RQD and J_n, K_b is calculated as

$$K_b = \frac{RQD}{J_n} \tag{4.67}$$

where $5 \le RQD \le 100$ and $1 \le J_n \le 5$

With the values of RQD ranging between 5 and 100, and those of J_n ranging between 1 and 5, the value of K_b ranges between 1 and 100 for rock.

If RQD data is unavailable, its value can be estimated with one or more of the following equations

$$RQD = (115 - 3.3J_c) \tag{4.68}$$

J_c is known as the joint count number, a factor representing the number of joints per m^3 of the material, which can either be measured or calculated with the equation

$$J_c = \left(\frac{3}{D}\right) + 3 \tag{4.69}$$

where D is the mean block diameter in meters. D can be calculated with the equation

$$D = (J_x \cdot J_y \cdot J_z)^{0.33} \qquad \text{for } D \ge 0.10 \text{ m} \tag{4.70}$$

Where J_x, $J_{y,}$, and J_z are average spacing of joint sets in meters measured in three mutually perpendicular directions x, y, and z. Joint set spacing can be determined by a fixed line survey (see, e.g., International Society for Rock Mechanics 1981, Geological Society of London 1977, Bell 1992). In essence, this technique entails measuring the spacing between joints in three orthogonal directions, and averaging the distances for each direction.

Other equations that can be used to calculate *RQD*, derived from those above, are

$$RQD = \left(105 - \frac{10}{D}\right) \tag{4.71}$$

and

$$RQD = \left(105 - \frac{10}{(J_x \cdot J_y \cdot J_z)^{0.33}}\right) \tag{4.72}$$

Cohesive and non-cohesive granular soil. K_b is set to one ($K_b = 1$) when indexing intact *cohesive* soils.

In the case of *non-cohesive*, granular soils (including silt, fine, medium, and coarse sands, and gravel and cobbles), the value of K_b is determined by means of the following equation

$$K_b = 1000\mathrm{D}^3 \tag{4.73}$$

where D is the characteristic particle diameter (m) of the bed material.

The characteristic particle diameter is equal to the median diameter of the armor layer, should that be present. If the boundary is not representative of an armor layer but an armor layer can potentially form during the scour process, then the characteristic diameter can be set equal to the D_{85} diameter of the bed material. The reason for this is that the median diameter of an armor layer is roughly equal to the D_{85} of the gradation of the underlying bed material.

Discontinuity/interparticle bond shear strength number (K_d)

The shear strength number K_d is calculated differently for rock and granular material. In the case of rock, the discontinuity shear strength number is determined as the ratio between two variables representing different characteristics of the surfaces that make up the discontinuity. In the case of granular material, K_d is equal to the tangent of the residual angle of friction of the material.

Rock. The discontinuity or interparticle bond shear strength number (K_d) is the parameter that represents the relative resistance offered by discontinuities in rock, determined as the ratio between joint wall roughness (J_r) and joint wall alteration (J_a)

$$K_d = \frac{J_r}{J_a} \tag{4.74}$$

J_r represents the degree of roughness of opposing faces of a rock discontinuity, and J_a represents the degree of alteration of the materials that form the faces of the discontinuity. Alteration relates to amendments of the rock surfaces, for example, weathering or the presence of cohesive material between the opposing faces of a joint. Values of J_r and J_a can be found in Tables 4.7 and 4.8.

The values of K_d calculated with the information in these tables change in sympathy with the relative degree of resistance offered by the joints. Increases in resistance are characterized by increases in the value of K_d. The shear strength of a discontinuity is directly proportional to the degree of roughness of opposing joint faces and inversely proportional to the degree of alteration.

Joint roughness is described by referring to both large and small-scale characteristics. The large-scale features are known as stepped, undulating, or planar; whereas the small-scale features are referred to as rough, smooth, or slickensided. Examples of planar and undulating joints are shown in Figs. 4.22 and 4.23, respectively. Figure 4.24 is a schematic presentation of conventional descriptions of joint roughness.

A planar, rough joint indicates that the large-scale feature is planar, but that the joint surfaces are rough. The concepts of closed, open, and

TABLE 4.7 Joint Roughness Number (J_r)

Joint separation	Condition of joint	Joint roughness number
Joints/fissures tight or closing during excavation	Stepped joints/fissures	4.0
	Rough or irregular, undulating	3.0
	Smooth undulating	2.0
	Slickensided undulating	1.5
	Rough or irregular, planar	1.5
	Smooth planar	1.0
	Slickensided planar	0.5
Joints/fissures open and remain open during excavation	Joints/fissures either open or containing relatively soft gouge of sufficient thickness to prevent joint/fissure wall contact upon excavation.	1.0
	Shattered or micro-shattered clays	1.0

SOURCE: From Kirsten 1982.

TABLE 4.8 Joint Alteration Number (J_a)

Description of gouge	Joint alteration number (J_a) for joint separation (mm)		
	1.0*	1.0–5.0†	5.0‡
Tightly healed, hard, non-softening impermeable filling	0.75	—	—
Unaltered joint walls, surface staining only	1.0	—	—
Slightly altered, non-softening, non-cohesive rock mineral or crushed rock filling	2.0	2.0	4.0
Non-softening, slightly clayey non-cohesive filling	3.0	6.0	10.0
Non-softening, strongly over-consolidated clay mineral filling, with or without crushed rock	3.0	6.0§	10.0
Softening or low friction clay mineral coatings and small quantities of swelling clays	4.0	8.0	13.0
Softening moderately over-consolidated clay mineral filling, with or without crushed rock	4.0	8.00§	13.0
Shattered or micro-shattered (swelling) clay gouge, with or without crushed rock	5.0	10.0§	18.0

NOTE:
*Joint walls effectively in contact.
†Joint walls come into contact after approximately 100-mm shear.
‡Joint walls do not come into contact at all upon shear.
§Also applies when crushed rock occurs in clay gouge without rock wall contact.
SOURCE: From Kirsten 1982.

Figure 4.22 Planar joints.

Figure 4.23 Undulating joints.

filled joints terminology used in Table 4.8 are illustrated in Fig. 4.22. The value of K_d that is calculated by means of Eq. (4.74) is roughly equal to the tangent of the residual angle of friction between the rock surfaces.

Cohesive and non-cohesive granular earth material. In granular materials the interparticle bond shear strength number is estimated by the following equation

$$K_d = \tan \phi \tag{4.75}$$

where ϕ is the residual friction angle of the granular earth material.

Relative ground structure number (J_s)

The relative ground structure number (J_s) represents the relative ability of earth material to resist erosion due to the structure of the ground (Table 4.9). This parameter is a function of the dip and dip direction of the least favorable discontinuity (most easily eroded) in the rock with respect to the direction of flow, and the shape of the material units. These two variables (orientation and shape) affect the ease by which the stream can penetrate the ground and dislodge individual material units. When assessing intact material, such as massive rock or fine-grained massive clay, or when assessing non-cohesive granular soils, the value of J_s is equal to 1.0.

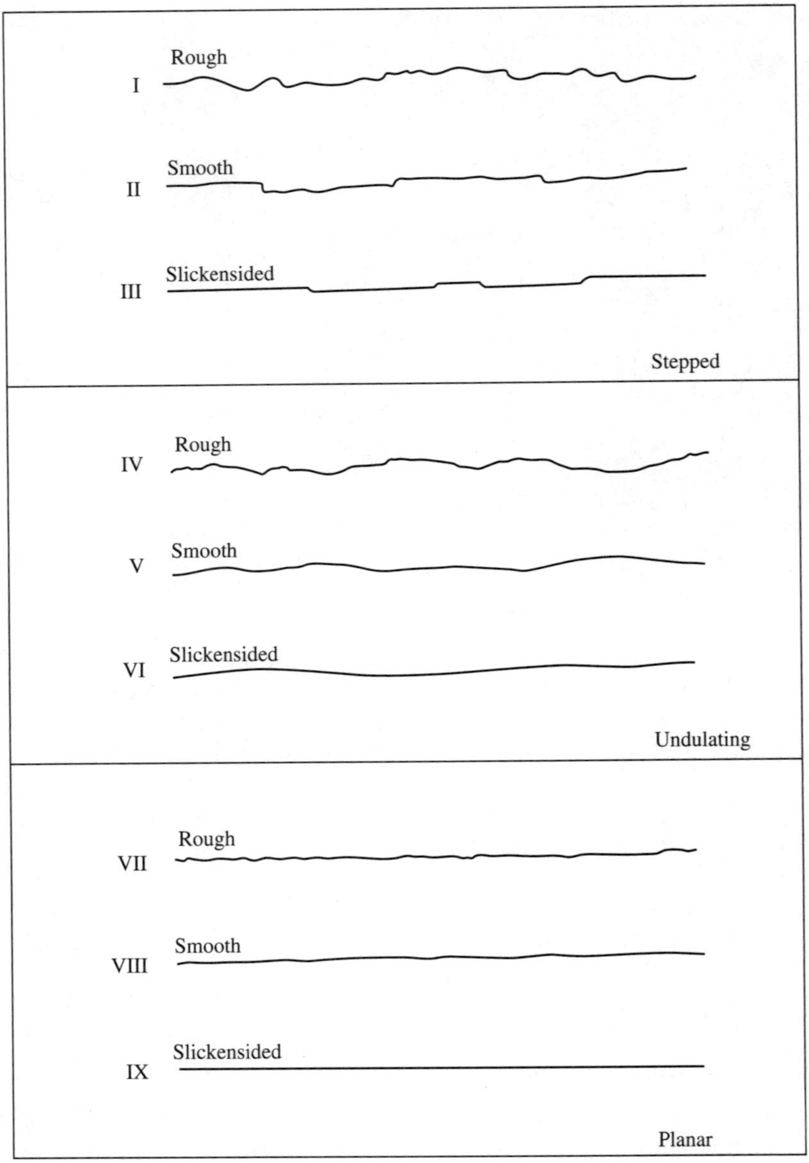

Figure 4.24 Schematic presentation of conventional descriptions of joint roughness.

The concepts of dip and dip direction of rock are illustrated in Fig. 4.25. This figure shows a perspective view of a block of rock with a slanting discontinuity. The line that is formed where the horizontal plane and the plane of the discontinuity intersect is known as the strike of the rock. The dip direction, measured in degrees azimuth, is the direction of a line

TABLE 4.9 Relative Ground Structure Number (J_s)

Dip direction of closer spaced joint set (degrees)	Dip angle of closer spaced joint set (degrees)	Ratio of joint spacing, *r*			
		1:1	1:2	1:4	1:8
180/0	Vertical 90	1.14	1.20	1.24	1.26
	89	0.78	0.71	0.65	0.61
In direction of	85	0.73	0.66	0.61	0.57
stream flow	80	0.67	0.60	0.55	0.52
	70	0.56	0.50	0.46	0.43
	60	0.50	0.46	0.42	0.40
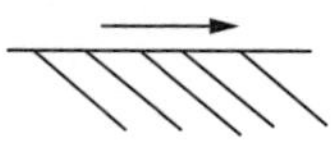	50	0.49	0.46	0.43	0.41
	40	0.53	0.49	0.46	0.45
	30	0.63	0.59	0.55	0.53
	20	0.84	0.77	0.71	0.67
	10	1.25	1.10	0.98	0.90
	5	1.39	1.23	1.09	1.01
	1	1.50	1.33	1.19	1.10
0/180	Horizontal 0	1.14	1.09	1.05	1.02
	−1	0.78	0.85	0.90	0.94
Against direction of	−5	0.73	0.79	0.84	0.88
stream flow	−10	0.67	0.72	0.78	0.81
	−20	0.56	0.62	0.66	0.69
	−30	0.50	0.55	0.58	0.60
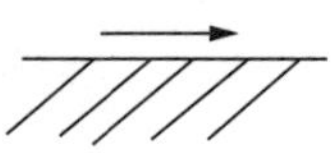	−40	0.49	0.52	0.55	0.57
	−50	0.53	0.56	0.59	0.61
	−60	0.63	0.68	0.71	0.73
	−70	0.84	0.91	0.97	1.01
	−80	1.26	1.41	1.53	1.61
	−85	1.39	1.55	1.69	1.77
	−89	1.50	1.68	1.82	1.91
180/0	Vertical −90	1.14	1.20	1.24	1.26

NOTES: 1. For intact material take $J_s = 1.0$.
2. For values of r greater than 8 take J_s as for $r = 8$.
SOURCE: From Kirsten 1982.

in the horizontal plane that is perpendicular to the strike and located in the vertical plane of the dip of the rock. The dip of the rock is the magnitude of the angle between the horizontal plane and the plane of the discontinuity, measured perpendicular to the strike. If the flow direction is roughly in the same direction as the dip direction, then the dip is said to be in the direction of the flow. If the flow direction is opposite to the dip direction, then the dip is said to be opposite to the direction of flow.

The shape of rock blocks is quantified by determining the joint spacing ratio (r), which is the quotient of the average spacing of the two most dominant high angle joint sets in the vertical plane (see Fig. 4.26). In cases where the value of r is greater than 8, use the values of J_s for $r = 8$.

Conceptually the function of relative ground structure number (J_s), incorporating shape and orientation, is as follows. If rock is dipped

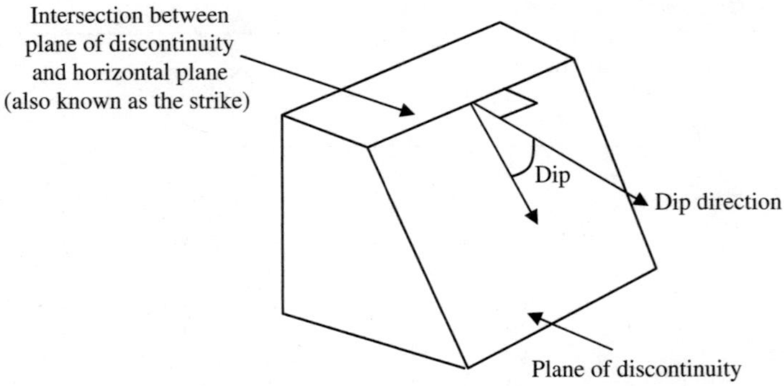

Figure 4.25 Definition sketch defining dip and dip direction of rock.

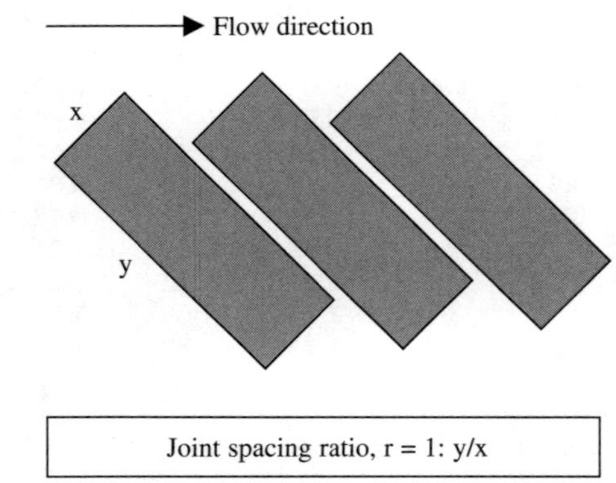

Figure 4.26 Determination of joint spacing ratio, *r*.

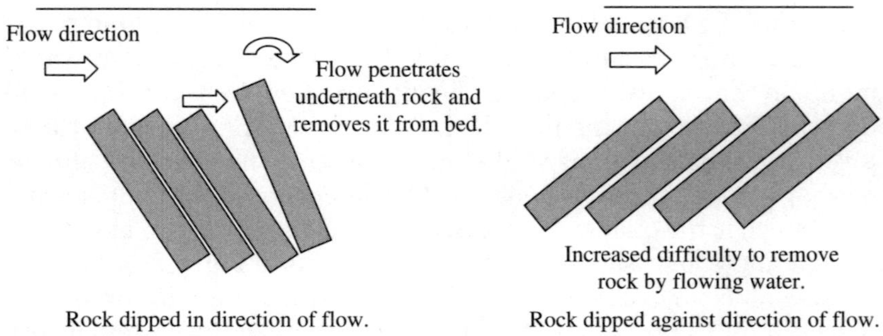

Figure 4.27 Influence of dip direction on scour resistance offered by rock.

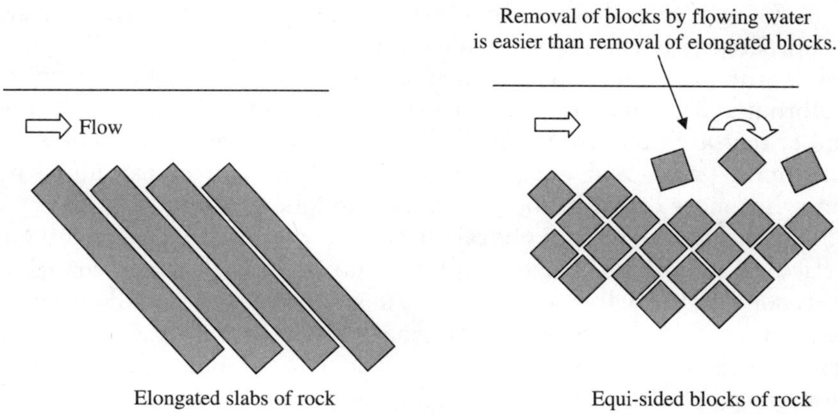

Figure 4.28 Influence of shape of rock blocks on scour resistance.

against the direction flow, it will be more difficult to scour the rock than when it is dipped in the direction of flow. When it is dipped in the direction of flow, it is easier for the flow to lift the rock, penetrate underneath, and remove it. Rock that is dipped against the direction of flow will be more difficult to dislodge (Fig. 4.27).

The shape of the rock, represented by the ratio *r*, impacts its erodibility in the following manner. Elongated rock will be more difficult to remove than equi-sided blocks of rock (Fig. 4.28). Therefore, large ratios of *r* represent rock that is more difficult to remove because it represents elongated rock shapes.

Vegetated Soils

Vegetated soils generally provide a greater resistance against erosion than non-vegetated soils. Establishment of the erosion threshold characterizing the erodibility index method developed by Annandale (1995) (see Fig. 4.18) incorporated analysis of vegetated soils that either eroded or not. A particular approach that was followed in estimating the erodibility index of vegetated soils formed part of the analysis determining the erosion threshold line. The same approach can be used to estimate the erosion resistance of vegetated soils for project work because quantification of the erodibility index and the use of the erosion threshold line in Fig 4.18 provides a means of determining the threshold stream power of vegetated soils.

In essence, the approach that was followed is based on the observation that the root architecture and root habit of plants play a significant role in determining the erosion resistance of vegetated soils. The *root architecture* of a plant describes the geometric characteristics of its roots. For example, the roots of a tree may be principally characterized by taproots, with some fibrous roots growing off the taproots. The root architecture of other plants, such as grass, may be more characterized by fibrous roots that grow very closely together in a clump. The latter kind of root architecture is desirable when using vegetation to protect soils against the erosive capacity of water. This will be explained in somewhat more detail in the paragraphs to follow.

The *root habit* of a plant characterizes the way the roots grow under particular conditions. For example, if plants grow in a non-cohesive sandy soil and the water source is deep down the roots would likely grow deeper down to reach the water. Alternatively, if the plants grow in, say, a clayey soil, one might find that the roots of a similar plant may not grow as deep, but may generally be located closer to the surface.

When considering the effect of vegetation in protecting soil against erosion it is useful to recall the factors used by the erodibility index method to characterize earth materials. These are mass strength, block-particle size, interparticle friction, and shape and orientation of the earth material. The mass strength of the earth material and its block or article size play greater roles than friction, shape, and orientation (this can be determined by comparing the relative values of the different parameters in the tables presented in the previous section).

It is therefore reasonable to expect that the greatest value will be gained by increasing the mass strength and block or particle size of an earth material that one wishes to modify. Realizing this, consider the modifying features that plants offer to improve the erosion resistance of earth material.

When roots develop within an earth material the increase in mass strength is relatively insignificant, but the potential to increase the "effective" particle size can be significant. From this point of view, it can be concluded that fibrous roots that grow in clumps would have the most useful root architecture. Such a root configuration binds the soil particles together. Although the increase in strength, resistance due to friction, shape and orientation added by the presence of the roots are not that great, the increase in effective particle size can be significant.

Therefore, when using the erodibility index method, the mass strength number M_s in the modified soil will not differ significantly from the virgin soil, nor would the interparticle shear strength number K_d and the shape and orientation number J_s. The only number that can changes significantly is the block size number K_b.

The effect that fibrous roots, growing in a clump, have on modifying earth material and increasing its resistance against the erosive capacity of water can be seen in Fig 4.29. This figure shows the condition on a floodplain in San Clemente, California, the day after it was inundated by a flood. The native soil on the floodplain consists predominantly of silt, with very little clay.

Figure 4.29 The effect of fibrous roots growing in a clump on increasing the erosion resistance of earth material.

The root clumps of the plants on the floodplain bind this fine textured soil together, forming larger effective particle sizes. This photograph is characteristic of the rest of the vegetation on the floodplain. What is noticeable in Fig. 4.29 is that the silt around the clump has eroded, but that the larger pseudo particle formed by the vegetation binding the soil did not.

Therefore, if the root architecture of the plants on a floodplain is known it is possible to estimate the size of the pseudo particle that will form once the plant is established. The particle size number K_b is then determined with the equation

$$K_b = 1000D^3$$

where D is the diameter of the root bulb bounded by the fine fibrous roots, measured in meters.

Once the values of the four index numbers have been assigned, the erodibility index of the vegetated soil can be calculated and the threshold stream power determined from Fig. 4.18.

Summary

This chapter presents typical properties of water and air that are relevant to assessing the erosive capacity of water. However, the principal focus is on earth and engineered earth material properties, determining their ability to resist the erosive capacity of water. In this regard, the earth and engineered earth materials are divided into two categories—physical and chemical gels.

Physical gels consist of elements (minerals) that occupy certain spaces and are connected to each other by mere touching. Examples of physical gels are non-cohesive soils and fractured rock formations. The principal properties of these materials determining their resistance to the erosive capacity of water are particle or block size, unit weight, and friction. Of these, the unit weight and particle/block size are the most important properties determining erosion resistance.

Chemical gels consist of elements (minerals) occupying locations in spaces that are connected to each other by fixed bonds. The fixed bonds are chemical in nature and as such provide more resistance than those of physical gels. Chemical gels include cohesive and cemented soils, concrete (an engineered earth material), and intact rock formations. Chemical gels are generally viewed as brittle materials that fail in either brittle fracture or fatigue failure.

Clays, although more malleable than other chemical gels such as rock, exhibits brittle fracture and fatigue failure characteristics. The failure mechanism of clays is complex. The reason for this is that the chemical forces determining its strength are sensitive to temperature, salinity, and pH of the ambient and interstitial water.

In order to understand the erosion of clays it is convenient to view erosion of these materials as a chemical process. This has been presented in some detail. It has been concluded that it is necessary to conduct tests on site-specific clay samples to understand each clay formation's unique erosion characteristics. When attempts are made to fully understand the erosion of a particular clay formation it is necessary to determine the activation energy required to initiate erosion, the number of bonds between individual clay particles, and the effects of pH, temperature, and salinity on erosion rate. These tests are discussed in Chap. 8.

The number of bonds of a clay can be determined by conducting rate of erosion tests under constant temperature. The activation energy of clay is determined by conducting rate of erosion test by varying the temperature of the pore- and ambient water. These tests are usually conducted in a Couette flow device. Determination of the effects of pH and salinity on erosion can also be investigated with the same device by varying these parameters under constant temperature.

The principal driving force leading to erosion of clays, as it is for any other earth material, is the effect of turbulent fluctuating pressures. One way to account for the effects of turbulent fluctuating pressures when analyzing scour is to use steam power as an indicator parameter. Alternatively, one can directly relate the magnitude of pressure fluctuations to the turbulent shear stress. Approaches for both options have been provided.

The fact that one can relate the relative magnitude of fluctuating pressures to turbulent shear stress does not mean that erosion by turbulent flow is a shear process. It is merely a convenient way to quantify the relative magnitude of fluctuating pressure.

Consideration of the various factors determining the rate of erosion of clays provides explanations for its behavior. It has been shown that both convex and concave relationships between rate of erosion and shear stress can be simulated using the equation developed to express the erosion properties of clays.

The failure characteristics of brittle materials like intact rock and engineered earth materials like concrete have been presented. This has been done in a practical way that allows direct consideration of the behavior of such materials when subjected to the erosive capacity of water. Methods to calculate the potential for brittle fracture and the material characteristics associated with that have been presented. The information in this chapter can be used to calculate the fracture toughness of the brittle material. If the stress intensity caused by the fluctuating pressures in turbulent flow exceeds the fracture toughness of the rock, it will fail in brittle fracture.

Material characteristics that can be used to calculate the rate of scour when the stress intensity caused by the fluctuating pressures does not exceed the fracture toughness of the rock were presented. In such cases, the brittle materials will fail in fatigue if the fluctuating pressures are applied long enough.

In addition to the material characteristics that can be used to directly calculate the potential for significant clay erosion, and the potential for brittle fracture and fatigue failure of chemical gels, and to calculate the erosion potential of physical gels, an empirical method that can be used to characterize both physical and chemical gels is presented in this chapter. This is known as the erodibility index method. This method provides a semi-empirical approach to accounting for the mass strength, block size, interparticle/block friction, and relative shape and orientation of any earth material. It indirectly accounts for the potential of brittle fracture and fatigue failure in chemical gels, and for the principal parameters affecting the erosion of physical gels. Research has indicated, as discussed in this chapter, that the concept of an erosion threshold is most probably not applicable to cohesive materials such as clay. It was stated that a "practical" threshold might be defined by allocating it a value equaling the value of the shear stress associated with a very low erosion rate. This is a somewhat qualitative assessment, which is dependent on the judgment of the individual testing the clay. In the case of the erodibility index method, such an erosion threshold was empirically established for clays. When using it, it is important to use the understanding acquired in this chapter to interpret the results of an analysis.

Lastly, material characteristics associated with the ability of vegetated soils to resist erosion were discussed. It has been pointed out that the major factors affecting the ability of vegetated soils to resist the erosive capacity of water are the root architecture and root habit of the plants. A fibrous root architecture binds the soil together, forming a larger "effective" particle size, which is the principal reason for the increased erosion resistance offered by vegetated soils.

Chapter

5

Erosive Capacity of Water

Introduction

This chapter describes the essential character of hydraulic action close to flow boundaries and presents practical methods for quantifying the relative magnitude of the erosive capacity of water resulting from such action. Insight into boundary flow processes is essential to advance understanding of scour and quantification of the erosive capacity of water is necessary to investigate the potential and determine the extent of scour.

Near-boundary processes

Chapter 3 explained that pressure fluctuations close to the boundary originate from near-bed processes associated with instabilities in the laminar sublayer. The near-bed processes lead to eddy formation and subsequent pressure fluctuations that follow the formation and breakup of hairpin vortices. Larger eddies originate from the central part of the hairpin vortices and smaller eddies from their sides. Additionally, flow from behind into the central part of the hairpin vortices cause high pressures on the bed. Turbulence production is another name for eddy formation.

The objective of this section is to gain more insight into turbulence production in flowing water, particularly in the near-bed region where direct interaction between the turbulent flowing water and the earth material occurs. This is done by mathematically investigating the distribution of the rate of energy supply and expenditure in a water column under turbulent flow conditions. The improved understanding leads to selection of reliable and consistent indicator parameters for quantifying the relative magnitude of the erosive capacity of water. Chapter 3 highlighted the need for selecting consistent indicator parameters.

The rate of energy supply and expenditure has units of power and is also known as stream power. Two forms of stream power exist in flowing water; available and applied stream power. The stream power that is made available to flowing water is the power that provides the impetus for the water to flow. The stream power that is applied to overcome resistance within the water and at its boundaries is known as the applied stream power.

To understand the distinction between available and applied stream power one can view the analog of a simply supported beam bending under its own weight. Its own weight is the load made available to provide the impetus for bending (the available force), whereas the internal stresses that develop within the beam during the course of bending give rise to the applied forces that overcome the internal resistance within the beam. In this case, as in the case of flowing water, the applied forces in the beam are converted from the available force, i.e., the weight of the beam.

Studying the spatial distribution of stream power in flowing water provides a useful way to understand the scour process. All translational energy in flowing water is eventually converted to heat and, once converted, cannot be regained. However, conversion from mechanical energy in turbulent flowing water into heat is not direct.

In the case of turbulent flow, the energy moves through an intermediate phase, in the form of turbulent kinetic energy, before it is finally converted to heat. Once the translational energy has been converted to turbulent kinetic energy during a given period of time, the macroturbulence thus generated leads to the development of smaller and smaller eddies. Once the eddies are small enough, the kinetic energy of turbulence is converted to heat by viscous action. The process of eddy formation that continually decreases in size is known as a cascade. The cascade results in the eventual dissipation of all the translational energy into heat, except for a small amount of energy that is required for formation of eddies when converting translational energy to kinetic energy of turbulence at the beginning of the process.

Our principal interest, as indicated above, is to quantify the kinetic energy of turbulence at the boundary prior to dissipation. The kinetic energy of turbulence at the boundary, prior to dissipation, leads to fluctuating pressures that are the prime drivers of incipient motion of earth materials and of the scour process.

Bakhmeteff and Allan (1946), who measured the distribution of energy loss in water flowing in natural open channels, found that only about 8 percent of the total available energy is lost in the main body of the flow while about 92 percent of the total available energy originates from this region. This points to an imbalance between the energy that is available in the main body of the flow and the amount that is dissipated in this same region. With only a small portion of the energy dissipated in the main body of the flow, it is obvious that the remainder of the energy is transferred to another part of the water body where it is dissipated.

Bakhmeteff and Allan found that the flow transmits about 90 percent of the energy from the main body of flow to the near-bed region. The latter generally occupies less than 15 percent of the total flow depth, directly adjacent to the bed (More often than not the near-bed region occupies about 11 to 12 percent of the total flow depth in open channel flow). The near-bed region should not be confused with the laminar sublayer, which is contained within its bounds. Nor should it be confused with the boundary layer, which can occupy the total flow depth in open channel flow, once fully developed. The pressure fluctuations resulting from the conversion of energy in the near-bed region to turbulent kinetic energy interact with the boundary and can lead to scour.

Schlichting and Gersten (2000) use the symbol y^+ to define the dimensions of the viscous (laminar) sublayer and the near-bed region:

$$y^+ = \frac{y}{\delta} \tag{5.1}$$

where y is the distance from the boundary and δ is the thickness of the wall layer.

The thickness of the wall layer is defined as

$$\delta = \frac{\nu}{u_*} = \frac{\nu}{\sqrt{\frac{\bar{\tau}_w}{\rho}}} \tag{5.2}$$

where ν = kinematic viscosity of the water

$\bar{\tau}_w$ = average wall shear stress

$\sqrt{\bar{\tau}_w/\rho} = u_*$ = shear velocity

The total thickness of the near-bed region is $y^+ = 70$. The viscous sublayer (also known as the laminar sublayer) and the buffer layer, the two components making up the near-bed region, are located within the following dimensions:

Viscous sublayer	$0 \le y^+ \le 5$
Buffer layer	$5 \le y^+ \le 70$

Chien and Wan (1999) considered the characteristics of the process of energy conversion by viewing shear stress τ acting on a small element of water moving at velocity u. Express the stream power in the water as the product of the two variables, i.e., τu (stream power per unit area). If one now wishes to discover how stream power is made available by the flowing water (the available stream power) and how it is used to overcome resistance (the applied stream power), the total stream power can be differentiated with respect to y, the incremental flow depth at right angles to the boundary, i.e.,

$$\frac{d}{dy}(\tau u) = u\frac{d\tau}{dy} + \tau\frac{du}{dy} \tag{5.3}$$

The first expression on the right hand side of the equation ($u d\tau/dy$) is the amount of stream power per unit volume that is made available by the flowing water, while the second term ($\tau du/dy$) is the applied stream power, or the rate of energy dissipation per unit volume of water. The applied stream power, as indicated previously, is also known as turbulence production (Schlichting and Gersten 2000).

The character of the two terms can be examined by first considering the balance of forces on an element of water in a two-dimensional flow situation. Consider the flow element in Fig. 5.1 with dimensions dy by dx and a thickness of one unit (at right angles to the paper). The flow velocity of this element is u, which changes as a function of depth. If the bed slope is s and uniform flow is assumed the gravity force acting on the body is

$$\gamma dxdys \tag{5.4}$$

The shear stress over the body varies from τ at the bottom of the element to $\tau + (d\tau/dy)dy$ at its top. The force balance on the element can

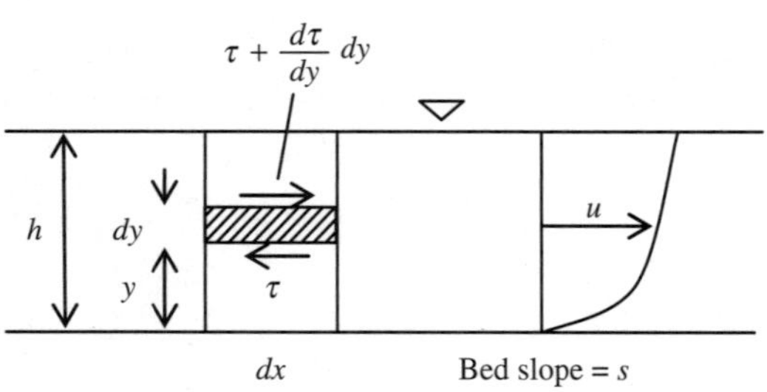

Figure 5.1 Forces acting on a water element in two-dimensional open channel flow, flowing from left to right.

therefore be expressed as

$$\left(\tau + \frac{d\tau}{dy}dy\right)dx - \tau dx + \gamma dxdys = 0 \tag{5.5}$$

from which follows

$$\frac{d\tau}{dy} = -\gamma s \tag{5.6}$$

If one now multiplies this equation with u, one gets

$$u\frac{d\tau}{dy} = -\gamma us \tag{5.7}$$

This is the *rate* by which potential energy per unit volume of water is made available. Recalling that energy expenditure per unit time (i.e., the rate of energy expenditure) is defined as power, it can be concluded that Eq. (5.7) represents the available stream power, i.e.,

$$P_{\text{available}} = -u\frac{d\tau}{dy} \tag{5.8}$$

where $P_{\text{available}}$ is the available stream power.

In order to understand the second term in Eq. (5.3) (i.e., $\tau\, du/dy$) consider the net rate of work required to deform an element of water at a distance y from the bed.

Figure 5.2 represents an element of water with dimensions dx by dy (and unit width at right angles to the paper) that deforms from the shape represented by the square with the solid lines to a parallelogram, represented by the dashed lines. The increase in velocity over the

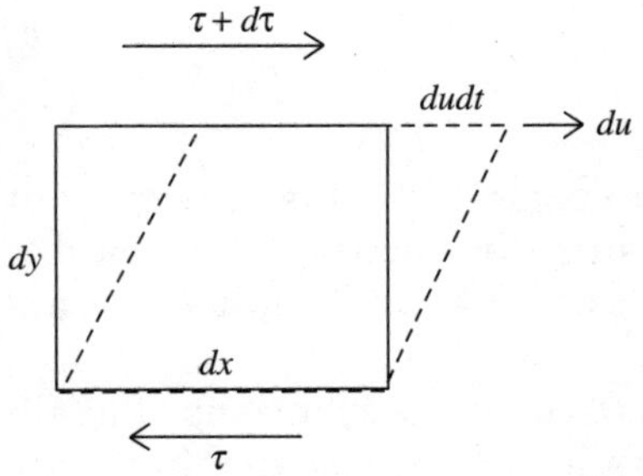

Figure 5.2 Deformation of an element of water resulting from shear stresses.

distance dy that occurs after a time dt is equal to du. The distance moved by the upper part of the element of water relative to the bottom part over a time dt is therefore equal to $dudt$.

Work done during a small time dt is equal to the product of the shear force and the distance moved. The net amount of work performed over the element during this time is therefore

$$(\tau + d\tau)dxdudt - \tau \cdot 0 = \tau \frac{du}{dy} dydxdt + d\tau dxdudt - 0 \tag{5.9}$$

By neglecting second order terms it follows from the above equation that the *energy applied* to overcome resistance in the body of flow and at the boundary per unit volume of water and over a unit period of time during the process of turbulence production can be expressed as

$$\tau \frac{du}{dy} \tag{5.10}$$

Again, by recalling that stream power is the rate of energy expenditure

$$P_{\text{applied}} = \tau \frac{du}{dy} \tag{5.11}$$

where P_{applied} is the applied stream power per unit volume of water (also known as turbulence production).

Determination of the amount of stream power applied to the boundary, which is the principal area of interest in scour technology, requires investigation into its spatial distribution. In order to do this a simplified equation representing the velocity distribution as a function of flow depth is used,

$$u = u_o \left(\frac{y}{h} \right)^m \tag{5.12}$$

where u_o is the flow velocity at the water surface and m is the exponent that varies between 0.10 and 0.25. By using this equation it is possible to develop typical distributions of the available and applied stream power (Fig. 5.3).

An interesting aspect of this figure is that the distributions of available and applied stream power differ significantly. The available stream power is concentrated in the main body of the flow, while the applied

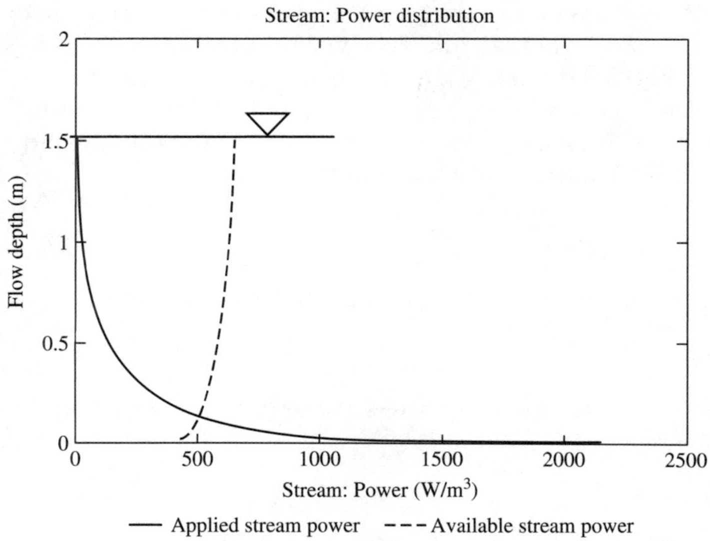

Figure 5.3 Available and applied stream power per unit volume distribution in open channel flow.

stream power is principally concentrated in the near-bed region. The reason for the latter is attributable to the steep velocity gradient at the boundary, which leads to turbulence production, i.e., eddy formation and consequent pressure fluctuations in the near-bed region.

Another interesting aspect of Eq. (5.3) is that the term $d(\tau u)/dy$ represents the algebraic sum of the available and applied stream power per unit volume of water, and in that sense provides an indication of how energy is transmitted to the boundary. One can show this by considering the mechanism of transmission of energy to the boundary.

Consider three elements of water A, B, and C (Fig. 5.4). The water flows from left to right and the dimensions of element A are dy by dx

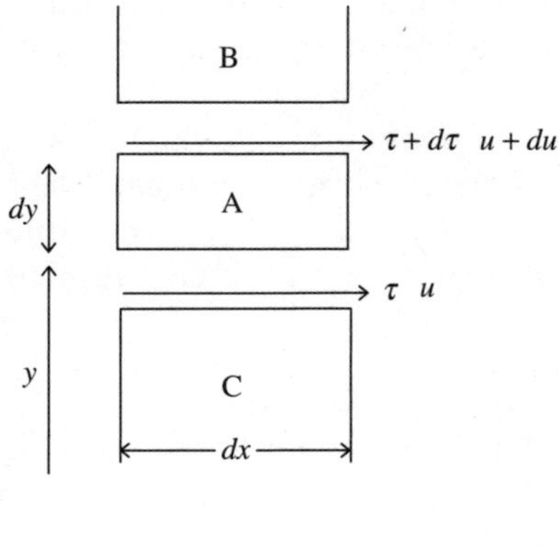

Figure 5.4 Mechanism for transmitting energy to the boundary.

(by a unit width rectangular to the page). The power received by body C from body A is τdxu, whereas that received by A from B is $(\tau + d\tau)dx$ $(u + du)$. From Fig. 5.3 it is known that the amount of energy per unit time received by A from B is less than the amount transmitted from A to C. One can therefore write an equation representing the rate of energy transmission to the boundary per unit time, i.e.,

$$\tau dxu - (\tau + d\tau)dx(u + du) \tag{5.13}$$

By neglecting second order terms it is possible to rewrite Eq. (5.13) as

$$-(\tau du + ud\tau)dx = -\frac{d}{dy}(\tau u)dydx \tag{5.14}$$

from which follows

$$P_{\text{transfer}} = -\frac{d}{dy}(\tau u) \tag{5.15}$$

where P_{transfer} is the amount of stream power per unit volume of water transmitted to the boundary from a depth y above the stream bed.

Plotting the spatial distribution of the stream power transmitted to the boundary on a common graph with the other two curves allows comparison of available, applied, and transferred stream power (Fig. 5.5). The transmission of power to the boundary is equal to the difference between available and applied stream power.

The negative values of the stream power transmission curve in the near-bed region (below line N) indicate a shortage of stream power in the near-bed region. The positive values of the transfer curve in the main body of the flow (also known as the core flow region) indicate that energy is transferred from this part of the flow to the near-bed region. This means that the energy applied to make flow possible per unit time and per unit volume of flow close to the bed (see the applied stream power curve) is transmitted to the boundary from the main body of the flow.

Additional insight into the relationship between these variables is gained by investigating the relationship of the integrated values. When integrating these variables it is found that the integral of the available stream power with respect to y is equal to the integral of the applied stream power with respect to y, i.e.,

$$\int_0^h ud\tau = \int_0^h \gamma usdy = \int_0^h \tau du \tag{5.16}$$

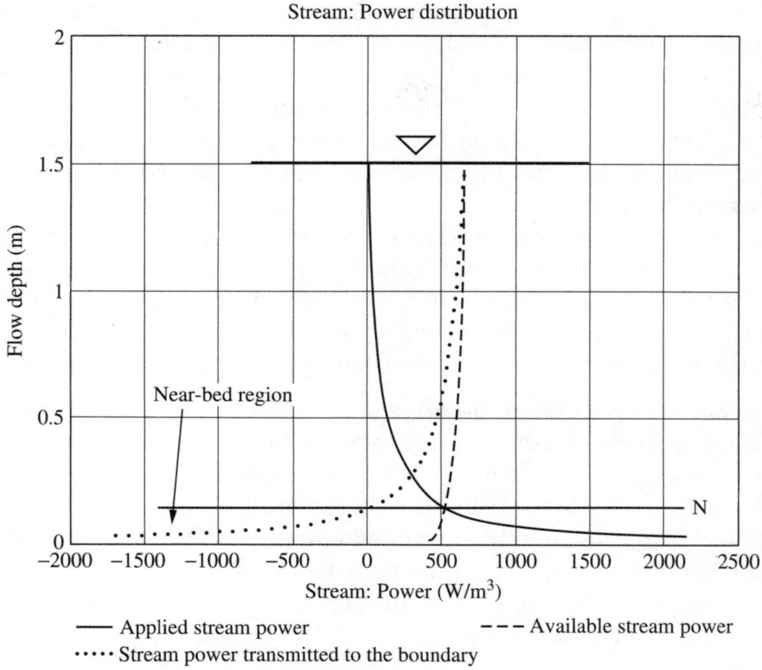

Figure 5.5 Available and applied stream power, and stream power transmitted to the boundary.

which means that

$$\int_0^h \gamma usdy + \int_0^h \tau du = 0 \tag{5.17}$$

Therefore, integration of the curve representing the transmission of stream power to the bed is also equal to zero, i.e.,

$$\int_0^h d(-\tau u) = 0 \tag{5.18}$$

The relationship above merely tells us how energy is transmitted to the bed, i.e., it comes from the main body of the flow (the positive values on the curve) and is transmitted to the near-bed region (the negative values on the curve). The algebraic sum of the positive and negative values is zero.

When it comes to understanding the flow processes relevant to considering incipient motion of earth materials and scour, the most important component that merits further study is the applied stream power ($\tau\, du/dy$)

in the near-bed region, i.e., the turbulence production. Chapter 3 explained that the pressure fluctuation that is the principal cause of scour originates with turbulence production. The curve representing applied stream power in Fig. 5.3 converges towards the bed without reaching a discrete value right at the bed, where $y = 0$.

The figure implies that the applied stream power at the bed approaches infinity, which is obviously not possible. This finding is mainly due to mathematical limitations and does not represent reality. However, quantification of the magnitude of applied stream power in the near-bed region is of critical importance when analyzing incipient motion and scour. The fluctuating pressures generated in this region are in direct contact with the earth material and determines whether scour will occur.

Turbulence production in the near-bed region. Schlichting and Gersten (2000) provide experimentally verified information on the universal distributions of energy supply, turbulence production, and direct dissipation during turbulent flow in the near-bed region. This information can be used to quantify the amount of applied power, i.e., the power generated by turbulence production, in the near-bed region.

The energy supply curve shown in Fig. 5.6 represents the term $-d(\tau u)/dy$. The direct dissipation curve in the figure represents the

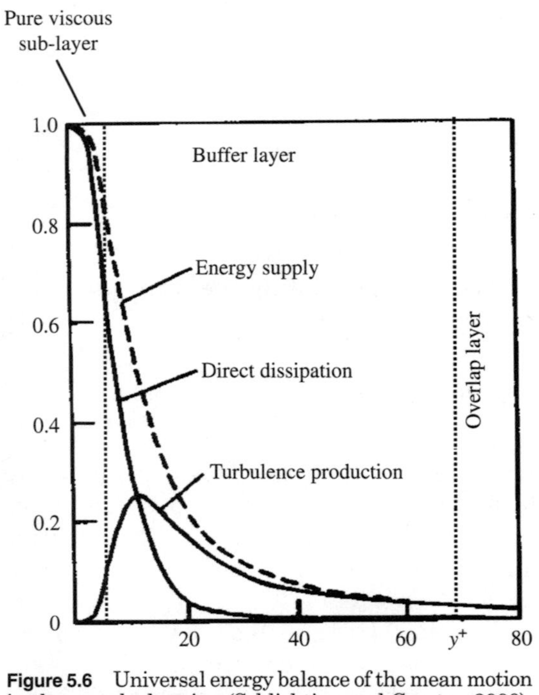

Figure 5.6 Universal energy balance of the mean motion in the near-bed region (Schlichting and Gersten 2000).

proportion of the energy supplied to the boundary that experiences viscous dissipation. The universal distribution of turbulence production $\tau du/dy$ in the near-bed region (the applied stream power) can also be seen in Fig. 5.6.

The figure shows the viscous sublayer and buffer layer that forms the near-boundary region, as well as the overlap layer. The latter layer lies between the near-bed region and the main body of the flow (also known as the core layer). The principal region of interest in scour analysis is the near-bed region, i.e., $y^+ \leq 70$.

Schlichting and Gersten (2000) indicate that the power distributions shown in Fig. 5.6 are universal and applicable to any kind of turbulent flow. The universal relationship for turbulence production allows derivation of an equation that can be used to quantify the amount of applied stream power in the near-bed region that affects the bed.

The ordinate of Fig. 5.6 is dimensionless, which allows Schlichting and Gersten to plot energy supply, direct (viscous) dissipation, and turbulence production on the same graph. The dimensionless scale for the turbulence production curve is defined by the term $\tau^+ du^+/dy^+$, which is the dimensionless turbulence production,

where $\tau^+ = \tau_t/\rho u_*^2$

τ_t = turbulent shear stress at the boundary

$u_* = \sqrt{\overline{\tau}_w/\rho}$ = shear velocity

$\overline{\tau}_w$ = average wall shear stress

$u^+ = \overline{u}/u_*$

$\overline{u}$ = average velocity

The curve representing dimensionless turbulence production in Fig. 5.6 is unique, and can be represented by the following equation:

$$\tau_t^+ \frac{du^+}{dy^+} = \frac{1}{-0.1117\times 10^2 + 0.6254y^+ + 0.9429\times 10^2/y^+} + 0.5976\times 10^{-3}\,y^+$$

Integration of this equation in the near-bed region ($0 \leq y^+ \leq 70$) leads to the following:

$$\int_0^{70} \tau_t^+ \frac{du^+}{dy^+} dy^+ = 7.853 \tag{5.19}$$

Written differently, this means that in the near-bed region

$$\tau^+ u^+ = 7.853 \tag{5.20}$$

Therefore, the value of the stream power in the near-bed region can be calculated as

$$\frac{\tau_t}{\rho u_*^2} \cdot \frac{\bar{u}}{u_*} = 7.853 \tag{5.21}$$

from which follows that the stream power due to turbulence production in the near-bed region (per unit area) can be expressed as

$$\tau_t \bar{u} = 7.853\left(\rho u_*^3\right) = 7.853\rho\left(\sqrt{\frac{\bar{\tau}_w}{\rho}}\right)^3 = 7.853\rho\left(\frac{\bar{\tau}_w}{\rho}\right)^{3/2} = 7.853\frac{\bar{\tau}_w^{3/2}}{\sqrt{\rho}} \tag{5.22}$$

The stream power generated by turbulence in the near-boundary region results in pressure fluctuations at the boundary, which is the principal factor leading to incipient motion and scour. Equation (5.22) can be used to calculate the proportion of the total amount of stream power that is applied to the bed.

By expressing the stream power generated by turbulence production in the near-boundary region as a proportion of the total available power it is possible to determine how much of the total available power is applied to the bed for differing flow conditions.

The ratio between the stream power applied to the bed and the total available stream power is

$$\frac{\tau_t \bar{u}}{\rho g q s_f} = \frac{7.853\rho\left(\frac{\tau_w}{\rho}\right)^{3/2}}{\rho g q s_f}$$

which can be rewritten as

$$= \frac{7.853\rho\left(\frac{\rho g y s_f}{\rho}\right)^{3/2}}{\rho g y \bar{u} s_f}$$

$$= 7.853\frac{\sqrt{g y s_f}}{\bar{u}} = 7.853\frac{u_*}{\bar{u}} \tag{5.23}$$

Equation (5.23) expresses the ratio between applied stream power at the boundary, due to turbulence production in the near-bed region, and the total available stream power.

This equation can also be rewritten in a form that allows estimation of the turbulent shear stress at the boundary, i.e.,

$$\tau_t = 7.853\left(\frac{u_*}{\bar{u}}\right)\rho g y s_f = 7.853\left(\frac{u_*}{\bar{u}}\right)\tau_w \tag{5.24}$$

The effect of flow condition on the stream power ratio, i.e., whether it is smooth turbulent, transition, or rough turbulent flow, can be determined by quantifying the average flow velocity $\bar{u}$. In this regard it becomes important to represent the effects of boundary roughness as accurately as is practically possible, which warrants a discussion on estimation of boundary roughness in hydraulic engineering.

The conventional approach in the United States, and some other locations around the world is to use the Manning equation to estimate average flow velocity. This equation is written as

$$\bar{u} = \frac{R^{2/3}S^{1/2}}{n} \tag{5.25}$$

where R = hydraulic radius = A/P
A = cross-sectional area of flow
P = wetted perimeter
n = a roughness coefficient known as Manning's n

In other parts of the world, mostly Europe, the average flow velocity is determined by making use of the Chezy equation, which is expressed as

$$\bar{u} = C\sqrt{RS} \tag{5.26}$$

where C is a roughness coefficient known as the Chezy coefficient.

The practical approaches used to estimate the values of these two respective roughness coefficients differ quite substantially. When values of Manning's n are assigned, a standard procedure is to consider photographs of different stream types and select a value based on that or previous experience of the engineer. The problem with selecting the values of Manning's n based on previous experience for most engineers lies with the fact that they have never had the chance to really evaluate how accurate their experience has been in the past. The previous experience of most engineers in this regard essentially means that they have selected values of Manning's n in the past based on what they thought were fit, and they are just repeating that process every time when they select a new value. By selecting these values on a continuous basis most engineers think that they are gaining experience, which in actual fact is not the case. Unless one has the opportunity to check the selection of these roughness values with actual flood events the experience that is gained is worthless.

This is a particular concern in the case of selecting values of Manning's n because these roughness values are sensitive to flow depth, a fact often not realized by most engineers using the approach. An example of how Manning's n changes as a function of hydraulic radius (average flow depth) for varying absolute roughness values is illustrated in Fig. 5.7. The selection of Manning's n does not have a physical basis, except that empirical testing and general experience provides an indication of what values should be selected for particular flow conditions.

The Chezy coefficient, on the other hand, is related to the absolute roughness k_s, the hydraulic radius, and the Reynolds number. The Chezy coefficient C can be expressed in terms of the Darcy friction coefficient f as (Henderson 1966)

$$C = \frac{\sqrt{8g}}{\sqrt{f}} \tag{5.27}$$

The value of the Darcy friction coefficient is determined from equations shown in Table 5.1 for varying values of absolute roughness, hydraulic radius, and Reynolds number. The selection of an appropriate equation for calculating the roughness coefficient is also dependent on the relationship between the absolute roughness and the dimension of the near-bed region.

In this regard it is useful to note that the dimensionless depth y^+ [Eq. (5.1)] can also be written as

$$y^+ = \frac{y}{\delta} = \frac{yu_*}{\nu}$$

Therefore, if one is interested in determining the ratio between the absolute roughness k_s and the dimensions of the near-bed region, the above equation can be rewritten as

$$\frac{k_s u_*}{\nu} \tag{5.28}$$

If $k_s u_*/\nu < 5$ the absolute roughness is so small that smooth turbulent flow will result. Alternatively, if $k_s u_*/\nu > 70$ the roughness is greater

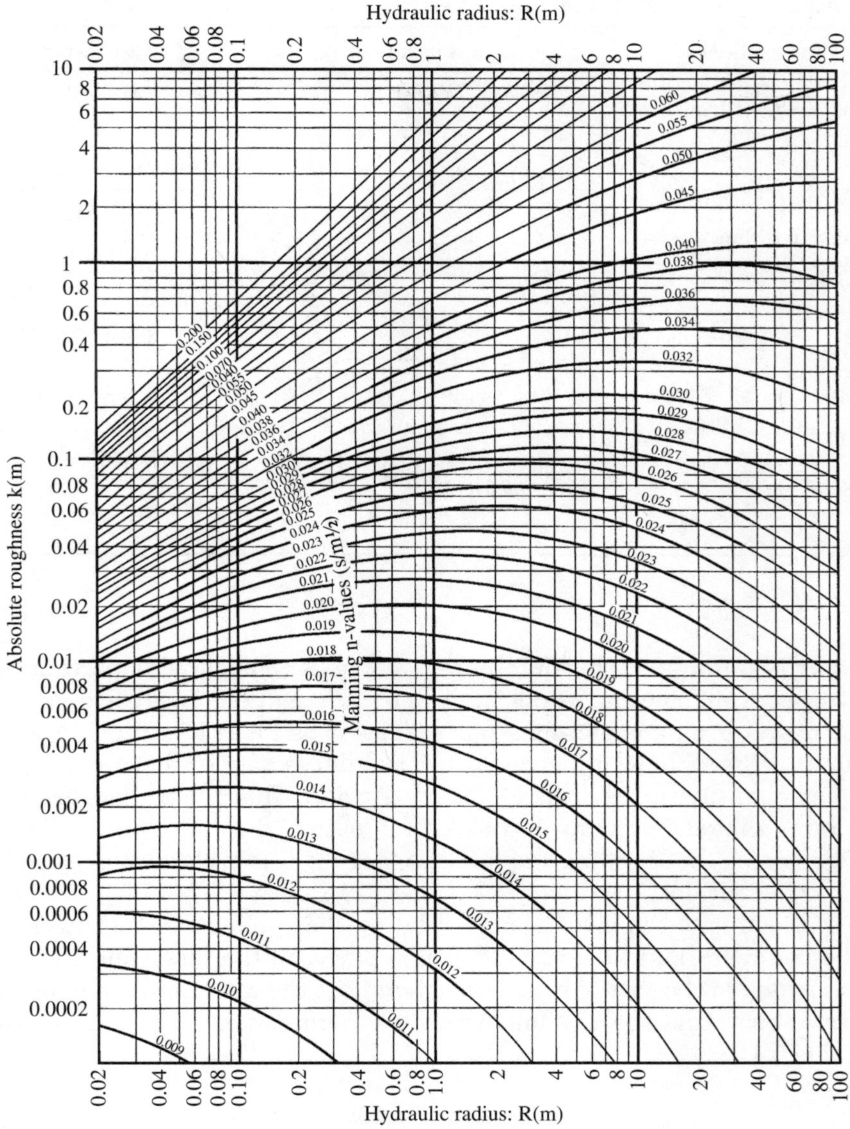

Figure 5.7 Hydraulic roughness estimates for open channel flow as a function of hydraulic radius for rough turbulent flow (Rooseboom et al., 2005).

than the thickness of the near-bed region and rough turbulent flow will result. In the range between these two limits transition flow will occur.

Examination of Table 5.1 shows that the selection of values for the Chezy roughness coefficient has a physical basis, which is lacking when

TABLE 5.1 Darcy Friction Coefficient for Varying Turbulent Flow Conditions

Flow condition	Roughness equation	Condition
Smooth turbulent	$f = \dfrac{0.316}{\mathrm{Re}^{1/4}}$	$\mathrm{Re} < 5000$ $\dfrac{u_* k_s}{\nu} < 5$
Smooth turbulent	$\dfrac{1}{\sqrt{f}} = 2.0\log_{10}\left(\dfrac{\mathrm{Re}\cdot\sqrt{f}}{2.51}\right)$	$\mathrm{Re} > 5000$ $\dfrac{u_* k_s}{\nu} < 5$
Transition	$\dfrac{1}{\sqrt{f}} = 2.0\log_{10}\left(\dfrac{k_s}{12R} + \dfrac{2.5}{\mathrm{Re}\sqrt{f}}\right)$	$2\cdot 10^3 < \mathrm{Re} < 10^6$ $5 < \dfrac{u_* k_s}{\nu} < 70$
Rough turbulent	$\dfrac{1}{\sqrt{f}} = 2.0\log_{10}\left(\dfrac{12R}{k_s}\right)$	$\mathrm{Re} > 2000$ $\dfrac{u_* k_s}{\nu} > 70$

using Manning's n. This difference may not be considered important in some civil engineering works, but when investigating the erosive capacity of water for conducting scour studies it becomes of critical importance. What follows illustrates the usefulness of the Chezy equation when quantifying the erosive capacity of water. This illustration demonstrates that the amount of information obtained from scour analyses when using the Manning's equation is lacking. It is preferable to use the Chezy equation.

Returning to the investigation relating the applied stream power at the boundary and the available steam power in the flow, consider the following: in the case of *rough turbulent* flow, the ratio between applied stream power at the bed and available stream power is written as (see Table 5.1)

$$\frac{\tau_t \bar{u}}{\rho g q s_f} = \frac{7.853 u_*}{\sqrt{8} \times 2 \times u_* \log_{10}\left(\frac{12y}{k_s}\right)} \tag{5.29}$$

from which follows

$$\frac{\tau_t \bar{u}}{\rho g q s_f} = \frac{7.853}{2\sqrt{8}} \cdot \frac{1}{\log_{10}\left(\frac{12y}{k_s}\right)} \tag{5.30}$$

Equation (5.30) indicates that the proportion of the total stream power that is applied to the bed is a function of the ratio between the absolute roughness k_s and the flow depth y (or the hydraulic radius if a wide channel is not investigated). The change in the proportion of the total available stream power applied to the bed as a function of dimensionless depth to absolute roughness ratio [using Eq. (5.30)] is shown in Fig. 5.8.

Figure 5.8 indicates that if the absolute roughness is approximately equal to the flow depth then almost all the available stream power under rough turbulent flow conditions is applied to the bed. As the flow depth increases relative to the absolute roughness the proportion of the total available stream power applied to the bed reduces and levels off at a ratio of approximately 0.4 for rough turbulent flow.

In the case of *smooth turbulent* flow when Re > 5,000 the ratio can be written as

$$\frac{\tau_t \bar{u}}{\rho g q s_f} = \frac{7.853}{2\sqrt{8}} \cdot \frac{1}{\log_{10}\left(\frac{\mathrm{Re}\sqrt{f}}{2.51}\right)} \tag{5.31}$$

Using a representative range of the product $\mathrm{Re}\sqrt{f}$ it is found that the proportion of the total available stream power that is applied to the

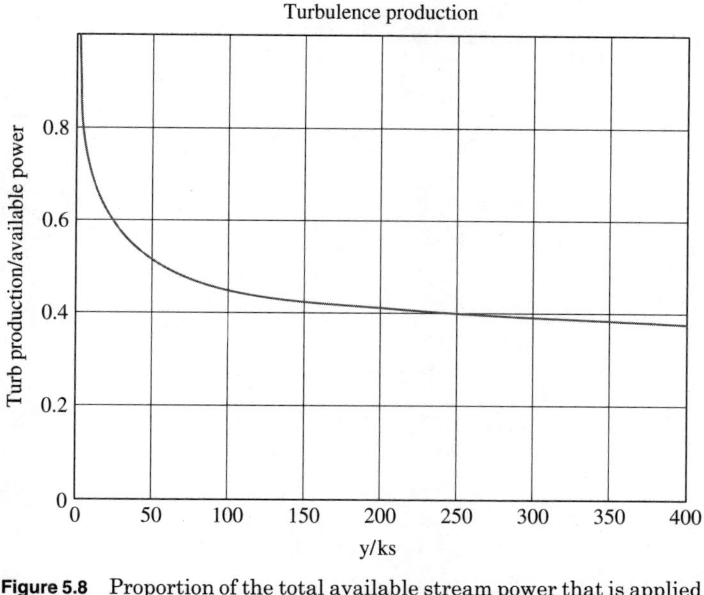

Figure 5.8 Proportion of the total available stream power that is applied to the bed as a result of turbulence production in the near-bed boundary layer as a function of flow depth-absolute roughness ratio for rough turbulent flow.

bed under smooth turbulent flow conditions (Fig. 5.9) is lower than the proportion applied to the bed during rough turbulent flow (Fig. 5.8). The ratio levels off at a value of approximately 0.25 for smooth turbulent flow conditions.

Plotting the dimensionless ratio of the proportion of the total available stream power that is applied to the bed for *transition turbulent* flow is somewhat more complex and is not presented here. However, by rewriting Eq. (5.23) for turbulent flow in the transition range, the following is found:

$$\frac{\tau_t \bar{u}}{\rho g q s_f} = \frac{7.853}{2\sqrt{8}} \cdot \frac{1}{\left(-\log_{10}\left(\frac{k_s}{12R} + \frac{2.51}{\mathrm{Re}\sqrt{f}}\right)\right)} \tag{5.32}$$

This ratio will be located between those for rough and smooth turbulent flow, and is expected to level off at values somewhere between 0.25 and 0.4.

Indicator parameter selection

A discussion in Chap. 3 indicates that the conventional indicator parameters used to quantify the relative magnitude of the erosive capacity of

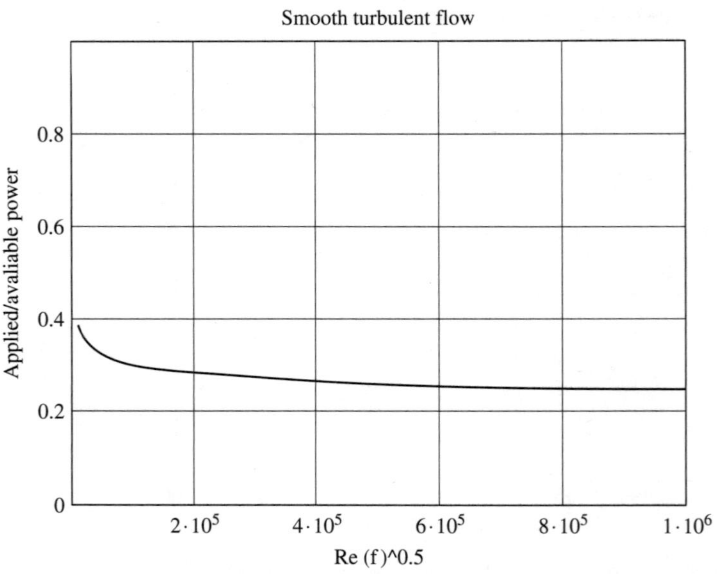

Figure 5.9 Proportion of the total available stream power that is applied to the bed as a result of turbulence production in the near-boundary layer for smooth turbulent flow.

water, i.e., wall shear stress, stream power, and average flow velocity lead to inconsistent trends when expressed as a function of hydraulic roughness. The equations normally used in practice to quantify these variables are

$$\tau_w = \gamma y s_f \tag{5.33}$$

$$P = \tau_w \bar{u} = \gamma y s_f \bar{u} = \gamma q s_f \tag{5.34}$$

The average flow velocity $\bar{u}$ is calculated by making use of either the Manning's or Chezy equations.

When expressing the values of these variables as a function of hydraulic roughness it is found that wall shear stress τ_w increases, stream power P remains constant, and the average flow velocity $\bar{u}$ decreases with increasing hydraulic roughness (see Chap. 3). The inconsistency in trends points to a practical problem when using these variables as indicator parameters for quantifying the relative magnitude of the erosive capacity of water.

The information presented in this chapter resolves this problem by showing that the relative magnitude of pressure fluctuations, the principal cause of scour, can be represented by the magnitude of turbulence production, i.e., stream power, at the boundary. The use of boundary layer theory shows that the stream power applied to the boundary is a function of flow depth, absolute roughness, flow velocity, and flow type. It is also a function of the wall shear stress.

Furthermore, the turbulent boundary shear stress is not equal but directly proportional to the wall shear stress. Fig. 5.10 shows the trends in turbulent boundary shear stress [Eq.(5.24)] and applied boundary stream power [Eq. (5.22)] as a function of hydraulic roughness for a constant unit discharge of $q = 1\ \text{m}^2/\text{s}$ and a channel slope of 0.001. This figure indicates that the trends in applied boundary stream power and the turbulent boundary shear stress are similar, although the trends in available stream power ($P_{\text{avail}} = \rho g \bar{u} y s$) and available shear stress (i.e., the wall shear stress $\tau_w = \rho g y s$) differ. Both the applied boundary stream power and the turbulent boundary shear stress increase as a function of hydraulic roughness.

Therefore, although it is prudent to emphasize that scour in turbulent flow is not a shear process but principally the result of turbulent pressure fluctuations; the trends in both wall shear stress and turbulent boundary shear stress are similar to that of the applied boundary stream power. This means that although applied boundary stream power is the preferred indicator parameter for quantifying the relative magnitude of the erosive capacity of water, one can also use shear stress as

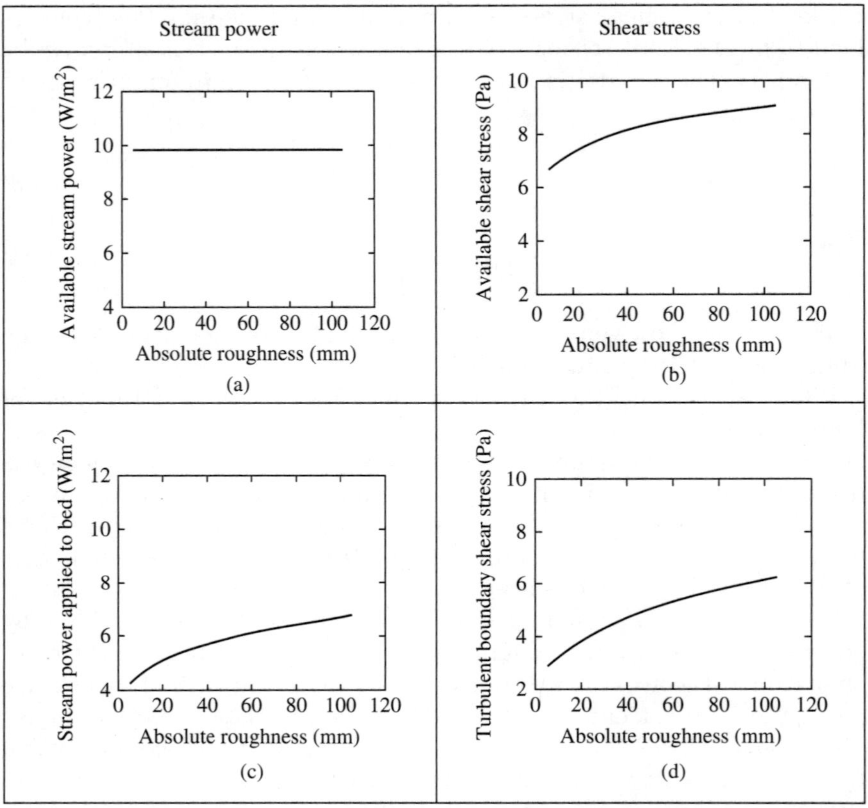

Figure 5.10 Comparison of trends in indicator parameters as a function of hydraulic roughness. (*a*) Available steam power (*b*) Available shear stress (*c*) Stream power applied to the boundary (*d*) Turbulent boundary shear stress.

an indicator parameter. Applied boundary stream power is preferred because it more closely represents the relative magnitude of pressure fluctuations at the boundary.

Summary

The explanations and mathematical derivations in this section are quite lengthy, so it is desirable to summarize the essential concepts. In essence we have distinguished between available and applied stream power. The available stream power is the rate at which energy is released to provide impetus for the water to flow, while the applied stream power is the rate at which energy is applied to overcome friction in the fluid and along its boundaries.

Another important conclusion is that the applied stream power is identical to turbulence production. This means that if one can quantify the turbulence production at the boundary (i.e., quantify the applied stream power) it should be possible to quantify the relative magnitude of pressure fluctuations. Pressure fluctuations in turbulent flow play a dominant role in the incipient motion of earth materials, as explained in Chap. 3 and shown in more detail in the following sections of this chapter.

The equation to quantify the magnitude of turbulence production at the boundary (i.e., the applied stream power) shows that the actual amount of the total available stream power that is applied to the boundary varies as a function of the boundary roughness, and the type of turbulent flow, i.e., smooth turbulent, transition, or rough turbulent flow.

Application of the equation indicates that the proportion of the total stream power applied to the boundary is larger in the case of rough turbulent flow than it is in the case of smooth turbulent flow. The calculations indicate that the amount of stream power applied to the boundary can range anywhere from about 25 percent to close to 100 percent. When the flow depth to absolute roughness ratio is close to one (i.e., very rough turbulent flow conditions), the amount of stream power applied to the boundary is close to 100 percent. When the flow depth to absolute roughness ratio is closer to 50 the amount of stream power applied to the boundary converges to about 40 percent (Fig. 5.8). The percentage of the total power applied to the bed under smooth turbulent flow converges to roughly 25 percent (Fig. 5.9).

Further confirmation of a relationship between stream power and pressure fluctuations at the boundary can be found in an empirical relationship between the relative magnitude of pressure fluctuations below a hydraulic jump and the rate of energy dissipation (Annandale 1995). Figure 5.11 shows a linear relationship between stream power and the relative magnitude of pressure fluctuations under a hydraulic jump, based on experimental data by Fiorotto and Rinaldo (1994). This correlation confirms a linear relationship between stream power (turbulence production, or rate of energy dissipation) and the relative magnitude of pressure fluctuations.

Therefore, from a practical engineering point of view it is concluded that stream power is a good indicator of the relative magnitude of pressure fluctuations, and therefore of the erosive capacity of water in turbulent flow. The actual stream power on the bed should in all cases be equal to or less than the total available stream power.

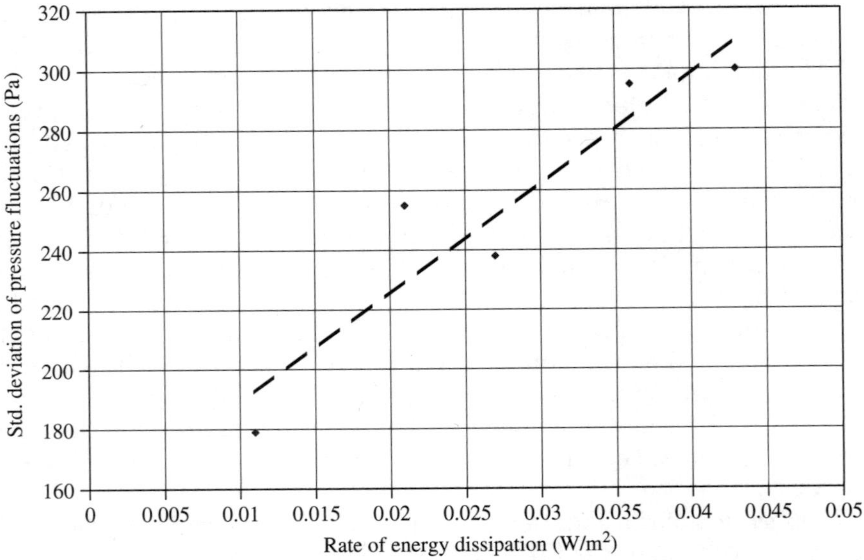

Figure 5.11 Relationship between the relative magnitude of pressure fluctuations and the rate of energy dissipation (stream power) in a hydraulic jump (Annandale, 1995).

Quantification of Erosive Capacity

The preferred methods in this book for determining the magnitude of the erosive capacity of turbulent flowing water are to quantify either

- The actual magnitude of fluctuating pressures by making use of published research, or
- The relative magnitude of fluctuating pressures by making use of stream power concepts

These approaches are followed because most scour problems dealt with by practicing engineers occur under turbulent flow conditions where pressure fluctuations are the driver leading to incipient motion. The approach followed in this book therefore differs from those followed in conventional practice where the use of shear stress and average flow velocity as indicator parameters for quantifying the relative magnitude of the erosive capacity of water is common.

Scour in turbulent flow is not a shear process. It is caused by turbulent, fluctuating pressures. Scour in laminar flow, not characterized by fluctuating pressures, is a shear process (see Chap. 3). Very few scour problems encountered by engineers are associated with laminar flow. It is therefore imperative to view and interpret scour in practical situations as a process resulting from turbulent pressure fluctuations.

We can divide methods for quantifying the erosive capacity of flowing water into two groups; structural hydraulics and environmental hydraulics. Structural hydraulics refers to the hydraulics of geometrically engineered structures such as internal erosion in embankment dams, flow around bridge piers, and flow in spillways, spillway chutes, stilling basins, plunge pools, gates, valves, and so forth. Environmental hydraulics refers to the hydraulic characteristics of flow in natural systems like natural river channels, wetlands, lakes, beaches, and the like. Although both structural and environmental hydraulics are obviously based on the same principles, each have specific, dominant characteristics as they relate to scour.

Structural hydraulics

As already indicated, the relative magnitude of the erosive capacity of water can be quantified by either direct quantification of the magnitudes of turbulent fluctuating pressures or by making use of stream power. Quantification of pressure fluctuations resulting from plunging jets can be accomplished by making use of work by Ervine and Falvey (1987), Ervine, et al. (1997), and Bollaert (2002).

A number of approaches exist for quantifying stream power representing various flow conditions. These include the stream power associated with plunging jets, jet diffusion in plunge pools, flow around bridge piers, flow through cracks in embankment dams (internal erosion of embankment dams), flow through fissures in dam foundations, flow through cracks in aprons used to protect dam foundations against fissure erosion, flow over drops, and hydraulic jumps.

The principal focus in what follows is on quantification of stream power. However, methods for quantifying the actual magnitude of fluctuating pressures under turbulent flow conditions are also presented. The latter is limited to the quantification of turbulent pressure fluctuations generated by plunging jets.

Plunging jet geometry. Plunging jets occur in various engineering applications, inlcuding overtopping dams, at the ends of spillway chutes, eminating from gates and valves, and the like. (see e.g., Figs. 5.12 and 5.13). Prior to developing equations for calculating the stream power of plunging jets it is appropriate to first present the known relationships between geometric properties of jets. This is followed by jet stream power equations.

The relevant jet components are jet issuance, the plunging jet, and the plunge pool. Figure 5.14 shows a jet discharging over a dam with issuance velocity V_i, issuance jet thickness D_i, and average issuance angle θ. As the jet plunges through the air its inner core contracts due

Figure 5.12 Plunging jet, aeration, and energy dissipation.

Figure 5.13 Jet discharging from a valve under pressure.

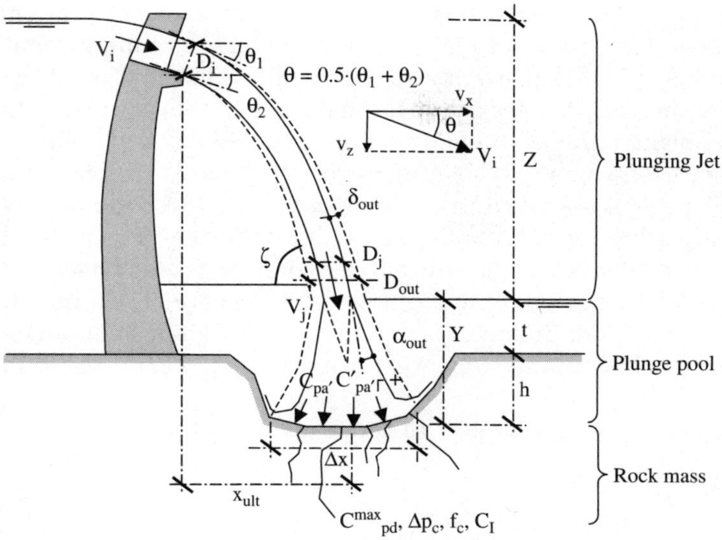

Figure 5.14 Nomenclature for a jet discharging over an ogee spillway and plunging into a pool (Bollaert 2002).

to the effects of gravity and the outer portion expands due to the effects of turbulence. Once the jet plunges into a pool, it experiences additional diffusion.

The issuance turbulence intensity, defined as

$$T_u = \frac{\sqrt{(v')^2}}{V} = \frac{v'}{V} \tag{5.35}$$

is an important parameter, at jet issuance, determining overall jet characteristics. The variable v' is the root mean square value of the fluctuating velocity, and V is the mean axial flow velocity of the jet.

Table 5.2 contains estimates of issuance turbulence intensity for use in practice.

TABLE 5.2 Typical Values of Issuance Turbulence Intensity T_u at Various Outlet Structure Types

Type of outlet structure	Turbulence intensity T_u
Free overfall	0.00–0.03
Ski jump outlet	0.03–0.05
Valve	0.03–0.08

SOURCE: From Bollaert 2002

When calculating the characteristics of plunging jets one is interested in its trajectory, trajectory length, breakup length, impingement angle, and spread. Knowledge of its trajectory allows calculation of the locations of impingement zones, while the spread is used to calculate the size of the footprint of the jet. Comparison of the trajectory and breakup lengths allows assessment of the state of the jet when it plunges into the pool. The state of the jet can be intact, undeveloped, or completely developed. An undeveloped jet is defined as a jet that has experienced reduction of its central core with air entrained in its frayed edges. By the time such a jet impinges onto a plunge pool the core is still intact, although its thickness or diameter is diminished. The angle of impingement is used to calculate the submerged trajectory length of the jet in the plunge pool.

The jet trajectory is calculated with the following equation:

$$z = x\tan\theta - \frac{x^2}{K_2[4(D_i + h_v)(\cos\theta)^2]} \tag{5.36}$$

where z = vertical distance
x = horizontal distance
θ = issuance angle
K_2 = coefficient allowing for the effects of air resistance on the jet trajectory
D_i = thickness of jet
$h_v = V_i^2/2g$
g = acceleration due to gravity

The value of K_2 is normally set to 0.75, but this is not necessarily representative of actual conditions. For example, when introducing air into a spillway chute to prevent cavitation, the flow velocity of the jet can increase and values of K_2 may be equal to 1.0 or higher. The reason for the increase in jet velocity in an aerated spillway chute is that the air reduces the friction on the spillway chute bed.

Very little air movement is generally associated with undeveloped, plunging turbulent jets. Most of the air interacting with a turbulent jet is sucked into the outer edges of the jet from the space around it and air resistance experienced by undeveloped turbulent jets is minimal (Ervine et al., 1997). The almost stationary mist surrounding large plunging jets, like those occurring at Niagara Falls or Victoria Falls, is evidence. The mist around such jets essentially "hangs" in the air around the jet and hardly moves.

When a jet is completely developed (Fig. 5.15) it no longer contains a core but essentially consists of blobs of water that disintegrate into finer

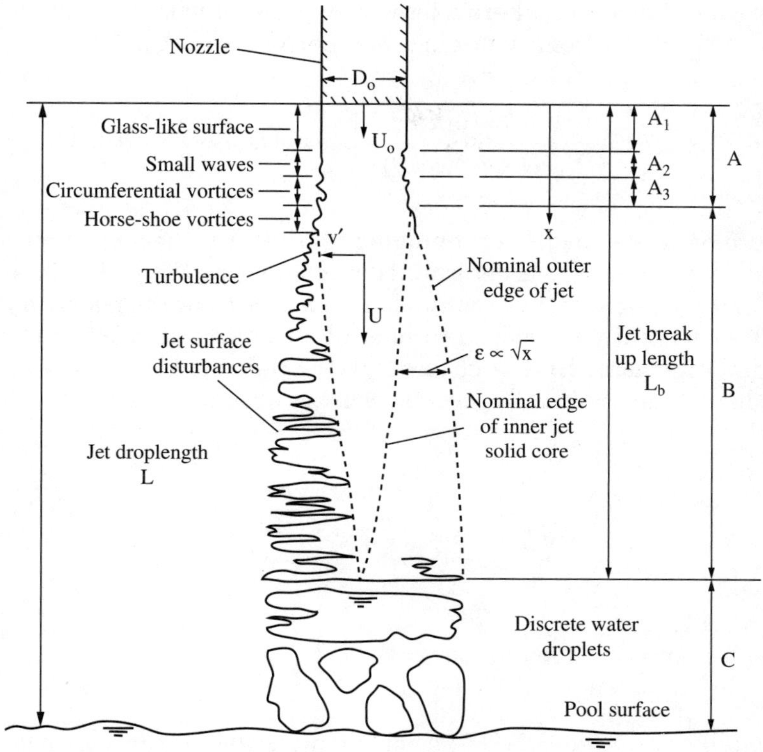

Figure 5.15 Jet characteristics (Ervine et al., 1997).

and finer drops. Individual blobs and drops of water slow down due to air drag and eventually reach terminal velocity. The latter occurs when the drag introduced by the air equals the weight of individual water globules or drops. Such interaction limits the erosive capacity of a fully developed jet.

By manipulating Eq. (5.36), it is possible to develop an equation that can be used to calculate the trajectory length of a plunging jet (L_j) as follows:

$$L_j = \int_0^x \sqrt{1 + \left[\tan\theta - \frac{2x}{K_2\left[4(D_i + h_v)(\cos\theta)^2\right]} \right]^2}\, dx \qquad (5.37)$$

This equation needs to be integrated numerically to solve for the jet trajectory length.

The horizontal distance where a jet will intersect a horizontal plane at a vertical distance z from an origin is also derived from Eq. (5.36), i.e.,

$$x = \left[\tan\theta + \sqrt{(\tan\theta)^2 - \frac{(-1)z}{K_2(D_i + h_v)(\cos\theta)^2}} \right] \times 2K_2(D_i + h_v)(\cos\theta)^2 \quad (5.38)$$

The state of development of a plunging jet (i.e., whether it is undeveloped or fully developed) is determined by comparing the jet breakup and trajectory lengths. If the breakup length is less than the trajectory length, the jet is fully developed (broken) by the time it reaches the point of impingement. Ervine et al. (1997) found the following relationship describing the breakup characteristics of round jets:

$$C = \frac{1}{\left(\sqrt{\frac{2L_b}{D_i Fr_i^2} + 1}\right)\left(\sqrt{\frac{2L_b}{D_i Fr_i^2} + 1} - 1\right)} \quad (5.39)$$

where $C = 1.14 T_u Fr_i^2$
Fr_i = issuance Froude number
L_b = breakup length of the jet

Additionally Ervine et al. (1997) used laboratory data to develop a jet breakup length equation for round jets, i.e.,

$$\frac{L_b}{D_i Fr_i^2} = \frac{1.05}{C^{0.82}} \quad (5.40)$$

Horeni (1953) proposed an equation to calculate the breakup length of rectangular jets, i.e.,

$$L_b = 6q^{0.32} \quad (5.41)$$

Other equations that can be used to estimate jet breakup length are summarized in Table 5.3. When applying the equation provided by Baron (1949) it is useful to recall that the Weber number is defined as

$$We = \frac{\rho U^2 L}{\sigma} \quad (5.42)$$

and that the surface tension of water $\sigma \approx 0.073\ N/m$.

Ervine and Falvey (1987) and Ervine et al. (1997) determined that the relationship between the outer spread of the jet δ_{out} and the distance

TABLE 5.3 Equations for Calculating Jet Breakup Length

Jet type	L_b	T_u	Authors
Circular jets	$1.7\frac{We}{(10^{-4}\,\mathrm{Re})^{5/8}}$	3%	Baron (1949)
Circular jets	$60Q^{0.39}$ $17.4Q^{0.31}$ $4.1Q^{0.20}$	0.3% 3.0% 8.0%	Ervine et al. (1980)
Circular jets	$50D_j$ to $100D_j$	3% to 8%	Ervine and Falvey (1987)

along the jet trajectory X is related to the turbulence intensity, i.e.,

$$\frac{\delta_{\text{out}}}{X} = 0.38T_u \tag{5.43}$$

The issuance turbulence intensity of jets can range from values as low as about 1 to 2 percent for smooth jets to somewhere between 5 and 8 percent for highly turbulent jets. This means that the outer jet spread in a turbulent jet is on the order of 3 to 4 percent (i.e., $\delta_{\text{out}}/X \approx$ 3–4 percent, equivalent of 1.7° to 2.3°). Ervine and co-researchers also determined that the inner contraction of a turbulent jet is about 15 to 20 percent of the outer spread, which means that $\delta_{\text{in}}/X \approx$ 0.5–1 percent, i.e., it ranges between 0.3° to 0.6° (Fig. 5.16).

The outer dimension of the jet (D_{out}) can therefore be calculated as

$$D_{\text{out}} = D_i + 2\frac{\delta_{\text{out}}}{X}L_j \tag{5.44}$$

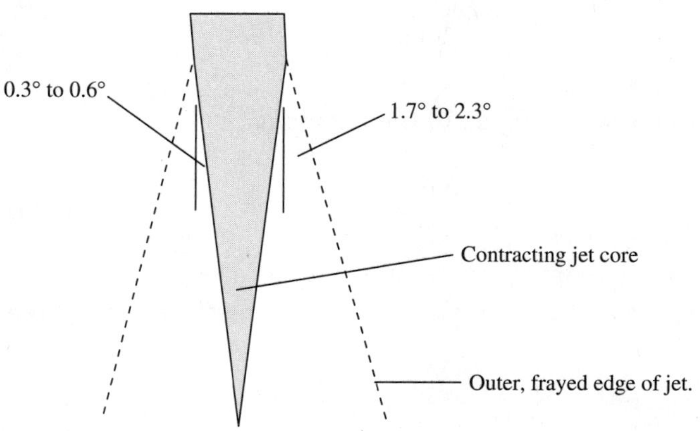

Figure 5.16 Inner contracting core and outer edge of plunging jet, showing inner and outer angles.

which is equivalent to

$$D_{\text{out}} = D_i + 2 \times 0.38(T_u L_j) \tag{5.45}$$

By making use of the continuity equation for round jets Ervine et al. (1997) showed that the diameter of the core of a round jet (D_j) can be expressed as

$$D_j = D_i \sqrt{\frac{V_i}{V_j}} \tag{5.46}$$

where the impact velocity (V_j) is expressed as

$$V_j = \sqrt{V_j^2 + 2gZ} \tag{5.47}$$

An equation to calculate the outer dimensions of a round jet have been proposed by Ervine et al. (1997) by making use of the core diameter of the jet D_j and an estimate of the jet spread ε, expressed as follows:

$$\varepsilon = \frac{1.14 T_u V_i^2}{g}\left[\sqrt{\frac{2L}{D_i Fr_i^2} + 1} - 1\right] \tag{5.48}$$

The outer dimension of the circular jet is calculated as

$$D_{\text{out}} = D_i \sqrt{\frac{V_i}{V_j}} + 2\varepsilon \tag{5.49}$$

The previous equation is subject to the following:

$$V_i \geq \frac{0.275}{T_u} \tag{5.50}$$

Castillo (1998) proposed an equation for calculating the spread of rectangular jets,

$$D_{\text{out}} = \frac{q}{\sqrt{2gZ}} + 4\varphi\sqrt{h_o}\left[\sqrt{Z} - \sqrt{h_o}\right] \tag{5.51}$$

where $\varphi = 1.07T_u$ for rectangular jets
h_o = overflow depth over a free-flowing ogee spillway
q = unit discharge

Equation (5.51) is applicable to nappe jets only, i.e., free-flowing jets over an ogee crest.

An expression for calculating the angle of impingement (ζ in degrees) is derived from Eq. (5.36), i.e.,

$$\zeta = \arctan\left[\tan\theta - \frac{x}{2K_2(D_i + h_v)(\cos\theta)^2}\right](-1) \tag{5.52}$$

If a flat jet discharges over an ogee spillway on the crest of, say, an arch dam prior to plunging through the air, its footprint is most likely rectangular (Fig. 5.17). However, when releasing it from a long, narrow chute the drag on the chute walls reduces the flow velocity at the edges. In such cases, the footprint of the jet assumes the shape of a horseshoe or inverted U (Figs. 5.18 and 5.19). The footprint shape and dimensions of a jet are important when calculating the impact conditions of a plunging jet.

Plunge pool—jet geometry changes. When a jet plunges into a pool it experiences additional diffusion. The outer boundary of the jet expands as the jet travels deeper into the pool. Ervine and Falvey (1987) and Ervine et al. (1997) studied the geometry of round jets in pools and found that the expansion angle of the outer jet boundary differs from the jet core contraction angle. If the core is still intact when the jet plunges into

Figure 5.17 Rectangular jet discharging over an ogee and forming a rectangular footprint.

Figure 5.18 Image of side of a plunging jet from a long, narrow chute.

the pool its angle of contraction is a function of the jet condition, i.e., whether it is a smooth laminar jet or a turbulent jet (Fig. 5.20).

The core angle of contraction for smooth, almost laminar round jets is about 5°, while that of smooth turbulent jets entraining only small amounts of air into the plunge pool is about 7° to 8°, and that of high

Figure 5.19 Footprint of a plunging jet discharging from a long, narrow spillway chute.

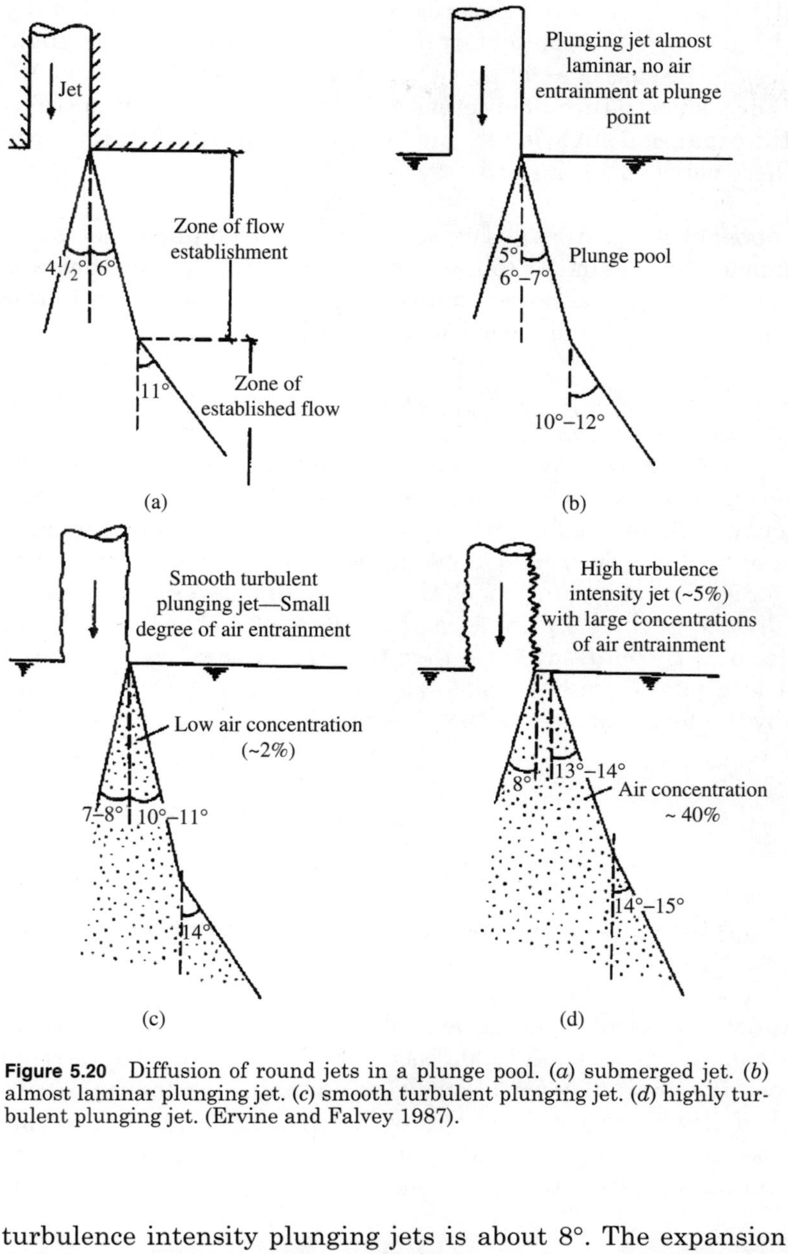

Figure 5.20 Diffusion of round jets in a plunge pool. (*a*) submerged jet. (*b*) almost laminar plunging jet. (*c*) smooth turbulent plunging jet. (*d*) highly turbulent plunging jet. (Ervine and Falvey 1987).

turbulence intensity plunging jets is about 8°. The expansion angle defining the outer boundary of jet flow in a plunge pool also varies as a function of jet conditions.

In the case of smooth, almost laminar round jets the expansion angle is about 6° to 7°. The same, for smooth turbulent round jets entraining small amounts of air into a plunge pool, is about 10° to 11°, while that of highly turbulent jets is on the order of 13° to 14°. Lower down in the pool where the jet is completely broken up, the expansion angle increases. In the case of an almost laminar jet it increases to about 10° to 12°. For a smooth turbulent jet with a minor amount of air entrainment, the expansion angle is 14°, and for a rough turbulent jet it ranges somewhere between 14° and 15°.

Stream power of plunging jets. The stream power of a plunging jet is calculated in a manner similar to that used when estimating hydropower potential. The amount of power that is available for causing scour after a jet has plunged through a verticle distance H is

$$P_{\text{jet}} = \gamma QH \tag{5.53}$$

where P_{jet} is the total stream power of the jet and Q is the total discharge.

The stream power per unit area is calculated by dividing the total stream power P_{jet} by the footprint area of the jet at the point of impact. For example, if it is desired to know the stream power per unit area at the water surface of a plunge pool, the total steam power at that elevation is divided by the footprint area at the same elevation. Or, if the jet impinges directly onto, say, rock, then the stream power per unit area at that location is similarly calculated by dividing the total stream power by the footprint area of the jet where it impinges onto the rock. Therefore,

$$p_{\text{jet}} = \frac{\gamma QH}{A} \tag{5.54}$$

where P_{jet} is the stream power per unit area and A is the footprint area of the jet.

Plunge pool diffusion of stream power. When a jet plunges into a pool its erosive capacity is affected by diffusion. Estimation of a jet's stream power per unit area at a given depth below the water surface can be estimated by following two optional approaches. The first approach simply scales the stream power by dividing it by estimated flow areas at various elevations below the water surface elevation of the plunge pool, i.e.,

$$p_{\text{pool}} = \frac{\gamma QH}{A_i} \tag{5.55}$$

where P_{pool} is the stream power per unit area at a particular depth below the water surface elevation of the plunge pool and A_i is the flow area of the jet at the desired depth below the water surface elevation. The flow area can be estimated by making use of the guidelines provided by Ervine and Falvey (1987); see previous section.

It should be noted that the value of H is the drop height of the jet where it impinges onto the plunge pool water surface. The product γQH is the total power at the plunge pool water surface elevation. From here onwards the power dissipates in the pool.

The second approach is to make use of average and fluctuating stream power decay coefficients, following an approach similar to that used when estimating average and fluctuating dynamic pressures (see section dealing with estimation of fluctuating pressures further on). The convention when estimating the average dynamic pressure in a plunge pool is to make use of an average dynamic pressure coefficient C_p. Once the magnitude of the average dynamic pressure coefficient is known the average dynamic pressure D_p can be calculated:

$$D_p = C_p \gamma \frac{V_j^2}{2g} = C_p \frac{1}{2} \rho V_j^2 \tag{5.56}$$

where V_j is the jet velocity at the water surface of the plunge pool.

Average stream power can be determined by multiplying the shear stress by the velocity. It means that a relationship between the average dynamic pressure coefficient C_p and average stream power decay coefficient C_{sp} can be developed. This can be done by first expressing the average dynamic pressure coefficient as

$$C_p = \frac{1/2 \rho V_z^2}{1/2 \rho V_j^2} \tag{5.57}$$

where V_z is the jet velocity at an elevation z in the plunge pool.

Should one now define the average stream power decay coefficient in a similar manner, i.e.,

$$C_{sp} = \frac{1/2 C_f \rho V_z^3}{1/2 C_f \rho V_j^3} = \frac{1/2 \rho V_z^3}{1/2 \rho V_j^3} \tag{5.58}$$

It can be shown that

$$C_{sp} = C_p \frac{V_z}{V_j} \tag{5.59}$$

The problem with this equation is that one does not necessarily know what the variable velocity V_z in the pool is. If this is known (by e.g., measuring it is physical model studies) the stream power adjustment factor can be suitably calculated. When such information is not known, or cannot be estimated with reasonable accuracy, then it is considered acceptable to use the dynamic pressure coefficient as an estimate of the value of the stream power decay coefficient. Justification for this recommendation can be found by comparing the theoretical values of the average dynamic pressure coefficient and the stream power decay coefficient.

The relationship can be determined by making use of equations to quantify the dynamic pressure of a submerged jet. Hanson et al. (2000) provide equations for calculating these pressures, i.e.,

$$D_p = \frac{1}{2}\rho V_j^2 \quad \text{if} \quad J \leq J_p \tag{5.60}$$

and

$$D_p = \frac{1}{2}\rho\left(\frac{J_p}{J}V_j\right)^2 \quad \text{if} \quad J > J_p \tag{5.61}$$

where J_p = the length of the core of the jet = K_jD_j
K_j = empirically determined factor = 6.3 for most jets
J = actual length of the jet

Using Eqs. (5.60) and (5.61) the average dynamic pressure coefficient can be reformulated as

$$C_p = \frac{D_p}{1/2\rho V_j^2} \tag{5.62}$$

The average stream power decay factor can then be written as

$$C_{sp} = C_p\frac{V_z}{V_j} = C_p\frac{J_p}{J} \quad \text{if} \quad J > J_p \tag{5.63}$$

In cases when $J \leq J_p$, then $C_{sp} = C_p = 1$.

The average dynamic pressure coefficient and the average stream power decay coefficient as a function of the dimensionless plunge pool depth (Y/D_j) is shown in Fig. 5.21.

Figure 5.21 shows that the value of the stream power decay coefficient is theoretically less than or equal to the average dynamic pressure coefficient for all dimensionless plunge pool depths. It is therefore considered

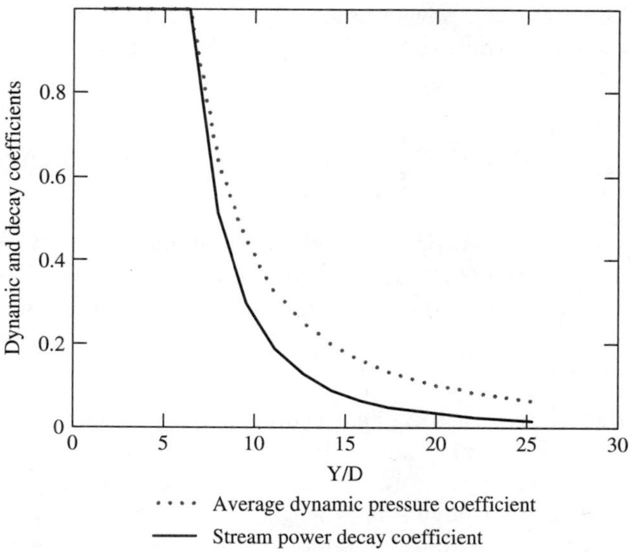

Figure 5.21 Comparison of average dynamic pressure coefficient and the average stream power decay coefficient in plunge pools as a function of dimensionless depth for an impinging jet with a solid core.

reasonable, and conservative, to use the average dynamic pressure coefficient to approximate the decay of stream power in a plunge pool if the varying jet velocity V_z is not known.

In summary, the average stream power per unit area at the water surface elevation of a plunge pool is calculated as

$$p_{\text{jet}} = \frac{\gamma QH}{A} \tag{5.64}$$

and the variation in average stream power within the pool as

$$p_{\text{jet}}\left(\frac{Y}{D}\right) = C_{\text{sp}}\left(\frac{Y}{D}\right)\frac{\gamma QH}{A} \approx C_p\left(\frac{Y}{D}\right)\frac{\gamma QH}{A} \tag{5.65}$$

where $p_{\text{jet}}(Y/D)$ is the average stream power per unit area as a function of Y/D and $C_{sp}(Y/D)$ is assumed to equal $C_p(Y/D)$ the average dynamic pressure coefficient as a function of Y/D.

The fluctuating portion of stream power per unit area around the mean is calculated as

$$p'_{\text{jet}}\left(\frac{Y}{D}\right) = C'_p\left(\frac{Y}{D}\right)\frac{\gamma QH}{A} \tag{5.66}$$

The variation in total stream power per unit area as a function of dimensionless pool depth is expressed as the sum of the mean and fluctuating portions, i.e.,

$$p_{\text{total}}\left(\frac{Y}{D}\right) = p_{\text{jet}}\left(\frac{Y}{D}\right) + p'_{\text{jet}}\left(\frac{Y}{D}\right) \tag{5.67}$$

The values of the average dynamic and fluctuating dynamic pressure coefficients can be determined from the section dealing with quantification of average and fluctuating dynamic pressures further on in this Chapter.

Stream power in stilling basins. Stilling basins downstream of dams dissipate energy by means of hydraulic jumps. In order to ensure that their linings can withstand imposed pressure fluctuations it is necessary to quantify either the actual or relative magnitudes of these pressures. Quantification of actual magnitudes of fluctuating pressures can be accomplished by making use of work by Fiorotto and Rinaldo (1992a and 1992b), Fiorotto and Salandin (2000), and Fiorotto and Tanda (1984).

The rate of energy dissipation (stream power) developed by a hydraulic jump can be calculated once the energy loss ΔE over a hydraulic jump is quantified. From Henderson (1966),

$$\Delta E = y_1 + \frac{q^2}{2gy_1^2} - \frac{y_1}{2}\left(\sqrt{1+8Fr_1^2}-1\right) - \frac{4q^2}{2gy_1^2\left(\sqrt{1+8Fr_1^2}-1\right)^2} \tag{5.68}$$

where ΔE = energy head loss over the jump

y_1 = upstream water depth

$Fr_1 = V_1/\sqrt{gy_1}$ = Froude number of flow upstream of the jump

The average stream power per unit area underneath a wide hydraulic jump is the product of the unit weight of water, unit discharge and the average energy loss per unit length of the jump, i.e.,

$$P_{\text{avaliable}} = \gamma q \frac{\Delta E}{L}$$

$$= \gamma \frac{q}{L}\left(y_1 + \frac{q^2}{2gy_1^2} - \frac{y_1}{2}\left(\sqrt{1+8Fr_1^2}-1\right) - \frac{4q^2}{2gy_1^2\left(\sqrt{1+8Fr_1^2}-1\right)^2}\right) \tag{5.69}$$

where L is the effective hydraulic jump length over which the energy is dissipated.

Information of how energy dissipation is distributed over the extent of a hydraulic jump is not currently readily available. It is most probably reasonable to expect that the spatial distribution of energy loss over a hydraulic jump follows a negative exponential shape (obviously depending on the type of jump), with the greatest amount of energy dissipation at the beginning of the jump and lower rates of dissipation further downstream. Should the actual distribution of energy head loss be known, it would be possible to calculate the distribution of the rate of energy dissipation over the total length of the jump. This would provide a realistic assessment of the spatial distribution of the erosive power under a hydraulic jump.

In the absence of appropriate data it is assumed, for purposes of design, that the energy is distributed over a unit length, i.e., $L = 1$ m when using the SI system. This is obviously a conservative approach, which can be improved if more information about the spatial distribution of energy loss along hydraulic jumps becomes available.

Internal erosion. Internal erosion occurs if an embankment contains cracks that will allow flow-through of water. For example, embankment dams in arid regions often experience desiccation that leads to the formation of transverse and longitudinal cracks (Fig. 5.22). In addition to cracks forming in embankments dams, the formation of earth fissures also poses a threat to such facilities. Regional ground water abstraction in Arizona leads to the development of ground subsidence over large areas. Figure 5.23 shows a foundation fissure in one of the dams in the general vicinity of Phoenix.

It is conceivable that water can flow through such cracks and fissures. If the erosive capacity of water exceeds the ability of the earth material to resist it, the features can increase in size and eventually lead to failure of the dam, its foundation, or both. Mathematical models calculating the relative magnitude of the erosive capacity of the water can be used to assess the erosion potential of cracks and fissures.

In what follows, equations for calculating the erosive capacity of water in foundation fissures are presented. A similar approach can be implemented to develop equations for flow through embankment cracks.

The total stream power through a fissure is expressed as

$$P = \gamma Q h_f \tag{5.70}$$

where γ = unit weight of water

Q = total discharge through the fissure

h_f = total head loss through the fissure

Figure 5.22 Desiccation cracks that form in embankment dams at regular close intervals in arid regions can lead to internal erosion of dams.

Figure 5.23 Earth fissure in a foundation of an earth embankment dam in the general vicinity of Phoenix, Arizona.

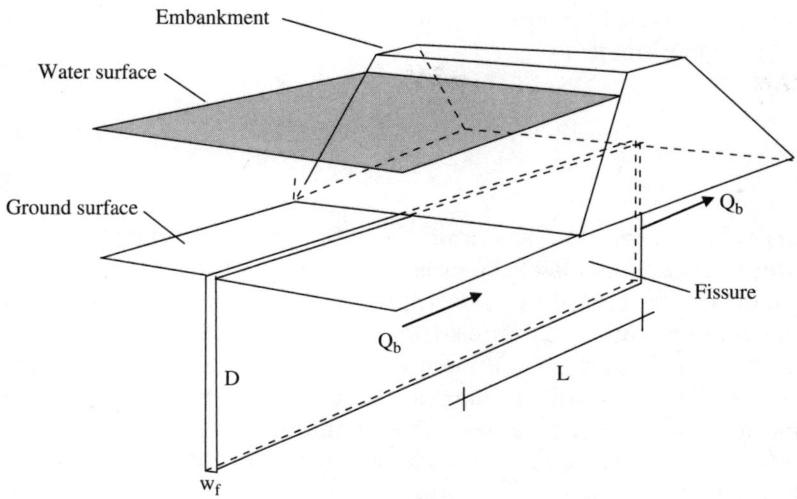

Figure 5.24 Embankment dam with fissure through foundation.

This expression can also be written as

$$P = \gamma h'_f \tag{5.71}$$

where $h'_f = Qh_f$ denotes the energy loss flux. In what follows equations representing the energy loss flux are developed.

Embankment with foundation fissure. Figure 5.24 depicts a fissure in the foundation underneath an embankment. The water surface elevation is at a depth H above the ground surface. Water flows through the fissure at a discharge rate Q_b. The total length of the fissure underneath the embankment is equal to L. The fissure width is w_f and its depth is D.

As water flows through the fissure, the energy head loss h_f is

$$h_f = f\frac{L(2D+2w_f)}{8(Dw_f)}\frac{Q_b^2}{g(Dw_f)^2} \tag{5.72}$$

where f is the Darcy friction factor (see Table 5.1).

The total applied stream power, which leads to erosion of the fissure, is expressed as

$$P_1 = \gamma Q_b h_f \tag{5.73}$$

Therefore, the total amount of power expended in the fissure underneath the embankment is

$$P_1 = \frac{f \rho L(2D + 2w_f)}{8(Dw_f)^3} \cdot Q_b^3 \tag{5.74}$$

Fissure with an apron. Lengthening of the flow path through a fissure can reduce the magnitude of the erosive capacity of the water and prevent the fissure from eroding. One approach of accomplishing this is to construct an apron on the ground surface upstream of the embankment.

Figure 5.25 depicts an embankment with an apron. The fissure extends from upstream of the apron, across and underneath the apron and embankment, and downstream of the embankment. The water surface elevation is at a depth H above the apron. Water flows through the fissure at a discharge rate Q_b. The total length of the fissure, underneath the apron and the embankment, is equal to $L_a + L$. The fissure width is w_f and its depth is D. As water flows through the fissure, the energy head loss h_f is

$$h_f = f \frac{(L_a + L)(2D + 2w_f)}{8(Dw_f)} \frac{Q_b^2}{g(Dw_f)^2} \tag{5.75}$$

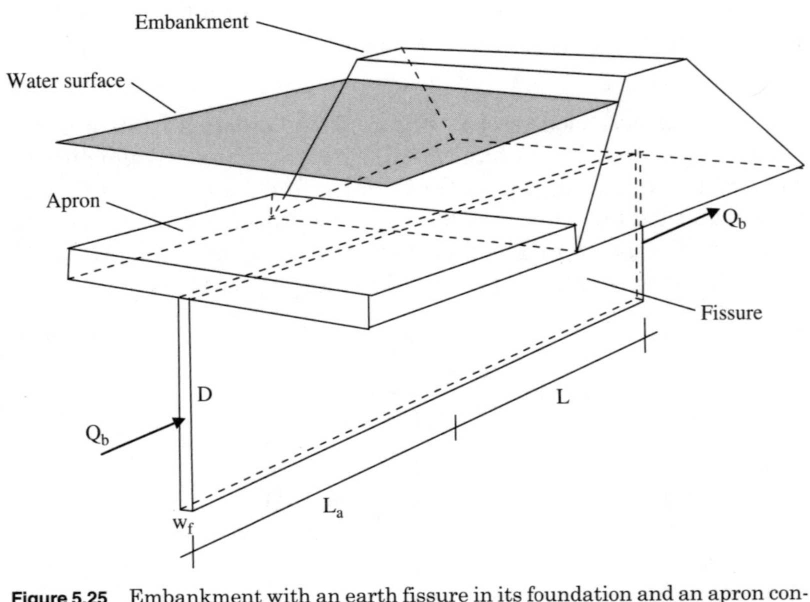

Figure 5.25 Embankment with an earth fissure in its foundation and an apron constructed upstream of the embankment to lengthen the flow path and reduce the erosive capacity of water in the fissure.

By making use of Eq. (5.75), it is possible to express the total amount of power expended through the fissure underneath the apron and embankment as

$$P_2 = \frac{f\rho(L_a + L)(2D + 2w_f)}{8(Dw_f)^3} \cdot Q_b^3 \tag{5.76}$$

Apron with crack. Economic construction of aprons can be accomplished by making use of soil cement, which is subject to cracking. If the cracks are very narrow, its effect can be beneficial in reducing the maximum erosive capacity in the earth fissure. Obviously, if the crack becomes too wide, assuming that it is located directly above the fissure, it will become ineffective in reducing the erosive capacity of the water in the fissure. In fact, if the crack is very wide the flow conditions revert to those shown in Fig. 5.24.

The simplified configuration of an apron with a crack is shown in Fig. 5.26. The presence of the fissure results in water flowing through the crack into the fissure, in addition to the water that discharges into the fissure from upstream.

It is considered reasonable to assume that the water discharges at a constant rate q_a through the fissure. The amount of water flowing into the fissure from upstream changes to Q'_b, which is less than Q_b. The total

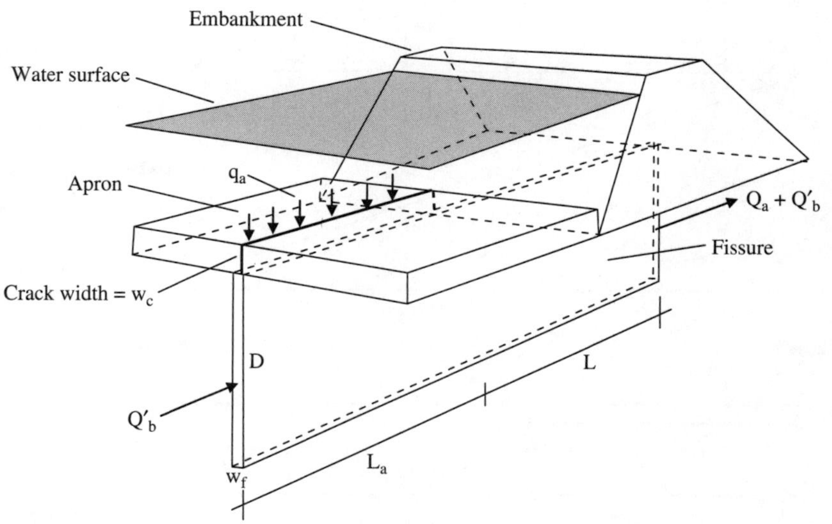

Figure 5.26 Embankment with a foundation fissure, upstream apron and crack in the apron located directly above the earth fissure.

outflow from the fissure, at the downstream toe, is equal to $Q_a + Q'_b$, where $Q_a = q_a L_a$.

From an overall mass balance point of view, the amount of water flowing into the fissure must equal the amount flowing out, i.e.,

$$Q'_b + q_a L_a = Q'_b + Q_a \tag{5.77}$$

Once the water flows through the fissure directly beneath the downstream edge of the apron its magnitude $(Q'_b + Q_a)$ remains constant from there onwards (Fig. 5.27). The distribution of discharge underneath the apron indicates that the erosive capacity of water in the fissure increases until it reaches a maximum at the upstream toe of the embankment.

It is necessary to set up a differential equation of flow through the fissure underneath the apron in order to develop an expression for calculating stream power in the fissure. This derivation commences with the Darcy equation, i.e.,

$$h_f = \frac{fL}{R_h}\frac{v^2}{2g}$$

where $R_h = 4A/P$ = hydraulic radius of an enclosed conduit
$A = Dw_f$ = cross-sectional area of the fissure
$P = (2D + 2w_f)$ = wetted perimeter of the fissure

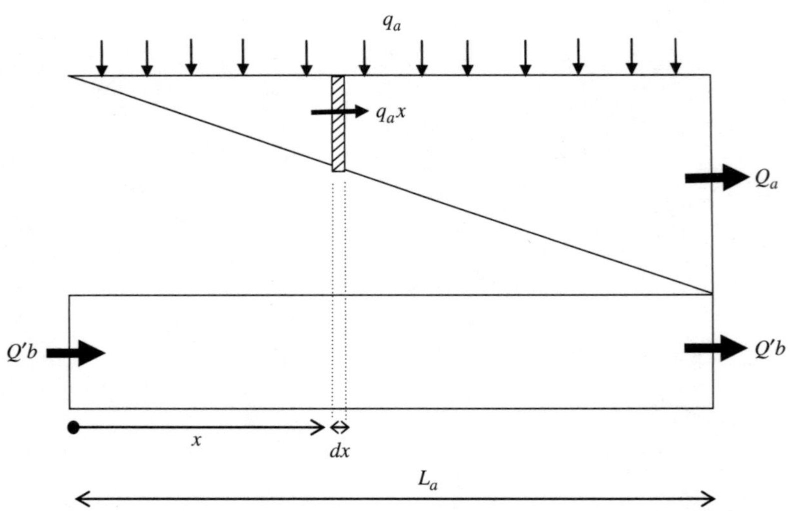

Figure 5.27 Distribution of discharge in the fissure reach directly underneath the apron.

One can rewrite this equation to express the energy slope over a short distance dx as

$$\frac{dh_f}{dx} = \frac{f}{R_h}\frac{v^2}{2g} \tag{5.78}$$

The velocity in the fissure can also be expressed as a function of x (Fig. 5.27)

$$v = \frac{Q'_b + q_a x}{w_f D} \tag{5.79}$$

from which follows

$$dh_f = \frac{f}{2gR_h w_f^2 D^2}(Q'_b + q_a x)^2 dx \tag{5.80}$$

Converting Eq. (5.80) to an expression representing *energy flux* by multiplying it with the discharge, i.e.,

$$dh'_f = \frac{f}{2gR_h w_f^2 D^2}(Q'_b + q_a x)^3 dx \tag{5.81}$$

where h'_f is the flux of energy loss.

The total energy loss flux in the fissure underneath the apron (excluding the losses through the crack at this stage) can therefore be expressed as

$$h'_f = \int_0^{L_a} \frac{f}{2gR_h w_f^2 D^2}(Q'_b + q_a x)^3 dx \tag{5.82}$$

solving for the integral results in

$$h'_f = \frac{f(2D + 2w_f)}{8gw_f^3 D^3}\left(Q_b'^3 L_a + \frac{3}{2}Q_b'^2 q_a L_a^2 + Q'_b q_a^2 L_a^3 + \frac{1}{4}q_a^3 L_a^4\right) \tag{5.83}$$

However, this is not the total energy loss flux over the distance L_a. The energy loss flux of the water flowing through the crack in the apron needs to be added. If the crack width is w_c then the velocity through the crack can be expressed as

$$v_c = \frac{q_a}{w_c} \tag{5.84}$$

And the energy head loss per unit length of crack is

$$h_{fc} = K\frac{v_c^2}{2g} = K\frac{q_a^2}{2gw_c^2} \tag{5.85}$$

where K is an energy loss coefficient.

The flux of energy loss per unit length of crack is

$$h'_{fc} = K\frac{v_c^2}{2g}\cdot q_a = K\frac{q_a^3}{2gw_c^2} \tag{5.86}$$

From which follows that the total energy loss flux through the crack can be expressed as

$$h'_{fc} = K\frac{q_a^3}{2gw_c^2}L_a \tag{5.87}$$

Therefore, the total flux of energy loss through the crack and fissure underneath the apron is

$$h'_{f\_\text{apron}} = \frac{f(2D+2w_f)}{8gw_f^3D^3}\left(Q_b'^3L_a + \frac{3}{2}Q_b'^2q_aL_a^2 + Q_b'\,q_a^2L_a^3 + \frac{1}{4}q_a^3L_a^4\right) + K\frac{q_a^3}{2gw_c^2}\cdot L_a \tag{5.88}$$

and the total flux of energy loss by the time the water reaches the downstream toe of the embankment is

$$h'_{f\_\text{total}} = \frac{f(2D+2w_f)}{8gw_f^3D^3}\left(Q_b'^3L_a + \frac{3}{2}Q_b'^2Q_aL_a + Q_b'Q_a^2L_a + \frac{1}{4}Q_a^3L_a\right) + K\frac{Q_a^3}{2gw_c^2L_a^3}\cdot L_a + \frac{fL(2D+2w_f)}{8gw_f^3D^3}\cdot(Q_b'+Q_a)^3 \tag{5.89}$$

With Eq. (5.89) representing the total flux of energy loss, the total power expended through the crack in the apron, along the fissure underneath the apron, and in the fissure underneath the embankment dam is

$$P_3 = \gamma h'_{f\_\text{total}} \tag{5.90}$$

Comparison. The formulation of equations to calculate the total amount of power expenditure for the three optional flow scenarios through the fissure and apron crack can become quite complex. However, when comparing the total amount of power expended in each of these scenarios one finds that they are similar. This is surprising at first sight. However, the value of using an apron to reduce erosion potential becomes clearer if one considers the spatial distribution of the power along the fissure.

The maximum discharge in the fissure is determined by the energy head between upstream and downstream conditions. This is roughly equal to the difference between the water surface elevation upstream of the dam and the water surface elevation at the downstream toe of the dam. It is therefore reasonable to conclude that

$$Q_b \approx Q'_b + Q_a \tag{5.91}$$

From which follows that the total power expenditure over the length of the fissure in each of the three scenarios is equal, i.e.,

$$\gamma Q_b h_f \approx \gamma (Q'_b + Q_a) h_f \tag{5.92}$$

However, the spatial distribution of the power expended over the length of the fissure is different for the three scenarios. In the case without an apron the applied stream power *per unit area* is

$$P = \frac{\gamma Q_b h_f}{2LD + 2Lw_f} \tag{5.93}$$

In the case of an apron without a crack the applied stream power *per unit area* is

$$P_a = \frac{\gamma Q_b h_f}{2(L_a + L)D + 2(L_a + L)w_f} \tag{5.94}$$

Comparison of Eqs. (5.93) and (5.94) indicates that the stream power per unit area in the fissure with an apron is lower than in the case without an apron, which explains the value of using an apron to reduce the possibility of fissure erosion.

The distribution of stream power per unit area, along a fissure in the presence of an apron with a *narrow* crack is more complex. The reason for this is that the discharge in the fissure underneath the crack and apron varies. Discharge on the upstream end of the fissure, below the apron is low and gradually increases until it reaches its maximum value right underneath the upstream toe of the embankment. From here onwards the flow in the fissure underneath the embankment remains constant.

The energy head difference at the downstream toe of the embankment can be divided into two components (see Fig. 5.28), i.e.,

$$h_f = h_{f1} + h_{f2} \tag{5.95}$$

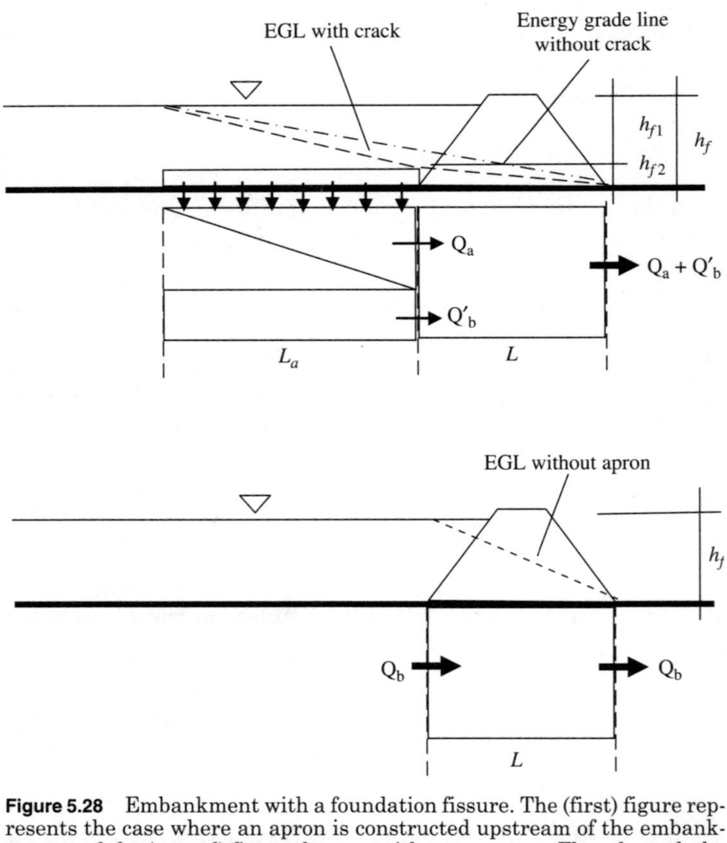

Figure 5.28 Embankment with a foundation fissure. The (first) figure represents the case where an apron is constructed upstream of the embankment, and the (second) figure the case without an apron. Flow through the fissure in the top figure can occur with or without a crack in the apron.

The head loss h_{f1} represents the combined energy head loss of flow through the crack in the apron and along the portion of the fissure underneath the apron. The head loss h_{f2} is representative of the energy loss of the flow in the fissure directly underneath the embankment.

An important observation is that the energy grade line slopes for the three cases differ (Fig. 5.28). The slope of the energy grade line for the case without an apron is the steepest. Addition of an apron on the upstream end of the embankment leads to a decrease in the energy grade line slope. Once a crack appears in the apron, the energy grade line can be split into two sections. In the reach demarcated by the apron the slope becomes steeper than for the case without a crack, while it becomes milder in the reach directly underneath the embankment.

The stream power expenditure per unit area in the fissure below the embankment for the case when a narrow crack exists in the apron can be expressed as

$$P'_a = \frac{\gamma(Q'_b + Q_a)h_{f1}}{2LD + 2Lw_f} \tag{5.96}$$

It can therefore be shown that

$$P > \mathrm{P}_a > P'_a \tag{5.97}$$

It is concluded that if a *narrow* crack forms in an apron upstream of the dam then the total amount of power that is available to scour the fissure is less than when the crack does not exist. This makes sense because a large amount of the energy is consumed by the flow through the *narrow* crack. The amount of power per unit area that remains after that for potentially causing erosion of the fissure is lower than when the narrow crack does not exist.

Spatial distribution of stream power. The equations for calculating stream power in the previous section provide an indication of the average stream power per unit area in the fissure. When simulating the formation of a breach (i.e., an embankment crack or fissure that widens as a function of time due to erosion) it is necessary to estimate the distribution of stream power on the top and bottom of the crack and on its sides.

Observations of dam breach failure indicate that the width of a breach is limited to a certain maximum, which can vary from case to case. Once a particular breach width has been reached is seems as if it remains stable while water is still discharging through the breach. Interpretation of this observation leads to the conclusion that the erosive capacity of the water acting on the sides of a crack or fissure gradually decreases with increasing breach width.

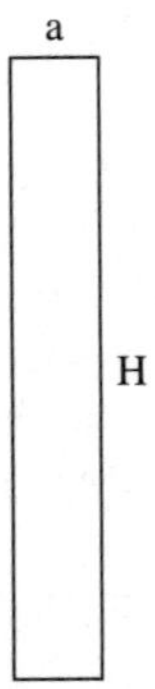

Figure 5.29 Schematic of a crack or fissure with dimensions a by H.

This has indeed been found to be true by Knight and Patel (1983) and Rhodes and Knight (1994), who conducted laboratory experiments to determine the distribution of wall shear stress in smooth closed ducts. Knight and Patel (1983) originally investigated the distribution for aspect ratios ranging between 1 and 10. Their findings were improved upon by Rhodes and Knight (1994), who extended the study to aspect ratios ranging from 1 to infinity. The work by Rhodes and Knight is more complete and used here to provide a means of estimating the distribution of erosive capacity of water in cracks and fissures.

Rhodes and Knight (1994) developed the following equation for estimating the percentage of the shear force applied to the walls:

$$\%SF_w = \frac{100}{1+\left(\dfrac{1+1.345\frac{H}{a}}{1+1.345\frac{a}{H}}\right)^{-1.057}} \tag{5.98}$$

The dimensions of the crack or fissure used to calculate the aspect ratio is shown in Fig. 5.29. The stress on the top and bottom, and on the sides of the crack can be calculated as follows (Knight and Patel 1983):

$$\frac{\overline{\tau}_w}{\tau} = 0.01 \times \%SF_w \left(1+\frac{a}{H}\right) \tag{5.99}$$

$$\frac{\overline{\tau}_b}{\tau} = \left(1-0.01\times\%S_w\right)\left(1+\frac{H}{a}\right) \tag{5.100}$$

where $\overline{\tau}_w$ = mean shear stress on the wall
$\overline{\tau}_b$ = mean shear stress on the top and bottom
ρ = density of water
τ = the total boundary shear stress

The distribution of shear stress on the top and bottom, and on the vertical walls as a function of aspect ratio is presented in Fig. 5.30. This figure shows that as the crack or fissure becomes wider (i.e., the ratio *H*/*a* decreases in value) the proportion of the erosive capacity of the water applied to the vertical walls decreases. When the crack or fissure is very narrow (i.e., the ratio of *H*/*a* is high) the proportion the erosive capacity on the walls is at its maximum, and the proportion of the erosive capacity on the top and bottom of the fissure is low.

The relationship between applied stream power at the boundary and wall shear stress leads to the conclusion that the spatial distribution of stream power in a crack or fissure is similar to that of shear stress. Annandale (2004) used these relationships to simulate widening in embankment cracks and foundation fissures.

Bridge piers. The complex flow patterns around bridge piers increase turbulence intensity and the erosive capacity of the water. The increase

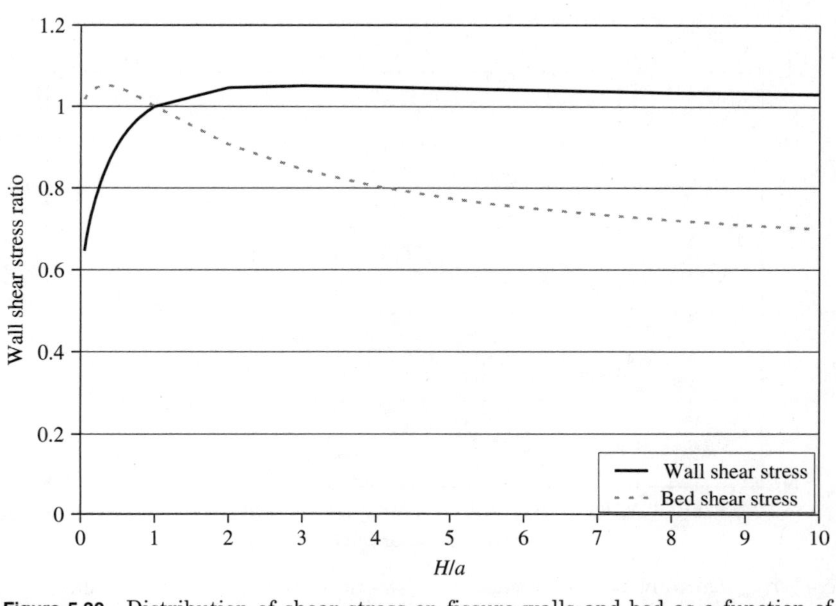

Figure 5.30 Distribution of shear stress on fissure walls and bed as a function of aspect ratio.

in erosive capacity causes scour around bridge piers, which can result in bridge failure. Research conducted by the Federal Highway Administration (FHWA) concluded that the erosive power of water around bridge piers decrease as scour holes increase in depth (see e.g., Smith et al., 1997). This finding has significant implications because earth material often increases in strength as a function of elevation below a riverbed. Concurrent decrease in the magnitude of the erosive power of water and increase in earth material strength causes scour holes around bridge piers to reach finite depths. The maximum scour depth occurs at the elevation where the erosive capacity of water is less than the erosive power required to cause scour of the earth material at that elevation.

Estimates of the magnitude of the erosive capacity of water as a function of scour depth can be made by means of graphs that are based on the results of the FHWA research (Figs. 5.31 and 5.32). Both figures show the change in stream power around bridge piers as scour holes increase in depth, one for round piers and the other for all pier shapes tested (round, square, and rectangular). The stream power is expressed in dimensionless form on the ordinate of the graphs as the ratio P/P_a. P_a is the magnitude of the stream power in the river upstream of the pier, and P is the magnitude of the stream power at the base of the scour hole as it increases in depth. The abscissa of both figures represents dimensionless scour depth. Figure 5.31 expresses dimensionless

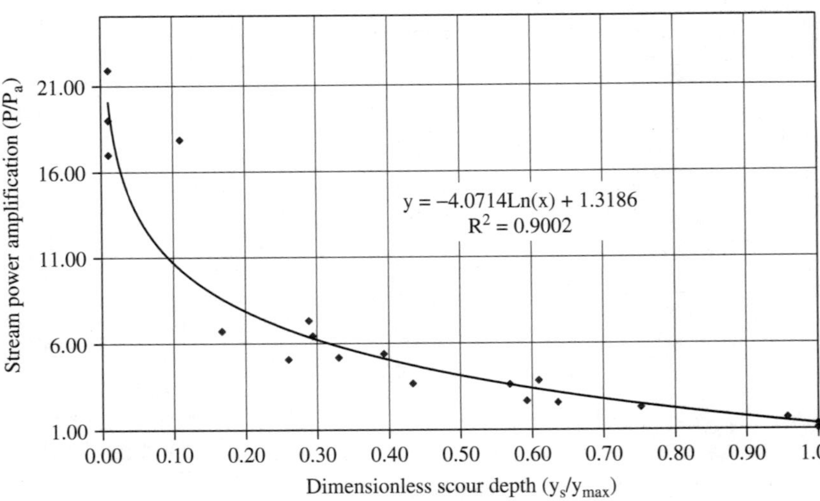

Figure 5.31 Change in stream dimensionless stream power as a function of dimensionless scour depth for round piers, expressed as a function of maximum possible scour depth.

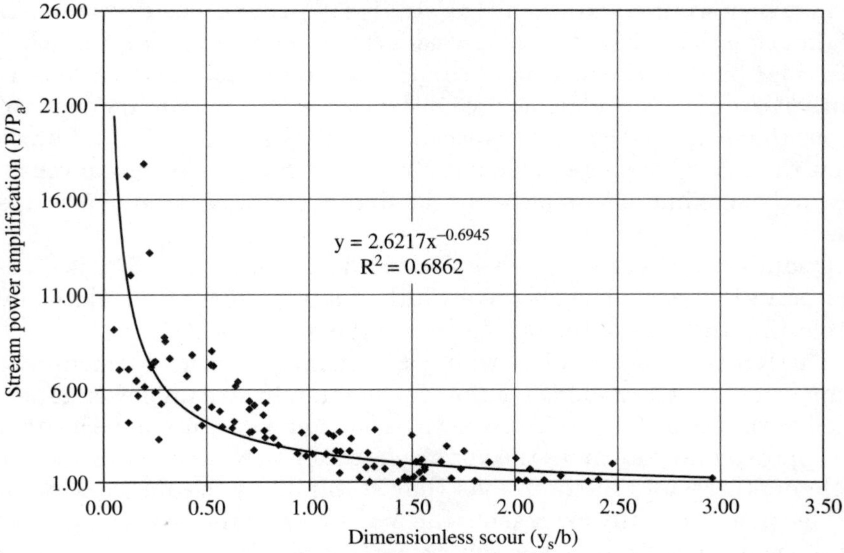

Figure 5.32 Change in dimensionless stream power as a function of dimensionless scour depth for square, round, and rectangular piers, expressed as a function of the effective bridge pier width *b*.

scour depth as the ratio $y_s/y_{\max}$, whereas Fig. 5.32 expresses it as the ratio y_s/b. The variable $y_{\max}$ represents the maximum scour depth that can develop around a bridge pier under given flow conditions, whereas y_s represents variable scour depth ($y_s \le y_{\max}$). The variable b represents the effective pier width in the direction of flow.

Quantification of the axes of Figs. 5.31 and 5.32 requires estimates of the approach stream power (P_a) and the maximum possible scour depth ($y_{\max}$) or the effective width of the pier (b). The magnitude of the approach stream power is calculated as [see Eq.(5.22)]

$$P_a = 7.853\rho\left(\frac{\tau_w}{\rho}\right)^{3/2} \tag{5.101}$$

where $\tau_w = \rho g y s$.

An estimate of the maximum possible scour depth that can occur around a given bridge pier ($y_{\max}$), assuming negligible scour resistance by the earth material, is required to make the abscissa of Fig. 5.31 dimensional. Such an estimate can be obtained by making use of the bridge pier scour equation in HEC-18 (FHWA 1993). The development of that equation is based on an envelope curve embracing a large number of bridge pier scour experiments and is considered to provide a conservative estimate of bridge pier scour depth, i.e., the maximum depth of scour that can reasonably be expected to occur. Use of Fig. 5.31 assumes that the scour hole depth calculated with the HEC-18 equation represents the maximum possible scour depth that can occur around a bridge pier.

Quantification of the effective pier width *b* for use in Fig. 5.32 is illustrated in Fig. 5.33. The effective width of a bride pier is calculated by projecting the area of the bridge pier in the direction of flow.

The figures relating dimensionless stream power as a function of dimensionless scour indicates that the maximum stream power generated at the base of a pier just prior to scour commencing can be as high as approximately 21 times the approach stream power. Increasing scour hole depth results in conditions that streamlines the flow around a bridge pier, gradually decreasing the magnitude of the erosive capacity of the water until it is so low that scour ceases.

Quantification of pressure fluctuations—plunge pools. The methods for quantifying the actual magnitude of pressure fluctuations in turbulent flowing water are presented for plunging jets only. The principal focus is on the formation of pressure fluctuations in plunge pools.

The dynamic pressures introduced into a plunge pool by a plunging jet play an important role in determining the potential and extent of rock and concrete scour. Rock scour can occur in unlined plunge pools, while scour of concrete can occur in concrete lined plunge pools.

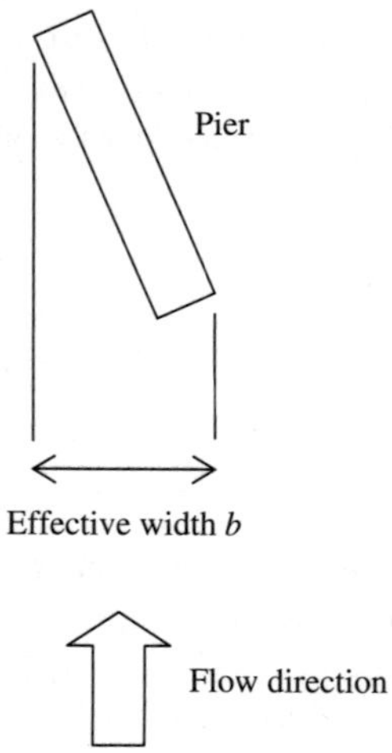

Figure 5.33 Definition sketch for determining effective width of bridge pier *b*.

It is necessary to distinguish between the dynamic pressures in the plunge pool itself and those within discontinuities (cracks, fissures, joints, and so on.) in a rock mass or concrete lining. The dynamic pressures in the plunge pool itself are dealt with in what follows, and the pressures that develop in discontinuities within the material mass are discussed in the next sub-section.

The dynamic pressure introduced into a pool by a plunging jet is a function of its issuance turbulence intensity, issuance jet diameter (round jets) or thickness (rectangular jets), jet trajectory length, and the water depth below the surface of the plunge pool. The dynamic pressure introduced into a pool consists of two components; the mean and fluctuating dynamic pressures. The total dynamic pressure is the sum of these two components, which, for analysis and design purposes is expressed as (Ervine et al., 1997)

$$P_{\max} = (C_{pa} + C'_{pa})\gamma\phi\frac{V_j^2}{2g} \tag{5.102}$$

where $P_{\max}$ = total dynamic pressure
C_{pa} = mean dynamic pressure coefficient
C'_{pa} = fluctuating dynamic pressure coefficient
ϕ = kinetic energy velocity coefficient (often assumed = 1)
γ = unit weight of water

Ervine et al. (1997) prepared a relationship between average dynamic pressure and dimensionless depth below the plunge pool water surface for round jets with a breakup length ratio of 0.5 (i.e., $L/L_b = 0.5$) shown in Fig. 5.34. The figure also shows the theoretical relationship for the average dynamic pressure coefficient [using Eqs. (5.60), (5.61), and (5.62)].

The Ervine et al. (1997) expression for the average dynamic pressure coefficient for round jets with a breakup length ratio (L/L_b) of 0.5 is

$$C_{pa} = 38.4(1-\alpha_i)\left(\frac{D_j}{Y}\right) \quad \text{if} \quad \frac{Y}{D_j} > 4$$
$$\text{and } C_{pa} = 0.875 \quad \text{if} \quad \frac{Y}{D_j} \le 4 \tag{5.103}$$

The air concentration α_i in this equation is calculated as

$$\alpha_i = \frac{\beta}{1+\beta} \tag{5.104}$$

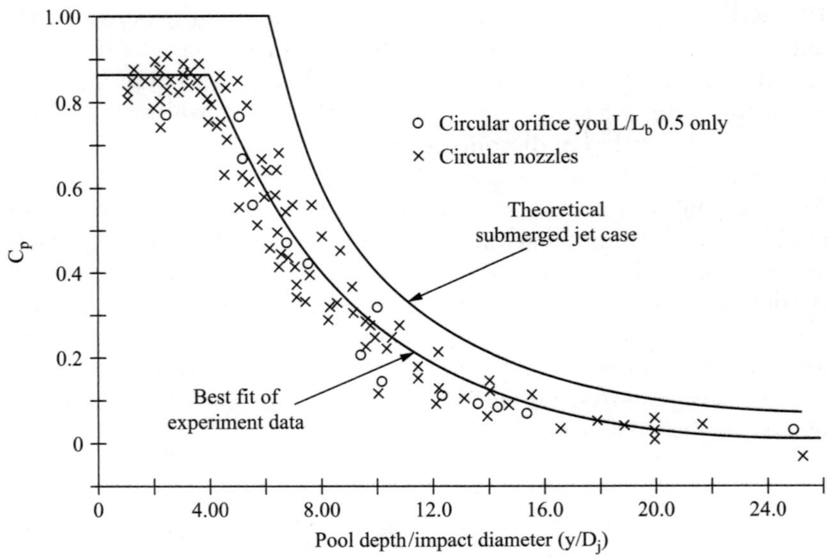

Figure 5.34 Variation of mean dynamic pressure coefficient (along jet center line) as a function of dimensionless pool depth for round jets (Ervine et al., 1997).

and the free air content as

$$\beta = \frac{q_a}{q} \tag{5.105}$$

where β = free air content
q = unit flow of water
q_a = unit flow of air

Unambiguous equations for calculating the air content in plunging jets are not currently available. The best equation at this stage is presented by Ervine (1998):

$$q_a = 0.00002(V_j - 1)^3 + 0.0003(V_j - 1)^2 + 0.0074(V_j - 1) - 0.0058 \tag{5.106}$$

This equation, strictly speaking, is only valid for rectangular jets with thickness exceeding 30 mm and velocities ranging between 1.5 and 15 m/s. The equation's accuracy is about +/– 30 percent.

Other equations to estimate the air content β are presented as follows:

For a rectangular plunging jet (Ervine & Elsawy 1975):

$$\beta \approx 0.13\sqrt{\frac{L}{D_j}} \tag{5.107}$$

For a circular plunging jet Ervine (1976):

$$\beta = K'\sqrt{\frac{L}{D_j}}\left(1-\frac{V_0}{V_j}\right) \tag{5.108}$$

where K' ranges between 0.2 and 0.4; V_0 (= 1 m/s) is the minimum plunging velocity leading to commencement of aeration.

As a practical check for estimated values of the air content β it is worth noting that Mason (1989) estimated that the maximum air content that could reasonably be expected to occur in water is on the order of about 65 to 70 percent. This estimate is roughly in agreement with measurements by van de Sande (1973), who claim to have measured air contents of up to 80 percent. It should be noted though that the wave celerity in water reaches a minimum if the air concentration is 50%. After that it becomes equal to the celerity of sound in air containing water, that is, it increases from 100 m/s to 300 m/s.

Dependence of the average dynamic pressure coefficient on the breakup length ratio of a jet has been studied by Castillo (2004) for rectangular jets. He related the coefficient to dimensionless plunge pool depth for varying breakup length ratios (Fig. 5.35). It is unclear why the average dynamic pressure coefficient presented by Castillo (2004) is greater than 1.0 for non-aerated jets. A maximum value of 1.0 is most probably more realistic.

An equation that can be used to calculate the values of the average dynamic pressure coefficient as a function of jet breakup length ratio and dimensionless plunge pool depth is

$$C_p = a e^{-b(Y/B)} \tag{5.109}$$

where B is the width (i.e., thickness) of a rectangular jet. The values of the parameters a and b as a function of jet breakup length ratio are presented in Table 5.4.

Relationships for quantifying the magnitude of the fluctuating dynamic pressure coefficient have been developed by Ervine et al. (1997), Castillo (2004), May and Willoughby (1991), and Bollaert (2002). Ervine et al. (Fig. 5.36) and Bollaert (Fig. 5.37) developed curves for round jets.

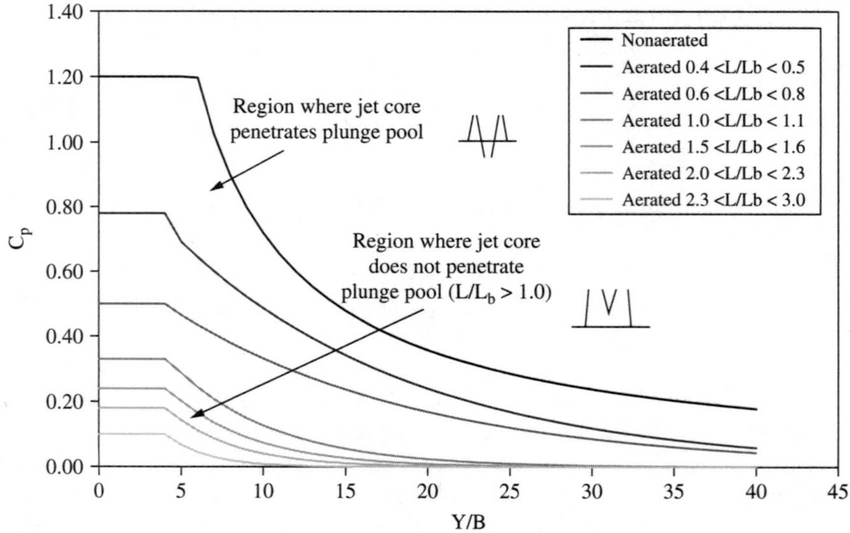

Figure 5.35 Average dynamic pressure coefficient as a function of dimensionless plunge pool depth for varying jet breakup length ratios (according to Castillo 2004).

The relationships developed by Castillo (Fig. 5.38) and May and Willoughby (Fig. 5.39) are for rectangular jets.

Bollaert (2002) developed an equation intended to quantify the fluctuating dynamic pressure coefficient as a function of the issuance turbulence intensity of the plunging jet:

$$C'_{pa} = a_1\left(\frac{Y}{D_j}\right)^3 + a_2\left(\frac{Y}{D_j}\right)^2 + a_3\left(\frac{Y}{D_j}\right) + a_4 \tag{5.110}$$

for $Y/D_j \le 20$.

TABLE 5.4 Parameters for Calculating Dynamic Pressure Coefficient as a Function of Jet Breakup Length Ratio

L/L_b	a	b	$C_p(Y/B < 4)$
0.4–0.5	0.98	0.070	0.78
0.5–0.6	0.92	0.079	0.69
0.6–0.8	0.65	0.067	0.5
1–1.10	0.65	0.163	0.33
1.1–1.3	0.65	0.185	0.31
1.5–1.6	0.55	0.200	0.24
1.8–1.9	0.55	0.250	0.20
2.2–2.3	0.50	0.250	0.18
2.3–3.0	0.50	0.400	0.10

SOURCE: From Castillo (2004)

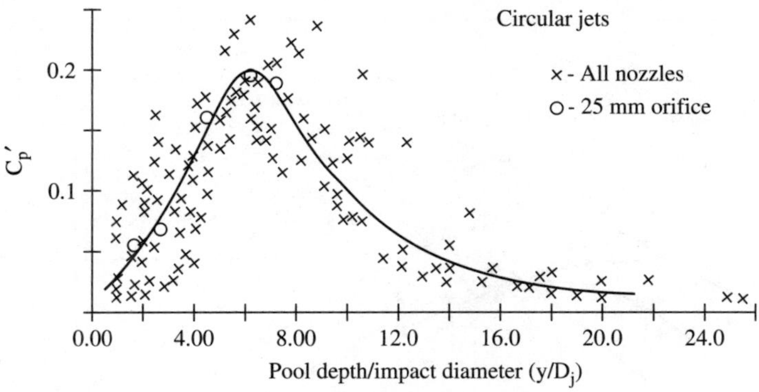

Figure 5.36 Variation of the fluctuating dynamic pressure coefficient as a function of dimensionless pool depth along the jet center line (Ervine et al., 1997).

For higher values of dimensionless depth, the C_p' value that corresponds to a ratio of 20 should be used as an interim rule until more information is available on its value beyond this depth.

The relationships between the issuance turbulence intensity T_u and the dimensionless coefficients a_i, are presented in Table 5.5.

The air content and degree of jet breakup affect the fluctuating dynamic pressure in a plunge pool (Ervine et al., 1997). The relationship between fluctuating dynamic pressure coefficient and jet breakup

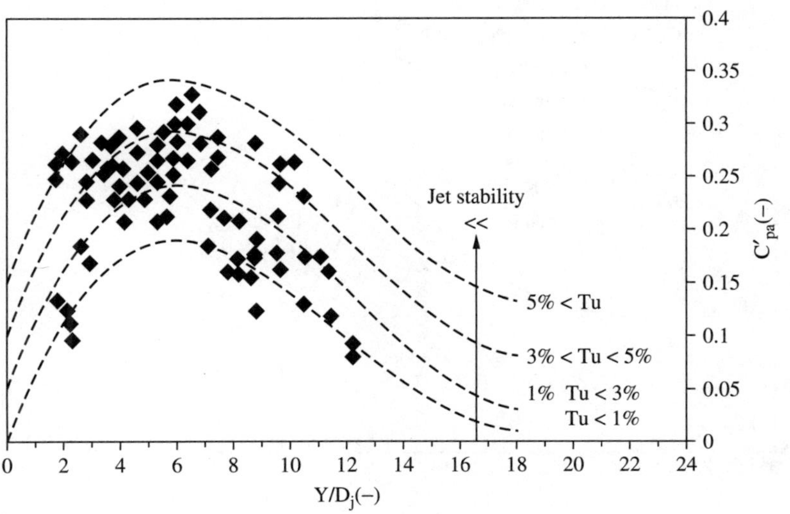

Figure 5.37 Fluctuating dynamic pressure coefficient as a function of dimensionless plunge pool depth and issuance turbulence intensity of plunging jets (Bollaert 2002).

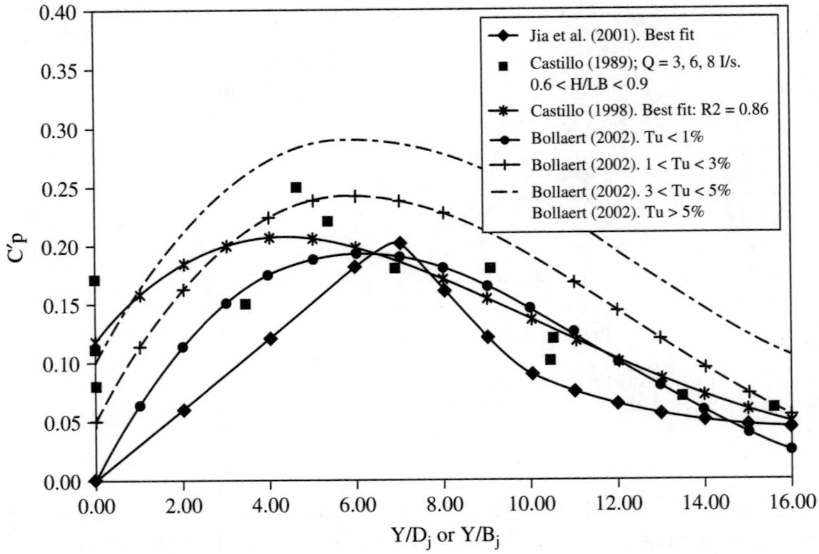

Figure 5.38 Fluctuating dynamic pressure coefficient for rectangular jets compared to Bollaert's relationships for round jets (Castillo 2004).

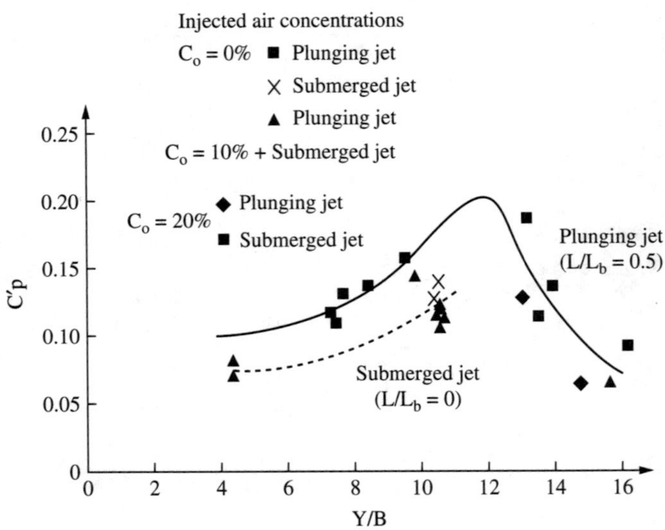

Figure 5.39 Fluctuating dynamic pressure coefficient as a function of dimensionless pool depth for rectangular jets (May and Willoughby 1991). *B* is the thickness of rectangular jet.

TABLE 5.5 Coefficient Values for Calculating the Fluctuating Dynamic Pressure Coefficient

T_u (%)	a_1	a_2	a_3	a_4	Type of jet
<1	0.00220	–0.0079	0.0716	0.00	Compact
1–3	0.00215	–0.0079	0.0716	0.050	Low turbulence
3–5	0.00215	–0.0079	0.0716	0.100	Moderate turbulence
>5	0.00215	–0.0079	0.0716	0.150	High turbulence

SOURCE: From Bollaert 2002

length ratio for round jets indicates that the coefficient reaches a maximum when the jet breakup length ratio is approximately 0.7 and reduces to approximately zero when the jet breakup length ratio is greater than 2.0 (Figs. 5.40 and 5.41).

Some analysis procedures use the difference between maximum and minimum fluctuating pressures to estimate the net uplift pressure over a unit of interest, such as a block of rock or a concrete slab. Relationships between the maximum and minimum fluctuating dynamic pressure coefficients as a function of dimensionless plunge pool depth, developed by Ervine et al. (1997) are presented in Fig. 5.42.

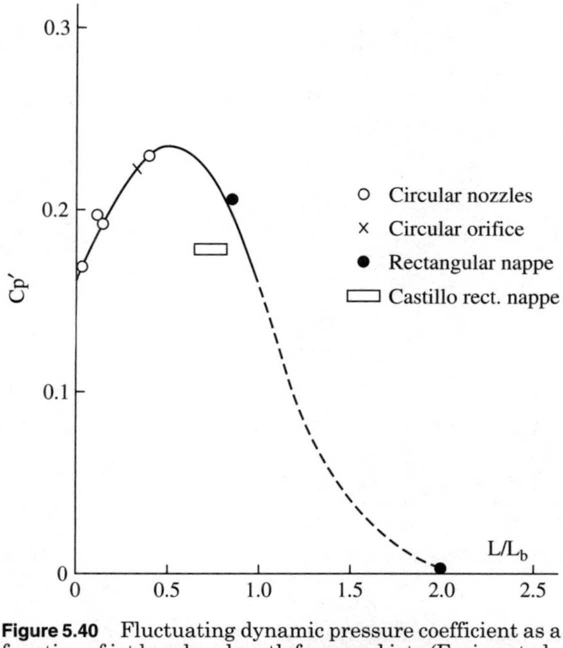

Figure 5.40 Fluctuating dynamic pressure coefficient as a function of jet breakup length for round jets (Ervine et al., 1997).

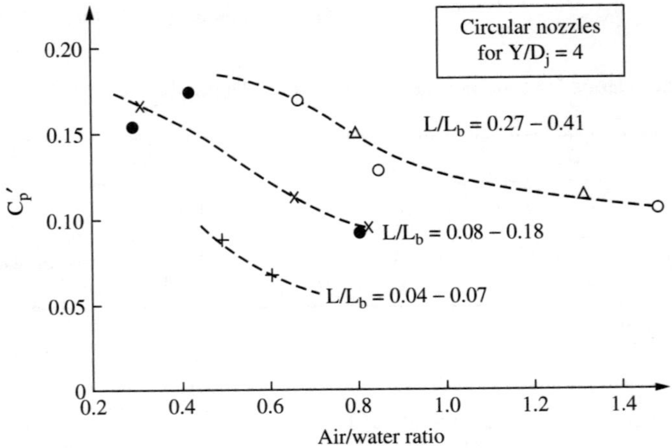

Figure 5.41 Fluctuating dynamic pressure coefficient as a function of jet breakup length and air content for round jets (Ervine et al., 1997).

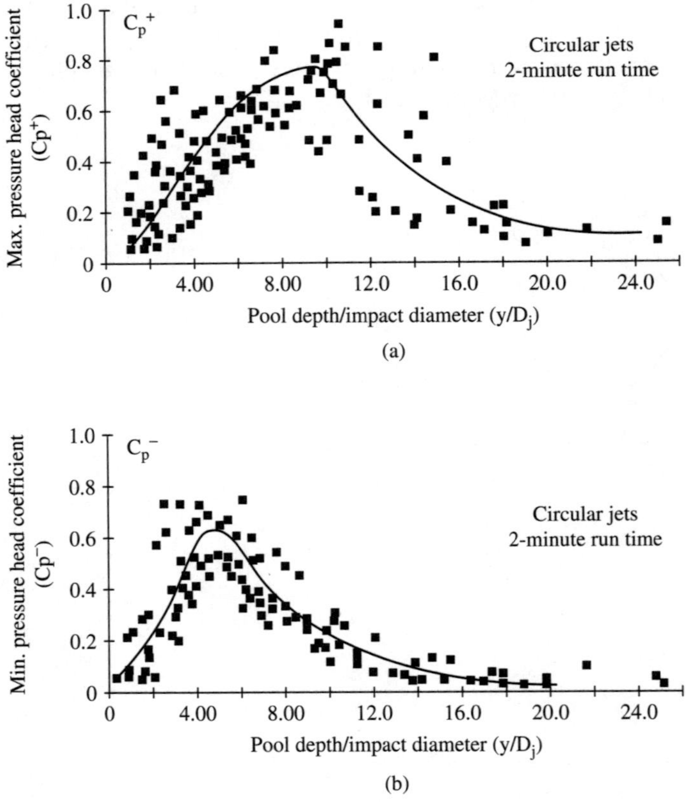

Figure 5.42 Maximum and minimum fluctuating dynamic pressure coefficients as a function of dimensionless plunge pool depth for round jets (Ervine et al., 1997).

The maximum fluctuating dynamic pressure coefficient reaches its highest value when the dimensionless pool depth is approximately 10 and the minimum fluctuating dynamic pressure coefficient occurs at a dimensionless depth of approximately 5. The maximum fluctuating dynamic pressure is added to the average dynamic pressure and the minimum fluctuating dynamic pressure is subtracted from the average dynamic pressure.

Pressure fluctuation in open- and close-ended discontinuities. Fluctuating pressures introduced into close-ended fissures and open-ended joints in rock or concrete masses by the average and dynamic fluctuating pressures originate from the interaction between plunging jets and the surrounding water in plunge pools. Research has shown that the extreme pressures in such discontinuities are affected by the presence of air, either free or dissolved. The presence of free air within the water in a joint can lead to resonance (Bollaert 2002).

When bubbles of free air in a plunge pool move close to or over the surface expressions of discontinuities, a sudden drop in pressure within a discontinuity can result in free air being drawn in. Additionally, free air can also come out of solution from the water within a discontinuity if the drop in pressure is large enough.

The presence of free air in water reduces its pressure wave celerity. What this means is that a pressure wave will travel slower through water containing free air, than through water containing no air. The pressure wave celerity in water with air contents in excess of 0.075 percent is lower than the pressure wave celerity in air. This phenomenon is mainly due to scatter and interference of the pressure wave, caused by the presence of the air bubbles in the water. An explanation of the physics of this phenomenon is found in Kafesaki et al. (2000) and Krokhin et al. (2003).

The pressure wave celerity in pure water is on the order of about 1000 m/s to 1440 m/s, depending on temperature. In an air/water mixture with an air content of about 1 percent, the celerity reduces to about 100 m/s. The significance of reducing the pressure wave celerity is that the natural frequency of such discontinuities changes significantly.

The natural frequency f_c of a close-ended fissure can be calculated with the equation

$$f_c = \frac{c}{4L} \tag{5.111}$$

and that of an open-ended joint with

$$f_c = \frac{c}{2L} \tag{5.112}$$

where c is the pressure wave celerity and L is the fissure length.

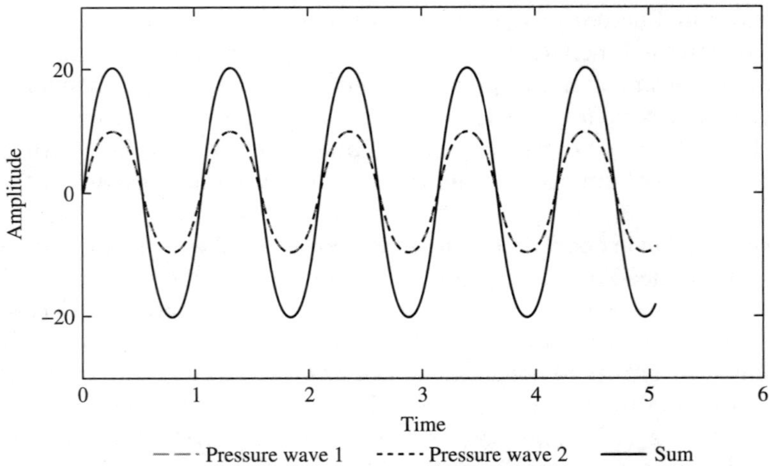

Figure 5.43 Resonance by adding two pressure waves that are in phase.

Therefore, if the length of a close-ended fissure equals 1 m and the pressure wave celerity of the air/water mixture in the fissure is 100 m/s, Eq. (5.111) indicates that its characteristic frequency is 25 Hz. With the frequency of pressure fluctuation in flowing water generally ranging between 2 and 25 Hz, it is conceivable that resonance can occur within such fissures.

In the case of plunging jets, Bollaert (2002) found that significant amounts of energy are contained in pressures fluctuating at 100 Hz, which is higher than normally found in channel flow. Therefore, even if the pressure wave celerity of the water is somewhat higher, say about 400 m/s, the characteristic frequency of a 1 m long fissure will be about 100 Hz. This can also lead to resonance.

In essence, it means that the frequency by which pressure fluctuations introduced into a close-ended discontinuities is reflected, is the same as the frequency by which they are introduced. This leads to an increase in pressure peaks experienced within a fissure if the two sets of waves are in phase. Figure 5.43 illustrates the resonance that occurs if two waves are in phase and have the same frequency. Pressure waves 1 and 2 in the figure have exactly the same amplitude and frequency and are therefore located on top of each other. Their sum is shown by the solid line, which reflects the resonance that occurs due to the coincident nature of these two pressure waves.

Chapter 4 provides the background for calculating the pressure wave celerity in a mixed fluid, like a mixture of water containing air, using the following equation:

$$c_{\text{mix}} = \sqrt{\frac{K_e}{\rho_e}} \qquad (5.113)$$

where K_e is the effective bulk modulus of the mixed fluid and ρ_e is the effective density of the mixed fluid.

The current state of the art as far as practical implementation of the effects of resonance on rock scour is concerned does not require exact knowledge of the pressure wave celerity in the air-water mixture. It merely requires an estimate, which for constant pressure (atmospheric pressure) and temperature (about 20°C) can be approximated using the following two equations (Bollaert 2002):

$$\rho_{\text{mix}} = \rho_{\text{air}} \cdot \beta + \rho_{\text{liq}} \cdot (1 - \beta) \tag{5.114}$$

where ρ_{air} is the density of air (assume 1.29 kg/m^3) and ρ_{liq} is the density of the liquid (assume 1000 kg/m^3 for water), and

$$c_{\text{mix}} = \sqrt{\frac{1}{\rho_{\text{mix}}} \cdot \frac{1}{\frac{(1-\beta)}{\rho_{\text{liq}} \cdot c_{\text{liq}}^2} + \frac{\beta}{\rho_{\text{air}} \cdot c_{\text{air}}^2}}} \tag{5.115}$$

where c_{liq} is the pressure wave celerity in the liquid (assume 1000 m/s for water) and c_{air} is the pressure wave celerity in air (assume 340 m/s). These two equations are valid for air contents β ranging between 0 and 50 percent.

Figure 5.44 is a plot of Eq. (5.115), which shows that the pressure wave celerity in water containing free air changes rapidly as a function of air

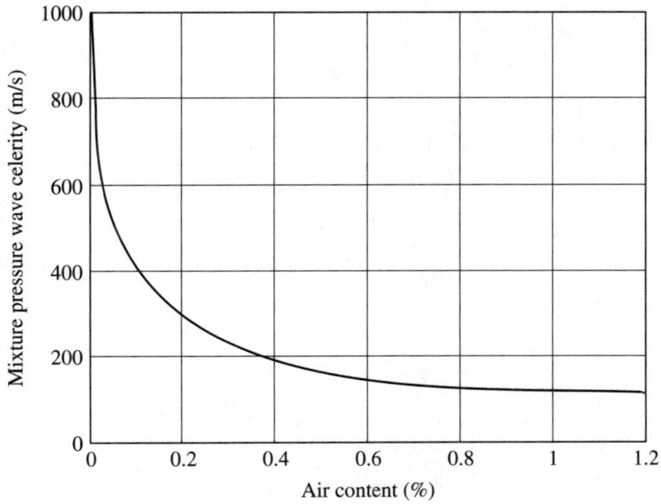

Figure 5.44 Change in pressure wave celerity in water containing air as a function of air content.

content. The pressure wave celerity in water changes from 1000 m/s, when it contains no air, to about 100 m/s when the air content is about 1 percent of free air by volume.

Open-ended discontinuities. Fluctuating pressures introduced into open-ended discontinuities of a material unit can experience amplification due to transients and resonance. View the hypothetical case where two fluctuating pressures of exactly the same magnitude are introduced into an open-ended discontinuity from opposite ends (Fig. 5.45). The pressure waves will travel through the water in the discontinuity and meet each other halfway, resulting in a net pressure that is equal to twice the original. If the water in the discontinuity contains free air, an additional possibility for resonance exists. This will increase the pressure even further.

Practical methods to account for both these phenomena (transients and resonance) are provided by Bollaert (2002). He found that the net uplift pressure underneath a unit of material with open-ended joints can be quantified with the equation

$$P_u = \gamma C_I \phi \frac{V_j^2}{2g} \tag{5.116}$$

where P_u is the *net* uplift pressure over an entity with open-ended joints impacted by a jet.

The *net dynamic impulsion* coefficient C_I has been expressed by Bollaert (2002) in terms of the dimensionless depth Y/D_j in a plunge pool (Fig. 5.46). Its magnitude can also be calculated with Eq. (5.117):

$$C_I = 0.0035\left(\frac{Y}{D_j}\right)^2 - 0.119\left(\frac{Y}{D_j}\right) + 1.2 \tag{5.117}$$

Examples applying these Eqs. [(5.116) and (5.117)] to calculate the possibility and extent of scour are presented in Chap. 7.

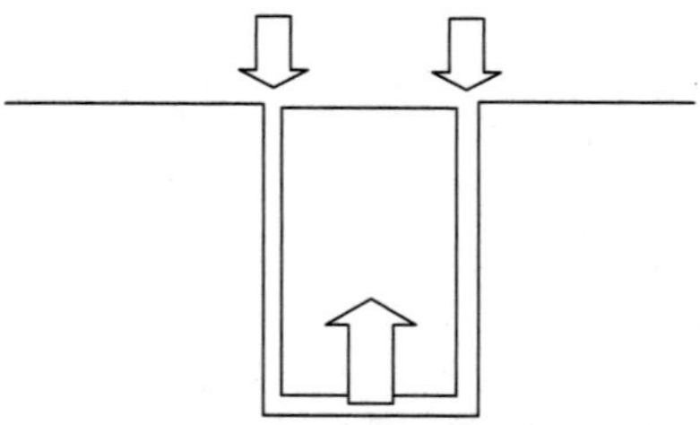

Figure 5.45 A material unit with open-ended joints subject to pressures introduced at the surface expression of the joints. The pressure increases underneath the material unit, due to transients and, possibly, resonance.

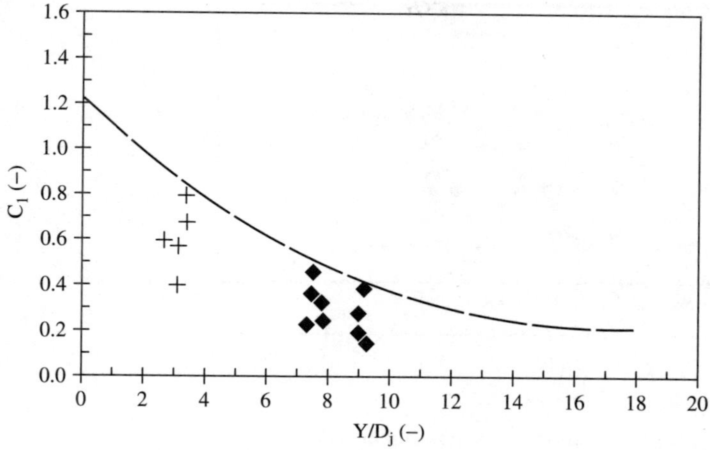

Figure 5.46 Dynamic Impulsion coefficient as a function of dimensionless depth (Bollaert 2002).

Close-ended fissures. The maximum pressure in a close-ended fissure (Fig. 5.47) is the sum of the mean and fluctuating dynamic pressures. The mean dynamic pressure will remain unchanged when transferred into the fissure, but the fluctuating dynamic pressure could be amplified by resonance if the natural frequency of the fissure is roughly equal to the frequency of introduced pressure fluctuations.

Bollaert (2002) developed an amplification factor (Γ^+) that is applied to the fluctuating dynamic pressure coefficient in Eq. (5.102) to account for the effects of resonance in close-ended fissures. The maximum pressure in a close-ended fissure is then expressed as

$$P_{\max} = \gamma\left(C_{pa} + \Gamma^{+}C'_{pa}\right)\frac{\phi V_j^2}{2g} \tag{5.118}$$

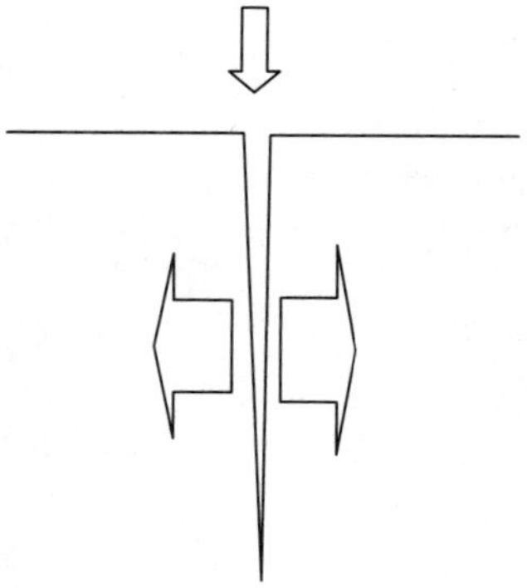

Figure 5.47 Introduction of pressure into a close-ended fissure, with resonance leading to amplification of the pressure within the fissure.

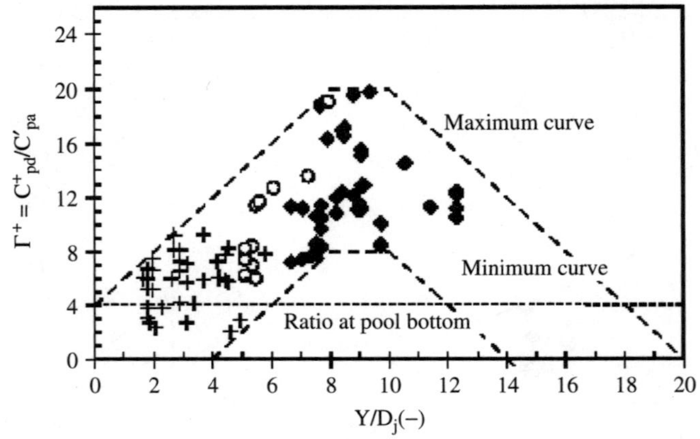

Figure 5.48 Amplification factor as a function of dimensionless depth (Bollaert 2002).

A relationship between the amplification factor and dimensionless depth Y/D_j, developed from experimental data, is shown in Fig. 5.48.

The factor can also be calculated as follows:

$$\Gamma^+ = 4 + 2Y/D_j \quad \text{for } Y/D_j < 8$$

$$\Gamma^+ = 20 \quad \text{for } 8 \le Y/D_j \le 10 \quad \text{curve of maximum values}$$

$$\Gamma^+ = 40 - 2Y/D_j \quad \text{for } 10 < Y/D_j$$

$$\Gamma^+ = -8 + 2Y/D_j \quad \text{for } Y/D_j < 8$$

$$\Gamma^+ = 8 \quad \text{for } 8 \le Y/D_j \le 10 \quad \text{curve of maximum values}$$

$$\Gamma^+ = 28 - 2Y/D_j \quad \text{for } 10 < Y/D_j$$

(5.119)

Environmental hydraulics

Open channel flow—straight reaches. Quantification of the stream power per unit area of the bed in straight open channel flow is calculated with the equation [see Eq. (5.22)]

$$P_{\text{channel}} = 7.853\rho\left(\frac{\tau_w}{\rho}\right)^{3/2} \tag{5.120}$$

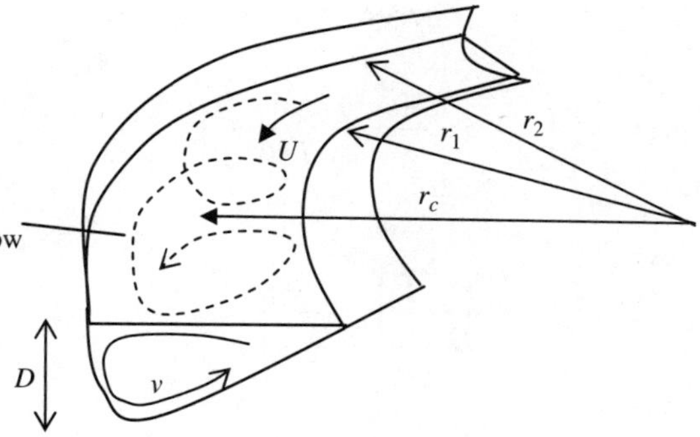

Figure 5.49 Flow around a bend, showing spiraling transverse flow and longitudinal flow.

Open channel flow—bends. When water flows around bends (Fig. 5.49) it leads to the development of a secondary flow, which occurs in the form of a helix. Equations that can be used to calculate the additional stream power in a river bend due to the presence of the transverse flow are found in Chang (1992). He showed that the general equation for power from transverse flow, neglecting the work done by upward and downward vertical velocities that cancel out, is

$$P'' = \int_{r_1}^{r_2} \int_0^D \rho v \frac{u^2}{r_c} dz dr \tag{5.121}$$

where ρ = density of water
v = transverse flow velocity
u = tangential flow velocity at a particular depth z
z = vertical axis
r = radial axis
D = maximum flow depth
r_1, r_2 = inner and outer radii of the flow region of concern
r_c = radius of the center line of the bend

When flow enters a bend the transverse velocity gradually increases, which can be calculated with the following difference equation developed by Chang (1992)

$$v_{j+1} = \left\{ v_j + F_1(f) \frac{U}{r_c} \exp[F_2(f)\Delta s]\Delta s \right\} \exp[-F_2(f)\Delta s] \tag{5.122}$$

where

$$F_1(f) = \left(\frac{f}{2}\right)^{1/2} \times \left[\frac{10}{3} - \frac{1}{\kappa} \times \frac{5}{9}\left(\frac{f}{2}\right)^{1/2}\right] \tag{5.123}$$

$$F_2(f) = \frac{\kappa}{D}\left(\frac{f}{2}\right)^{1/2} \cdot \frac{m}{1+m} \tag{5.124}$$

$m = \kappa(8/f)^{1/2}$
f = friction factor
κ = von Karman's coefficient = 0.41
U = depth-averaged flow velocity in the longitudinal direction
Δs = incremental distance along the centerline of the channel

Once the transverse velocity is known as a function of distance along the center line of the channel, the transverse stream power can be expressed as

$$P_j'' = \rho v_j \int_{r_1}^{r_2} \int_0^D \frac{u^2}{r_c} dz dr \tag{5.125}$$

where j is the incremental section number along the bend.

The relationship between the depth averaged velocity U and the variable velocity u at a distance z above the bed is (Chang 1992)

$$\frac{u}{U} = \frac{1+m}{m}\left(\frac{z}{D}\right)^{1/m} \tag{5.126}$$

By replacing u in Eq. (5.125) with the expression in Eq. (5.126)

$$P_j'' = \frac{\rho v_j}{r_c}\left(r_2 - r_1\right)\int_0^D U^2 \left(\frac{1+m}{m}\right)^2 \left(\frac{z}{D}\right)^{2/m} dz \tag{5.127}$$

which can be written as

$$P_j'' = \frac{\rho v_j}{r_c} \Delta r U^2 \left(\frac{1+m}{m}\right)^2 \frac{1}{1+\frac{2}{m}} D \tag{5.128}$$

This equation is used to calculate the secondary stream power resulting from transverse flow at different locations j along the bend.

The total stream power in the bend is then approximated as

$$P_{\text{total}_j} = \gamma Q s_f + P_j'' \tag{5.129}$$

at varying locations j around the bend. The reader should note that this expression reflects the *total stream power* around the bend, not the power per unit area as previously formulated. The stream power per unit area is approximated by dividing the total steam power calculated from Eq. (5.129) by the wetted bed area of the river between two sections around the bend.

Hydraulic jumps. Hydraulic jumps, at times, occur naturally in creeks and rivers and can cause erosion of the channel bed. The stream power of hydraulic jumps can be calculated in the same manner as that quantified for stilling basins [Eq. (5.69)].

Headcuts. A headcut is usually characterized by upstream erosion of a channel bed that is initiated by water flowing over a sudden drop in the channel bed. By making use of the momentum principle it is possible to develop equations that correspond reasonably well with measurements of energy loss over a drop. Such equations were developed during the early part of the previous century (Moore 1941) and were subsequently found to be quite useful for developing equations to calculate the rate of energy dissipation (stream power) at the base of a headcut (Annandale 1995).

Figure 5.50 shows an aerated jet plunging over a headcut with unit discharge q, critical depth y_c, and drop height Δz. The velocity of the jet increases as it plunges downwards, until it reaches a value of V_m at A, which is at the same elevation as the surface of the water trapped

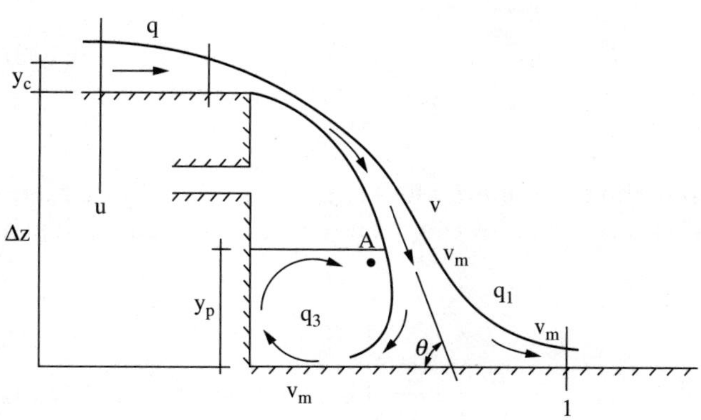

Figure 5.50 Plunging jet over a headcut.

between the face of the headcut and the jet. Moore (1941) showed that the elevation of the water trapped between the headcut face and the jet is determined by momentum transfer from the jet into the backroller and that venting of the jet has little effect on this phenomenon.

As the jet impinges onto the downstream bed of the stream at an angle θ it splits into two, part of it flowing upstream to form a backroller (unit discharge of backroller is q_3) and the rest flowing downstream. The unit discharge flowing in a downstream direction (q_1) is equal to the unit discharge q once equilibrium is reached. The discharge q_3 from the roller feeds back into the jet at A, with the same amount of water discharging back into it at the point of impingement. The discharge in the jet at the elevation of point A is therefore equal to $q + q_3$, leading to a local widening of the jet. It can be shown (Moore 1941) that the ratio between the flows is

$$\frac{q_1}{q_3} = \frac{(1+\cos\theta)}{(1-\cos\theta)} \tag{5.130}$$

By applying the momentum equation Henderson (1966) shows that

$$V_m = \frac{V}{2}(1+\cos\theta) \tag{5.131}$$

and that

$$\cos\theta = \frac{1.06}{\sqrt{\frac{\Delta z}{y_c}+\frac{3}{2}}} \tag{5.132}$$

Expressing the total energy head loss as

$$\Delta E = \Delta z + \frac{3}{2}y_c - y_1 - \frac{V_m^2}{2g} \tag{5.133}$$

where y_1 is the downstream depth.

It can be shown that the total energy loss can be expressed in dimensionless form solely as a function of the drop height and critical depth at the drop, i.e.,

$$\frac{\Delta E}{y_c} = \frac{\Delta z}{y_c} + \frac{3}{2} - \frac{y_1}{y_c} - \frac{1}{4}\left(\frac{3}{2}+\frac{\Delta z}{y_c}\right)\left[1+\frac{1.06}{\sqrt{\frac{\Delta z}{y_c}+\frac{3}{2}}}\right]^2 \tag{5.134}$$

With this estimate of energy head loss at the base of a headcut known, it is now possible to estimate the total rate of energy dissipation per unit width of flow (and thus the stream power per unit width of flow) at the point of impingement (impact) at the base of a headcut:

$$SP_{\text{impact}} = \gamma q y_c \left[\left(\frac{\Delta z}{y_c} + \frac{3}{2} - \frac{y_1}{y_c} - \frac{1}{4}\left(\frac{3}{2} + \frac{\Delta z}{y_c} \right) \right) \left(1 + \frac{1.06}{\sqrt{\frac{\Delta z}{y_c} + \frac{3}{2}}} \right)^2 \right] \tag{5.135}$$

If the thickness of the jet (D_j in meters) at the point of impact is known (in the direction of flow), it is possible to estimate the stream power per unit area by dividing SP_{impact} by D_j. This provides an estimate of the stream power per unit area, i.e., in W/m^2.

By using an equation derived by Henderson (1966) to calculate the portion of the energy loss in the backroller it is possible to calculate its rate of energy dissipation. This is the power per unit width of flow that will interact with the face of a headcut

$$SP_{\text{backroller}} = \gamma q y_c \left[\frac{1}{4}\left(\frac{3}{2} + \frac{\Delta z}{y_c} \right) \left(1 + \frac{1.06}{\sqrt{\frac{\Delta z}{y_c} + \frac{3}{2}}} \right)^2 \right] \tag{5.136}$$

The stream power per unit area on the face of the headcut can therefore be determined by dividing Eq. (5.136) by the depth of the pool y_p that forms behind the jet. This can be calculated with an equation developed by Chamani and Beirami (2002),

$$\frac{y_p}{y_c} = \sqrt{\left(\frac{y_u}{y_c} \right)^2 + 2Fr_u^2 \left(\frac{y_c}{y_u} \right) - \left(2Fr_u^2 + 1 \right)} \tag{5.137}$$

where y_u = upstream water depth
Fr_u = upstream Froude number

The authors tested this equation for both super and subcritical flow in the reaches upstream of the drop. The best agreement with experimental results was found for subcritical flow. The correlations between experimental data and pool depth for sub- and super-critical flow are shown in Fig. 5.51. Chamani and Beirami (2002) also compared energy loss over drops for upstream super- and subcritical flow (Fig. 5.52).

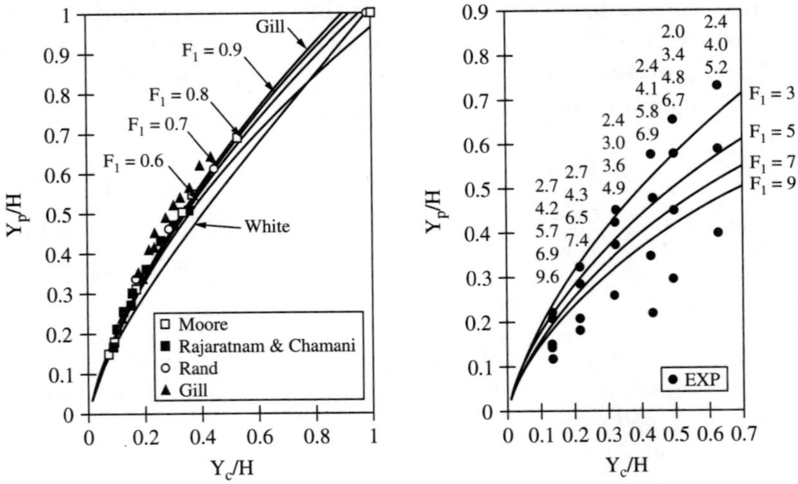

Figure 5.51 Correlation between experimental data and Eq. (5.137) for determining upstream pool depth for sub and supercritical flow (Chamani and Beirami 2002).

Knickpoints. A knickpoint is defined as a sudden slope change in a channel bed. As water flows over the slope change, the erosive capacity of the water increases locally. Estimation of the magnitude of the stream power per unit area of flow at a knickpoint in a channel is based on the same arguments that lead to development of the equations to quantify

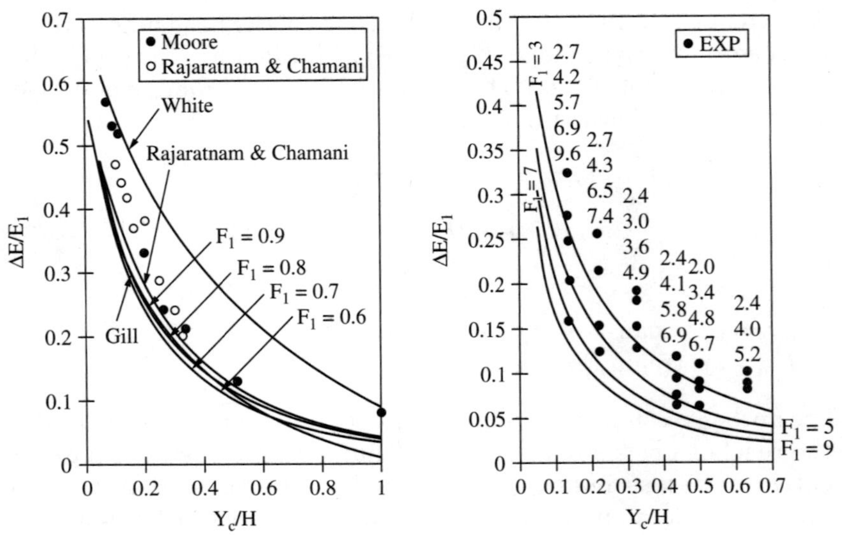

Figure 5.52 Comparison between experimental and theoretical estimates of energy loss for sub- and supercritical flow upstream of a drop (Chamani and Beirami 2002).

stream power at a headcut (see previous section). The proposed equation for calculating the stream power per unit area at a change in slope is

$$SP_{\text{knickpoint}} = \gamma \frac{q}{L}\left[\frac{1-\cos(\theta-\alpha)}{1+\cos(\theta-\alpha)} K_1 \frac{V_g^2}{2g} + LS_f\right] \tag{5.138}$$

where $SP_{\text{knickpoint}}$ = stream power over the knickpoint

V_g = velocity of water over the knickpoint

L = length of transition zone 2 (in the absence of more detailed information, assume a unit length)

S_f = average energy slope of water discharging over the knickpoint

K_1 = coefficient allowing for the non-hydrostatic head in the zone just downstream of the knickpoint (assume = 1)

q = unit flow.

A sketch defining the variables is shown in Fig. 5.53.

When water flows over a change in slope the stream lines are curved and the pressure distribution in the water is no longer hydrostatic. Incorporation of the factor K_1 is principally aimed at recognizing this phenomenon, but its practical implications are most probably very limited. The value of S_f follows from engineering judgment, and can be estimated as roughly the average of the up- and downstream energy slopes.

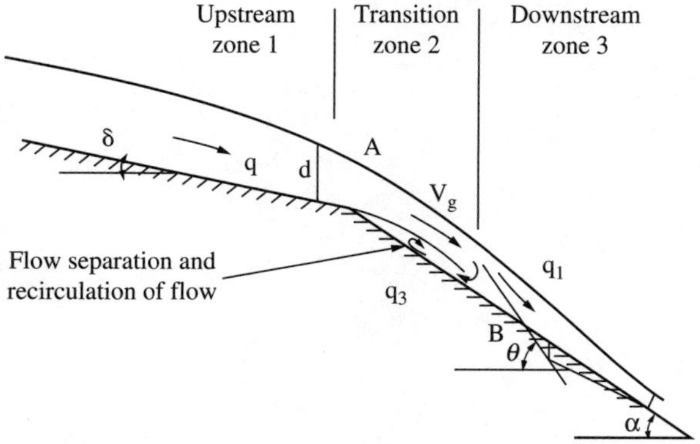

Figure 5.53 Flow over a knickpoint (slope change).

Chapter 6

Scour Thresholds

Introduction

This chapter solely focuses on providing scour thresholds for all earth and engineered earth materials. A scour threshold is a relationship between the erosive capacity of water and the relative ability of an earth or engineered earth material to resist it at the point of incipient motion. The following relationship is valid at the threshold:

$$P = f(K) \tag{6.1}$$

where P is the relative magnitude of the erosive capacity of the water and $f(K)$ is an expression that defines the relative ability of earth or engineered earth materials to resist the erosive capacity of water.

If $P > f(K)$ the earth or engineered earth material will experience scour. In cases when $P < f(K)$ the earth material will not experience scour.

In what follows scour thresholds for physical and chemical gels are presented. The physical gels under consideration include non-cohesive granular earth material and jointed rock masses. The chemical gels considered include intact rock and intact cohesive granular material, like clay.

After presenting threshold conditions for each particular gel, a method that can be used to determine scour thresholds for both gel types follows. This method is known as the erodibility index method (Annandale 1995), which is a semi-empirical approach that has been proven to work reliably in practice.

Physical Gels

The principal physical gels of interest to engineers investigating scour are non-cohesive granular material (like sand, gravel, cobble, and riprap) and jointed rock. The former has been studied quite extensively for many years, while methods to assess the erodibility of jointed rock are more recent.

The reader will note that some of the methods cited in the sections that follow uses all three indicator parameters most commonly used in practice, i.e., shear stress, flow velocity, and stream power. When the reader uses these methods it should be done against the background information provided in Chaps. 3 and 5, and the results should be assessed using the general decision-making process outlined in Chap. 2. Remember that flow velocity is an inconsistent indicatior of the relative magnitude of the erosive capacity of water (see Chaps. 3 and 5).

Non-cohesive granular material

Shields (1936). The erosion threshold relationship most often used in engineering is known as the Shields diagram (Shields 1936). This diagram (Fig. 6.1) relates dimensionless shear stress and the particle Reynolds number. Dimensionless shear stress is expressed as

$$\theta = \frac{\tau}{(\gamma_s - \gamma)d} \tag{6.2}$$

where θ = Shields parameter
d = representative particle diameter
γ_s = unit weight of the sediment
γ = unit weight of water

The shear stress, numerator of the Shields parameter, can also be expressed as

$$\tau = \rho u_*^2 \tag{6.3}$$

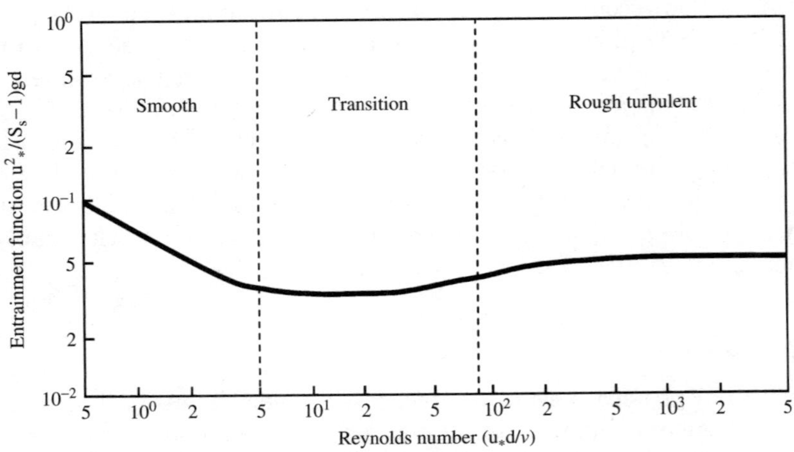

Figure 6.1 Shields diagram to determine conditions of incipient motion for non-cohesive granular material (Shields 1936).

From which follows that the Shields parameter (also known as the entrainment parameter) can be expressed as

$$\theta = \frac{u_*^2}{(S_s - 1)gd} \tag{6.4}$$

where $S_s = \gamma_s/\gamma$ and g is the acceleration due to gravity.

The particle Reynolds number Re_* is expressed as

$$\mathrm{Re}_* = \frac{u_* d}{\nu} \tag{6.5}$$

where $u_* = \sqrt{\tau/\rho}$ is the shear velocity and ν is the kinematic viscosity of the water.

When interpreting the general character of the Shields diagram it is useful to refer to the discussion about the near-bed region presented in Chap. 5. From those discussions it follows that

$$\frac{u_*}{\nu} = \frac{1}{\delta}$$

where δ is the wall layer thickness. The particle Reynolds number therefore also represents the ratio between the particle diameter and the wall layer thickness, i.e., $\mathrm{Re}_* = d/\delta$.

By making use of the work by Schlichting and Gersten (2000) it is shown in Chap. 5 that the near-bed region exists within the range $0 \le \mathrm{Re}_* \le 70$. This region contains the laminar sublayer, which lies within the region $0 \le \mathrm{Re}_* \le 5$, and the buffer layer, which lies in the region $5 < \mathrm{Re}_* \le 70$. When the diameter of the non-cohesive sediment particles, i.e., the absolute roughness of the bed, is less than the thickness of the laminar sublayer and the Reynolds number exceeds 2000 (i.e., $\mathrm{Re} > 2000$) smooth turbulent flow conditions exist. When absolute roughness on the bed increases and penetrates the buffer layer, the flow conditions are classified as transition flow. When the absolute roughness is larger than that, the flow becomes rough turbulent.

Flow condition	Particle Reynolds number
Smooth turbulent	$\frac{u_* d}{\nu} < 5$
Transition	$5 < \frac{u_* d}{\nu} < 70$
Rough turbulent	$\frac{u_* d}{\nu} > 70$

For smooth turbulent flow it can therefore be concluded that the noncohesive sediment particles are so small that they are contained within the laminar sublayer of the flow. Although the flow above the near-bed layer may be turbulent, the conditions within the laminar sublayer are laminar. The laminar flow in this sublayer therefore attempts to drag the small elements along with it as an assembly of particles. More effort is required to do that than when flow interacts with individual particles, as is the case under rough turbulent flow conditions. The difference in effort required to cause incipient motion under laminar and turbulent flow conditions respectively has been dealt with in quite a lot of detail in Chap. 3.

The Shields diagram confirms this behavior, i.e., the entrainment function increases when the particle Reynolds number is less than five. It can therefore be concluded that the sediment particles themselves experience laminar flow conditions when the flow is classified as smooth turbulent. This phenomenon can for example be seen on a beach where the wind might be strong enough to suspend larger sand particles, but spots of very fine sand may be observed that do not move at all. These particles are located in regions along the boundary that are covered by a viscous sublayer. It has been shown in Chap. 3 that the maximum value of the Shields parameter in this region is on the order of about 0.4.

The buffer layer in the near-bed region exists over the range $5 \leq Re_* \leq 70$. When the absolute roughness of the bed penetrates this region the flow conditions change to transition flow. Within this region the Shields parameter reaches a minimum value of 0.037. It has been shown theoretically in Chap. 3 that this low value can be explained by considering the relationship between the area occupied by low-pressure fluctuations and the diameter of sediment particles, and by balancing lift forces and the resistance offered by individual particles.

As the diameter of the sediment particles increases further, the flow at the boundary converts to rough turbulent flow because the individual elements start to shed eddies. In this region ($Re_* > 70$) the Shields parameter increases slightly and then stabilizes at a value of approximately 0.05 to 0.06. One of the reasons why the Shields parameter starts to increase when Re_* increases beyond a value of 30 is that the area occupied by low-pressure spots becomes smaller than the average diameter of sediment particles and more effort is required to remove particles from the bed (see Chap. 3).

Yang (1973). Yang (1973) developed a graph to determine conditions of incipient motion that relates the particle Reynolds number to dimensionless critical velocity. The origins of his relationship are found in unit stream power theory (Yang 1973). The threshold relationship is expressed mathematically as follows:

$$\frac{V_{cr}}{\omega} = \frac{2.5}{\log(\text{Re}_*) - 0.06} + 0.66 \qquad \text{for } 1.2 < \text{Re}_* < 70 \tag{6.6}$$

and

$$\frac{V_{cr}}{\omega} = 2.05 \qquad \text{for } \text{Re}_* \geq 70 \tag{6.7}$$

where V_{cr} is the critical average flow velocity, beyond which a particle will move and ω is the settling velocity of the sediment particle.

This relationship (Fig. 6.2) also shows the difference in conditions leading up to incipient motion for smooth, transition, and completely rough conditions right at the boundary. As before, smooth conditions exist when $\text{Re}_* < 5$, transition conditions exist when $5 < \text{Re}_* < 70$, and completely rough conditions exist when $\text{Re}_* \geq 70$.

Figure 6.2 confirms that the magnitude of the erosive capacity of water required for incipient motion under smooth flow conditions is larger than that required when the flow is rough turbulent. It also shows that the dimensionless parameter V_{cr}/ω assumes a constant value when the flow becomes rough turbulent.

Gessler (1965). Gessler (1965) developed a technique to assess the incipient motion of mixtures of non-cohesive material. It is has been found that application of this method in practice is very useful for predicting the grain size distributions of stable armor layers and for estimating general degradation of rivers.

Gessler, who investigated the Shields diagram in detail, showed for any particle size that

$$q(d) = p(d) = 0.5 \tag{6.8}$$

when

$$\tau_c(d) = \bar{\tau} \tag{6.9}$$

This means that the probability $q(d)$ that a particle with diameter d will *not move* is equal to the probability $p(d)$ that the same particle *will move* when the average shear stress $\bar{\tau}$ is equal to the critical shear stress $\tau_c(d)$ associated with a particle of diameter d. The critical shear stress is the stress that is expected to result in incipient motion of the sediment. He found that this is independent of whether the bed consists of a mixture of non-cohesive particles of different grain sizes, or whether

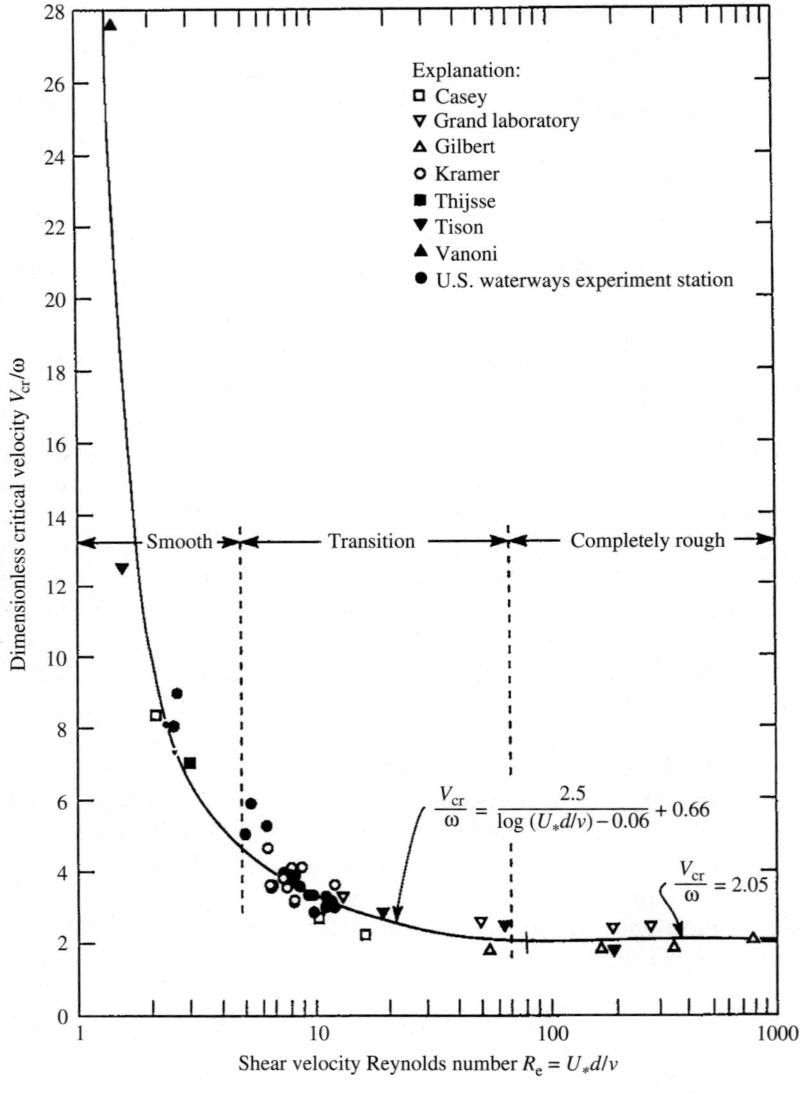

Figure 6.2 Incipient motion of non-cohesive granular material (Yang 1973).

it consists of uniform grains (similar to conditions used by Shields in developing his diagram). By making use of experimental data he developed a relationship between the probability $q(d)$ that a sediment particle of given diameter will *not move* as a function of dimensionless critical shear stress $(\tau_c/\bar{\tau})$ (Fig. 6.3), where $\bar{\tau}$ is the average shear stress on the bed.

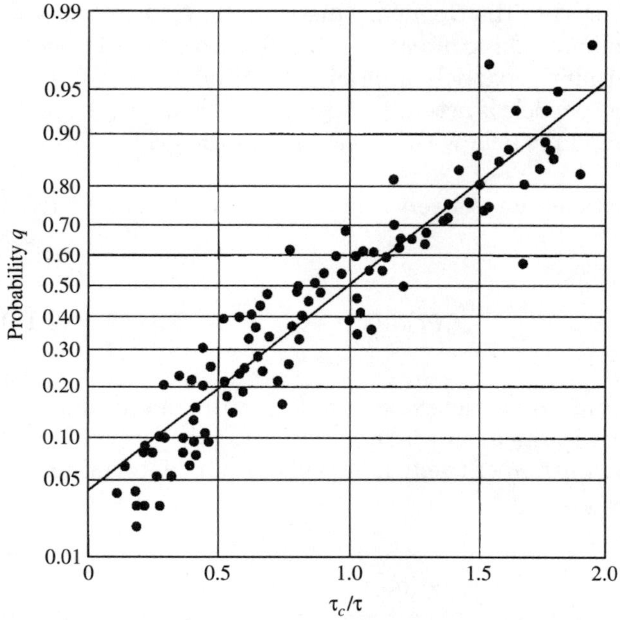

Figure 6.3 Probability that non-cohesive sediment grains will not move as a function of dimensionless critical shear stress (Gessler 1965).

When $\tau_c/\bar{\tau} = 1.0$, i.e., when the average shear stress equals the critical shear stress for that particle, then the probability of not moving is 0.50. Similarly, from Fig. 6.3 it can be deduced that the probability that a particle will not move if the average shear stress is equal to half the critical shear stress (i.e., $\tau_c/\bar{\tau} = 2.0$) is 95 percent; or the probability that a particle will not move when the average shear stress is twice the critical shear stress (i.e., $\tau_c/\bar{\tau} = 0.5$) is about 20 percent.

By making use of a probabilistic approach Gessler defines the grain size distribution of the original bed as

$$P_o(d) = \int p_o(d)dd \tag{6.10}$$

where p_o is the frequency function of the original grain size distribution.

When water flows over a bed with this distribution of particles then the frequency function defining the armor layer, i.e., the material that remains behind after some of it has been removed, is defined as

$$P_a(d) = k_1 q(d) p_o(d) \tag{6.11}$$

where k_1 is a constant.

Equation (6.11) states that the frequency distribution $P_a(d)$ of the sediment particles that remain behind after water has flown over it is the product of the probability that the particle will remain behind [i.e., $q(d)$] times the probability that the particle is actually present in the bed [i.e., the probability of encountering a particle with diameter d in the original gradation, i.e., $p_o(d)$].

Once the finer particles have been removed, it can be stated from basic probability theory that

$$\int_{d_{\min}}^{d_{\max}} p_a(d)dd = 1 \tag{6.12}$$

where $d_{\min}$ is the minimum particle size and $d_{\max}$ is the maximum particle size that remains in the armor layer.

The grain size distribution of the armor layer can be determined as follows:

$$P_a(d) = \frac{\int_{d_{\min}}^{d} q(d)p_o(d)dd}{\int_{d_{\min}}^{d_{\max}} q(d)p_o(d)dd} \tag{6.13}$$

and the grains size distribution of the material that will be removed is defined as

$$P_a(d) = \frac{\int_{d_{\min}}^{d} (1-q(d))p_o(d)dd}{\int_{d_{\min}}^{d_{\max}} (1-q(d))p_o(d)dd} \tag{6.14}$$

The distribution thus determined is what remains behind after the finer particles have been removed.

Jointed rock masses

In order to develop a relationship that can be used to calculate whether a block of rock can be removed from its matrix by pulsating forces it is necessary to set up a force balance representing the pulsating forces and integrate over the pulse period, Δt, as shown in the following equation (Bollaert 2002):

$$\int_0^{\Delta t} (F_{\text{up}} - F_{\text{down}} - W_g - F_{s1} - F_{s2})dt = F_{\Delta t} = mv_{\Delta t} \tag{6.15}$$

where F_{up} = total upward impulse caused by the transient pressure in the joint
F_{down} = total downward impulse caused by the fluctuating pressures on top of the rock block
W_g = submerged weight of the block of rock
F_{s1} and F_{s2} = instantaneous shear forces generated on the sides of the block of rock during the pulse period Δt
$F_{\Delta t}$ = net impulse
m = mass of the rock block
$v_{\Delta t}$ = average velocity attained by the mass of rock during the time period Δt

One can express the height through which a block of rock will be lifted by the pulses imposed on it during this short pulse period Δt as follows:

$$h = \frac{v_{\Delta t}^2}{2g} \tag{6.16}$$

It can be shown that the amount of power required to lift the rock block through this distance is

$$\frac{E}{\Delta t} = \frac{\left(\int_0^{\Delta t} (F_{up} - F_{down} - W_g - F_{s1} - F_{s2})dt\right) \cdot h}{\Delta t} = F_{\Delta t} \cdot \frac{h}{\Delta t} = F_{\Delta t} \cdot v_{\Delta t} \tag{6.17}$$

Stream power ($F_{\Delta t} v_{\Delta t}$) is considered a desirable indicator parameter for use in practice to quantify the relative magnitude of the erosive capacity of water when investigating scour of rock.

In order to determine whether such a block of rock will be removed from its matrix it is necessary to use Eqs. (6.15) and (6.16) to develop an expression that can be used to calculate the height h through which the rock will be lifted under given flow conditions.

Consider Eq. (6.15) and express the total shear force F_{sh} as the sum of all the shear forces acting on the rock block, i.e.,

$$F_{sh} = F_{s1} + F_{s2} + ... = \sum F_{si} \tag{6.18}$$

The expression representing the impulse forces on the rock block can therefore be expressed as

$$I_{\Delta t\_impulse} = \int_0^{\Delta t} (F_{up} - F_{down} - W_g - F_{sh})dt \tag{6.19}$$

where $I_{\Delta t\_impulse}$ is the net impulse on the block of rock.

The net impulse can now be quantified if each of the individual elements in the integral on the right is quantified.

Bollaert (2002) quantified the first portion of the equation, i.e., $\int_0^{\Delta t}(F_{\text{up}} - F_{\text{down}})dt$ for a jet impinging at a velocity v_j into a plunge pool. By using the definition of the average dynamic pressure coefficient on the bed of the plunge pool, we get

$$C_p = \frac{p/\gamma}{\phi v_j^2/2g} \tag{6.20}$$

and defining a time coefficient as

$$C_T = \frac{\Delta t}{2L/c} \tag{6.21}$$

where L = length of the open-ended discontinuity around a rock block in the plunge pool boundary
c = pressure wave celerity of the water
$2L/c$ = characteristic resonance frequency of the open-ended discontinuity

Using the concept that an impulse is the product of force and time, Bollaert (2002) defined a dynamic impulsion coefficient as the product of these two coefficients, i.e.,

$$C_I = C_P \cdot C_T \tag{6.22}$$

from which follows

$$C_I = \frac{p/\gamma}{\phi v_j^2/2g} \cdot \frac{\Delta t}{2L/c} \tag{6.23}$$

By manipulating this equation, it can be shown that

$$p \cdot \Delta t = \gamma C_I \left(\frac{\phi v_j^2}{g} \frac{L}{c} \right) \tag{6.24}$$

The product on the left hand side of the equation above is the impulsion per unit area, i.e.,

$$p \cdot \Delta t = \frac{I_{\Delta t\_\text{impulse}}}{A} = \frac{1}{A}\int_0^{\Delta t}(F_{\text{up}} - F_{\text{down}})dt \tag{6.25}$$

where A is the surface area of the block impacted by the impulse force.

From this follows that the net effect of the fluctuating pressures of an impinging jet on a plunge pool bottom (excluding the effects of the weight of the block and the friction forces around it) can be expressed as

$$I_{\Delta t\_\text{impulse}} = \int_0^{2L/c} (F_{\text{up}} - F_{\text{down}})dt = C_I A \phi \gamma \frac{v_j^2 L}{gc} \tag{6.26}$$

What remains is to quantify the magnitude of the weight of the rock block and the friction forces on its sides. The submerged weight can be calculated as

$$W_g = (\gamma_s - \gamma)V_b \tag{6.27}$$

where V_b is the volume of the block of rock.

The resistance offered by the weight of the block over the time of the impulse is

$$\int_0^{2L/c} W_g dt = \int_0^{2L/c} (\gamma_s - \gamma)V_b dt = (\gamma_s - \gamma)V_b \frac{2L}{c} \tag{6.28}$$

and the implulse resistance offered by the friction from the surroundings on the rock block can be expressed as

$$\int_0^{2L/c} F_{\text{sh}} dt = F_{\text{sh}} \frac{2L}{c} \tag{6.29}$$

Equation (6.19) can now be written as

$$I_{\Delta t\_\text{impulse}} = m v_{\Delta t} = \rho_s V_b v_{\Delta t} = \rho_s A z_b v_{\Delta t} = \int_0^{2L/c} (F_{\text{up}} - F_{\text{down}} - W_g - F_{\text{sh}})dt \tag{6.30}$$

where z_b denotes the vertical height of the rock block, assuming it is prismatic.

Using Eqs.(6.26), (6.28), and (6.29) in Eq. (6.30) it now follows that

$$v_{\Delta t} = \frac{1}{\rho_s A z_b} \cdot \left[C_I A \phi \gamma \frac{v_j^2 L}{gc} - (\gamma_s - \gamma) A z_b \frac{2L}{c} - F_{\text{sh}} \frac{2L}{c} \right] \tag{6.31}$$

The vertical distance h through which the block will be lifted if a net uplift force exists can now be determined using Eqs. (6.16) and (6.31).

If

$$C_I A\phi\gamma \frac{v_j^2 L}{gc} \geq (\gamma_s - \gamma) A z_b \frac{2L}{c} + F_{\text{sh}} \frac{2L}{c}$$

then,

$$h = \frac{1}{2g\rho_s^2 A^2 z_b^2} \cdot \left[C_I A\phi\gamma \frac{v_j^2 L}{gc} - (\gamma_s - \gamma) A z_b \frac{2L}{c} - F_{\text{sh}} \frac{2L}{c} \right]^2 \qquad (6.32)$$

$h = 0$ otherwise

The net dynamic impulsion coefficient C_I for *jets impinging into plunge pools* was determined by Bollaert (2002), who expressed it in terms of the dimensionless depth Y/D_j in a plunge pool, where Y is the plunge pool depth and D_j is the diameter of a round jet at the water surface elevation of the plunge pool. This relationship is shown in Fig. 6.4, and can also be calculated with Eq. (6.33).

$$C_I = 0.0035 \cdot \left(\frac{Y}{D_j} \right)^2 - 0.119 \cdot \left(\frac{Y}{D_j} \right) + 1.2 \qquad (6.33)$$

Equation (6.32) can be used to calculate removal of rock blocks for other flow conditions as well, as long as dynamic expulsion coefficients C_I for those particular flow conditions are known. The dynamic expulsion

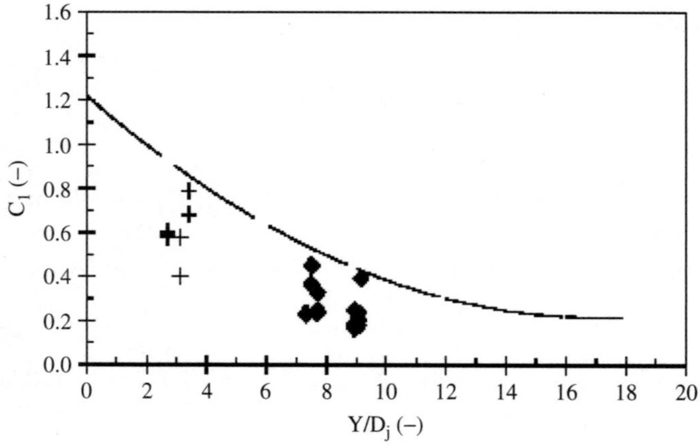

Figure 6.4 Dynamic impulsion coefficient as a function of dimensionless depth as the result of jets impinging into plunge pools (Bollaert 2002).

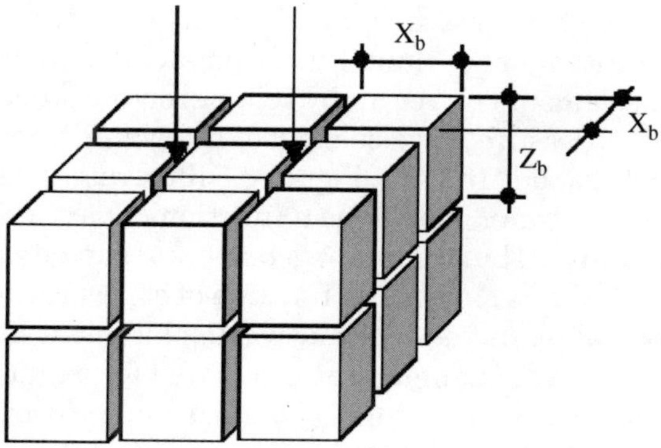

Figure 6.5 Simplified rock geometry for assessing rock scour by the dynamic impulsion method (modified from Bollaert 2002).

coefficient in Fig. 6.4 is applicable to plunge pool beds impacted by impinging jets only. By referring to Fig. 6.5 this equation can be rewritten as

$$h_{\text{up}} = \left[2\frac{(x_b + 2z_b)}{c} \right]^2 \cdot \frac{1}{2g \cdot x_b^4 \cdot z_b^2 \cdot \rho_s^2} \cdot \left[C_I \cdot \phi \cdot \gamma \cdot \frac{V_j^2}{2g} \cdot x_b^2 - (\gamma_s - \gamma) \cdot x_b^2 \cdot z_b - F_{\text{sh}} \right]^2 \tag{6.34}$$

Criteria for determining if rock blocks from a jointed rock mass will experience incipient motion, i.e., if scour will commence, have been proposed by Bollaert (2002) (Table 6.1).

Inspection of Eq. (6.34) shows that the expulsion of a rock block is inversely proportional to the square of the pressure wave celerity c. It has

TABLE 6.1 Proposed Criteria to Assess Rock Scour Potential by Dynamic Impulsion

$\frac{h_{\text{up}}}{z_b} \le 0.1$	Rock block remains in place.
$0.1 < \frac{h_{\text{up}}}{z_b} < 0.5$	Rock block vibrates and most likely remains in place.
$0.5 \le \frac{h_{\text{up}}}{z_b} < 1.0$	Rock block vibrates and is likely to be removed, depending of ambient flow conditions.
$\frac{h_{\text{up}}}{z_b} \ge 1.0$	Rock block is definitely removed from its matrix.

SOURCE: From Bollaert 2002.

been shown in Chaps. 4 and 5 that the pressure wave celerity of water is very sensitive to the free air content of the fluid. Even a small amount of free air, with an air content (i.e., amount of free air by volume) on the order of 1 percent or so, reduces the pressure wave celerity of the water from about 1000 m/s for pure water to about 100 m/s. Figure 6.6 illustrates the sensitivity of rock block removal from a matrix as a function of varying pressure wave celerity of the water. The rock block geometry assumed is shown in Fig. 6.5 and the sensitivity is shown as a function of aspect ratio z_b/x_b (assuming a square block in plan) and the relative displacement in the vertical, expressed as h_{up}/z_b. This example was calculated for expulsion of rock blocks due to the impact of an impinging jet into a plunge pool. More detail on applying this method is presented in Chaps. 7 and 9.

Figure 6.6 shows that the removal of rock blocks for this particular case is sensitive to the amount of air in the water. When the air content is close to zero (i.e., the pressure wave celerity = 1000 m/s), the graph shows that the rock block will not be removed. On the other hand, if the air content of the water is about 0.017 percent (the pressure wave celerity is about 600 m/s) blocks with aspect ratios of about 1 starts to vibrate, and when the air content reaches 1 percent (the pressure wave celerity is approximately = 100 m/s), it is likely that rock blocks with aspect ratios less than about 2 will definitely be removed from the rock matrix.

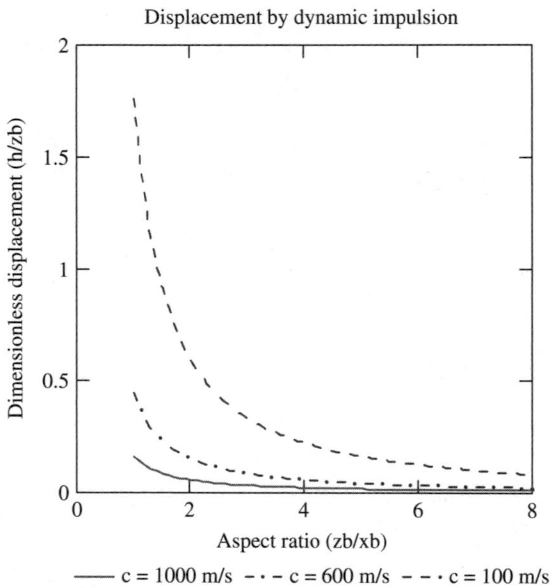

Figure 6.6 Displacement of a block of rock by dynamic impulsion for changes in the value of pressure wave celerity (c) and aspect ratio of the rock.

Figure 6.6 is presented for purposes of illustration. Similar analyses should be conducted in practice to determine the sensitivity of rock removal to the amount of free air in the water.

Keyblock theory

The removal of individual rock blocks could be affected by their relationship to other blocks of rock. If the removal of a certain rock block leads to other rock blocks being dislodged with more ease, it is appropriate to refer to a keyblock effect. An illustration of the concept is shown in Fig. 6.7*a*, which illustrates a plunge pool in jointed rock. If the keyblock is removed, as shown in *b*, it leads to the removal of the next keyblock *c*, and the next *d*, and so on. This cascading removal of rock blocks is known as the keyblock effect.

Keyblock theory introduced by Goodman and Shi (1985) has been used in design and analysis of various rock structures (dam foundations, cut slopes, and underground excavations). These methods are well established

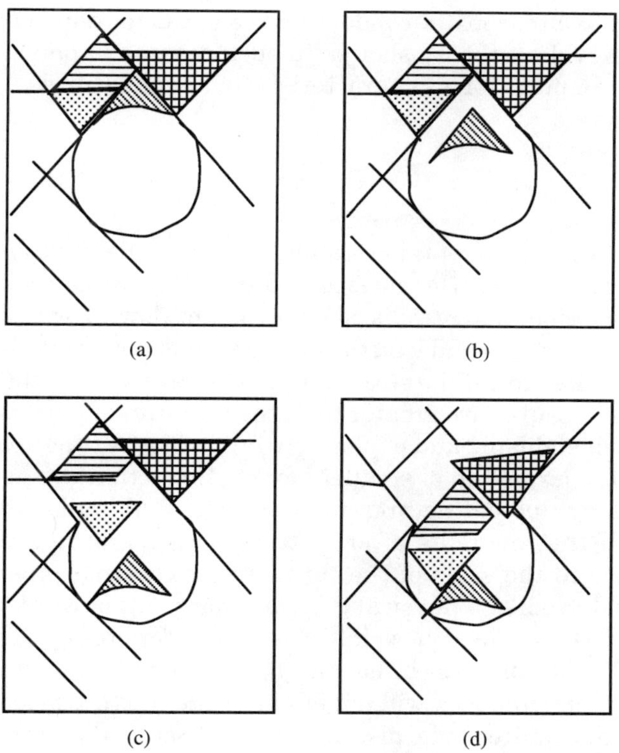

Figure 6.7 Removal of a keyblock and how it affects removal of other blocks of rock.

in the field of rock engineering and have been proven to be effective in evaluating the three-dimensional geologic structure relative to the rock free face. A free face is an excavation surface, natural rock slope, or the rock surface in a scour hole.

By applying the methods of Keyblock theory it is possible to locate potentially unstable rock blocks (removable blocks) and evaluate the stability of such blocks given the shear strength of the joint planes that define the blocks and the water forces acting on the blocks. The definition of removability is based on the relative orientations of the joint planes and free faces along a block pyramid. In scour analysis the free face could continually change, i.e., as blocks are dislodged and removed by dynamic hydraulic forces new free faces are formed. The new free faces have the potential to create additional removable blocks, i.e., blocks that are initially not removable become removable as the removable blocks are dislodged and removed from the scour hole. The progressive failure of removable blocks into a scour hole is a problem that can be solved using three-dimensional discontinuous deformation analysis (3D-DDA). This method is rooted in block theory and has been developed by Shi (1999, 2003).

Application of Keyblock theory is a complex process and may be considered as one of the approaches to conducting scour calculations on important projects. Keyblock theory should, in such cases, be applied with the estimates of impulse forces characteristic of turbulent flow.

Chemical Gels

Rock

Large continuous masses of rock contain irregularities, principally close-ended fissures that can start at the surface and extend into the rock mass. The pressure fluctuations introduced by turbulent flowing water enters such close-ended fissures and lead to development of stress intensities at the tips of the fissures. If the stress intensities developed at the tips of the close-ended fissures are greater than the fracture toughness of the rock mass it will suddenly and explosively fail in brittle fracture mode. This leads to the formation of smaller blocks of rock that can be more easily removed by fluctuating pressures in a manner similar to non-cohesive soils. If the remnants after brittle fracture failure are small enough relative to the scale of the turbulent fluctuating pressures the rock will behave like a physical gel from that point onwards.

If the pressure fluctuations introduced into close-ended fissures by the turbulent flow do not result in stress intensities that exceed the fracture toughness of the rock, then the rock will not fail in brittle fracture mode but might fail in fatigue failure mode, also known as sub-critical failure. When turbulent fluctuating pressures transferred into the close-ended fissures in the rock continue to fluctuate over a long enough period of time the rock will eventually fail in fatigue. The length of time that is required for fatigue failure to occur is determined by the frequency and magnitude of the pressure fluctuations and the strength characteristics of the rock. When the rock fails in sub-critical mode it can break up into smaller pieces. If the pieces of rock that remain behind after fatigue failure are small enough, the rock blocks will be removed by the turbulent fluctuating pressures in the water in the same manner as non-cohesive soils. The classification of the rock can change from a chemical gel to that of a physical gel after the occurrence of fatigue failure or brittle fracture.

Brittle fracture. The stress intensity K that develops in close-ended fissures is a function of the pressure in the fissure, a characteristic dimension and the shape of the fissure. It can be expressed as

$$K = \sigma_{\text{water}} \sqrt{\pi a} f \tag{6.35}$$

where σ_{water} = stress introduced by turbulent fluctuating pressures in a close-ended fissure
a = characteristic dimension of the fissure (usually the extent of the fissure into the rock mass from the surface of the rock)
f = a factor that accounts for the shape of the close-ended fissure

In order to determine whether the material will fail in brittle fracture, the stress intensity is compared to the fracture toughness of the material. The fracture toughness, K_I, is a material property. If the stress intensity is greater than the fracture toughness the material will fail in brittle fracture, i.e., in an explosive manner. Therefore, brittle fracture occurs if,

$$K > K_I \tag{6.36}$$

The fracture toughness of rock can be estimated with the following equations (Bollaert 2002):

$$K_{I,\,\text{insitu},T} = (0.105 \text{ to } 0.132)\ T + (0.054\ \sigma_i) + 0.5276 \tag{6.37}$$

$$\text{K}_{I,\,\text{insitu,UCS}} = (0.008 \text{ to } 0.010)\ \text{UCS} + (0.054\ \sigma_i) + 0.42 \tag{6.38}$$

where T = tensile strength of the rock (MPa)
UCS = unconfined compressive strength of the rock (MPa)
σ_i = confining stresses in the rock (MPa)

Examples illustrating how conditions for brittle fracture are determined are presented in Chaps. 7 and 9. It is important to note that Eqs. (6.37) and (6.38) provide approximations of the fracture toughness of a particular brittle material. It is therefore important to perform sensitivity analyses when executing scour assessments in order to gain insight into potential rock scour behavior and gather enough information for implementation of the decision-making process outlined in Chap. 2. One of the problems when applying this approach is that very few, if any, rock testing facilities currently exist in the world that can perform tests to determine the actual fracture toughness of a particular rock sample of interest.

Subcritical failure (fatigue). When the stress intensity does not exceed the fracture toughness of the material it is possible for the material to fail in fatigue, also known as subcritical failure. An equation that can be used to calculate the length of time it will take for the material to fail in fatigue was developed by Paris et al. (1961).

$$\frac{dL}{dN} = C(\Delta K)^m \tag{6.39}$$

where N = number of cycles of the fluctuating pressure that will lead to fatigue failure

C, m = material properties

ΔK = range of stress intensities introduced to the material by the fluctuating pressures

L = distance of crack growth required for the material to fail

Directions to determine values for C and m are presented in Chap. 4. Examples to calculate the time to failure are found in Chap. 8.

Cohesive granular earth material

Threshold conditions for cohesive granular materials like clay are quite difficult to define. Considerable effort has been put forward in Chap. 4 explaining the erosion characteristics of clay. A conclusion drawn from those discussions is that the concept of a threshold value of the erosive capacity of water where incipient motion commences in clays may not be valid.

A number of experiments have shown that erosion commences almost immediately when water starts to flow over a cohesive material like clay. Some researchers concluded that the rate of erosion for cohesive materials is most probably a more representative indicator of the ability of such materials to resist the erosive capacity of water than a threshold value of the erosive capacity (critical shear stress or critical stream power). This implies that clay with high erosion resistance will be characterized by a low rate of erosion and vice versa. This could be true when considering the fact that the erodibility of such materials appear to be more sensitive to changes in temperature, pH, and salt concentration than what they are to changes in the erosive capacity of water.

Hanson and Simon (2001) conducted extensive in situ testing of cohesive streambed soils in the loess area of the midwestern United States. They used the vertical jet tester (VJT) (see Chap. 8) to test these materials and based their estimates of critical shear stress and rate of erosion on the following equation:

$$\varepsilon = k_d(\tau - \tau_c) \tag{6.40}$$

where ε = rate of erosion (m/s)
k_d = rate of erosion coefficient (m^3/N-s)
τ_c = critical shear stress (Pa)
τ = shear stress applied to the sample by the testing equipment (Pa)

This equation presumes the existence of a critical shear stress.

Hanson and Simon (2001) performed 83 tests in western Iowa, eastern Nebraska, and the Yolabusha River Basin in Mississippi and found a wide range of critical shear stress and rate of erosion coefficient values. Histograms summarizing their results are shown in Fig. 6.8. From the summary provided by the authors it appears as if field assessment can

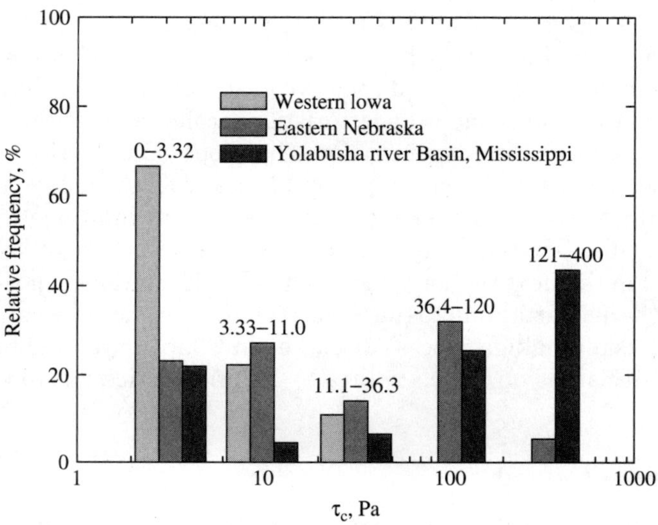

(a)

Figure 6.8 Threshold shear stress and erosion rate coefficient histograms for 83 in situ tests on cohesive stream bed material in the midwestern United States (Hanson and Simon 2001).

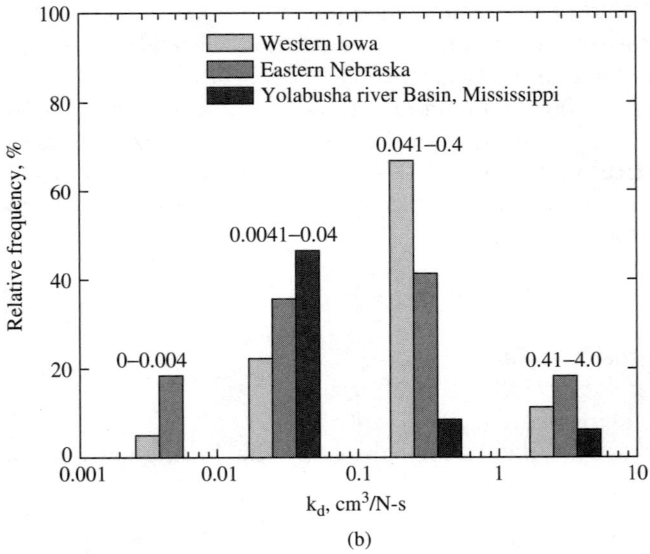

Figure 6.8 (*Continued*)

result in quite high values of critical shear stress. This contradicts observations made by others and indicates that the concept of critical shear stress might well be a valid concept for cohesive materials. No general agreement on this aspect currently exists in the profession.

This information, combined with the discussion provided in Chap. 4, emphasizes the complexity of erosion of cohesive granular materials like clay. It is concluded that the erosion characteristics of cohesive materials are complex and that various approaches should be adopted in practice to assess it. These include using testing equipment like the VJT, the hole erosion test (HET), the erosion function apparatus (EFA) and the Couette Flow Device (CFD) discussed in Chap. 8, and using the erodibility index method, discussed in the next section. Once as much information as possible has been gathered pertaining to the erosion characteristics of a particular clay the decision-making process discussed in Chap. 2 are used to make defensible decisions on which values to use in practice for analyzing scour.

The Erodibility Index Method

In the early 1990s a number of researchers analyzed field and laboratory scour and erosion data using a geomechanical index developed by Kirsten (1982). The interest in using this index to quantify the relative ability of earth material to resist the erosive capacity of water was stimulated by a discussion between Moore and Kirsten at the ASTM meeting on rock classification systems for engineering purposes (Kirkaldie 1988), when Moore suggested using Kirsten's index in this manner. The analyses by various researchers led to the development of essentially four threshold relationships, all using Kirsten's index (dubbed the "erodibility index") to quantify the relative ability of earth materials to resist the erosive capacity of water and stream power to quantify the relative magnitude of the erosive capacity of water.

The value that was added by following this approach is that it became possible to consider the erodibility of earth materials ranging from non-cohesive soils to cohesive and vegetated soils, and even rock and engineered earth materials like concrete. The pragmatic character of the erodibility index makes it possible to use either field characterization or laboratory data to index earth materials and quantify their relative ability to resist the erosive capacity of water. Guidance on quantifying this index is provided in Chap. 4.

Research was conducted by Temple and Moore (1994), Annandale (1995), van Schalkwyk et al. (1995), and Kirsten et al. (1996). Most of the data used by these researchers originated with the Agricultural Research Service of the U.S. Department of Agriculture, with some South African data added by Annandale, van Schalkwyk, and Kirsten. Additionally Annandale and Kirsten used published data on the incipient motion of non-cohesive granular material, and Kirsten also used data on jet cutting of intact rock materials. The total amount of data that was available was quite significant. Up to 137 field observations of events that either scoured or not were available from the data base of the U.S. Department of Agriculture. This information was collected by the Agricultural Research Service over a period of ten years under a program that monitored the performance of emergency spillways of the department's dams.

The fact that four groups of researchers independently worked on the analysis of field and laboratory data to establish an erosion threshold for all earth materials makes it possible to compare their findings and use conclusions made from this comparison to recommend an erosion threshold for use in practice. The threshold relationships proposed by each group of researchers are first presented and discussed, whereafter the threshold relationships are compared and a recommendation made for use in practice. Comparison of the four erosion threshold relationships emphasizes the amount of uncertainty inherent in scour technology. When applying this method, as all other methods in this field of specialization, the reader is reminded of the general approach to decision making outlined in Chap. 2. One can never merely use the result from a calculation for implementation in practice without further subjective and objective reasoning.

Temple and Moore (1994)

Temple and Moore (1994) used the product of unit discharge q and drop height H to represent the relative magnitude of steam power and correlated that with calculated values of the erodibility index using the field data collected by the U.S. Department of Agriculture. The unit discharge is expressed in units of $m^3/s/m$ (i.e., m^2/s) and the drop height H is expressed in m. The units of their indicator parameter of the relative magnitude of the erosive capacity of water are therefore m^3/s.

Selection of an abbreviated form of an expression for stream power limits the use of the threshold relationship to assessing headcutting only. For example, it is not possible to use this threshold relationship to determine the erodibility of a channel bed under channel flow conditions or when, say, a hydraulic jump forms in the channel. The relationship has been developed strictly for use in the SITES computer program developed by the U.S. Department of Agriculture to simulate head cut erosion. For this reason Temple and Moore (1994) called their approach the "headcut erodibility index method." This name is somewhat confusing, because the index that they use is exactly the same as Kirsten's index (with a few minor, but insignificant modifications). It is only the limitation introduced by the way the relative magnitude of the stream power is calculated that limits its use to head cut assessment.

However, if one were to multiply the product of qH with the unit weight of water and make the assumption that the thickness of the jet is approximately 1 m, it is possible to express the ordinate of their threshold relationship in terms of power per unit area, i.e., kW/m^2. By doing this it is possible to compare their threshold relationship with that of the other researchers. However, when such comparisons are made it is necessary to recall the limitations of these assumptions.

The threshold relationship developed by Temple and Moore (1994) is shown in Fig. 6.9. It contains two types of data points, those that experienced erosion and those that did not. The threshold line should lie between the regions on the graph representing events that experienced scour and those that did not. It appears as if the number of non-scour events that they used is insufficient (only five non-scour events are shown in the graph) to clearly demarcate the zone between scour and non-scour events. They principally used the bottom of the scour events to locate their threshold line.

van Schalkwyk et al. (1995)

van Schalkwyk et al. (1995) conducted research over a number of years with funding from the Water Research Commission in South Africa. They used data from 18 South African dams for which overtopping and rock scour information was available and the data collected by the U.S. Department of Agriculture to analyze scour of rock.

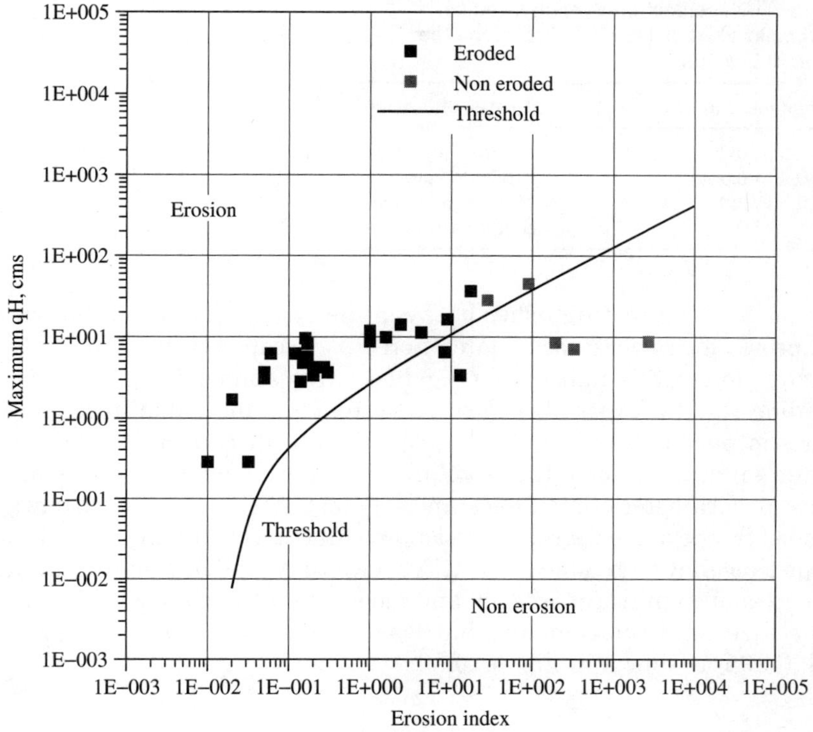

Figure 6.9 Erosion threshold for rock and vegetated soils (Temple and Moore 1994).

They also conducted laboratory experiments to better understand the scour of rock.

In their analysis they quantified stream power per unit area for both jet impact and for channel flow. They made some simplifying assumptions, principally that the thickness of an impinging jet at the point of impact can be calculated as the drop height divided by 3. The reasoning that led to this assumption is unknown. In essence they calculated the impact area of a rectangular jet as $A = bH/3$, which resulted in their equation for the stream power of plunging jets, i.e.,

$$P_{\text{vSchalkwyk}} = \frac{\gamma QH}{A} = \frac{3\rho gQH}{bH} = 3\rho gq \tag{6.41}$$

They calculated the stream power in channel flow using a conventional approach, i.e.,

$$P = \tau v$$

Additionally they also classified the degree of erosion using the criteria shown in Table 6.2. This assisted them in developing a threshold

TABLE 6.2 Classification System Used by van Schalkwyk et al. (1995) to Determine the Degree of Erosion

Depth of erosion (m)	Degree of erosion
0 to 0.2	No erosion
0.2 to 0.5	Little erosion
0.5 to 2.0	Moderate erosion
> 2.0	Significant erosion

relationship that distinguishes between the degree of erosion that can be expected for varying rock quality and erosive capacity. In their analysis they plotted this information against various indicator parameters including the RMR rock classification system and individual rock characteristic parameters like mass strength and discontinuity volume. Kirsten's index, renamed the erodibility index, was found to be the only indicator parameter of rock resistance against erosion that consistently provided trends of the degree of erosion as a function of stream power. This means that they were able to identify zones that provided an indication of no erosion, significant erosion, and moderate to little erosion of rock.

The threshold relationship that was developed by van Schalkwyk et al. (1995) is shown in Fig. 6.10.

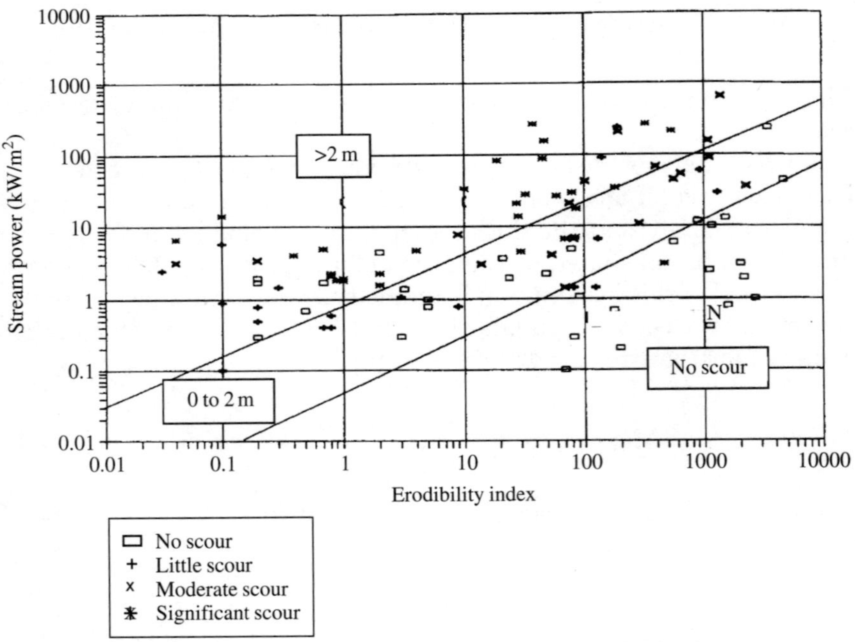

Figure 6.10 Erosion threshold for rock formations (van Schalkwyk et al., 1995).

Kirsten et al. (1996)

Kirsten et al. (1996) also analyzed the data set collected by the U.S. Department of Agriculture, some of the South African dams, published data on incipient motion of non-cohesive sediment and data on cutting of intact materials with hydraulic jets. They found two threshold curves, based purely on the data collected by the U.S. Department of Agriculture and the South African data, and another threshold based on the comprehensive data set (Figs. 6.11 and 6.12).

The two curves differ somewhat. The development of a single relationship spanning all data, i.e., from non-cohesive silt material at the low end to hard intact materials subject to cutting by hydraulic jets at the high end, appears somewhat ambitious. What is evident though is that it is possible to develop a threshold relationship for a wide variety of materials by relating the erodibility index and stream power.

Annandale (1995)

Annandale (1995) used the data collected by the U.S. Department of Agriculture, and scour data from Bartlett Dam, Arizona, and four South African dams provided by van Schalkwyk, as well as published data on incipient motion of non-cohesive earth materials to develop his threshold relationship. Although the threshold data seem to plot on a continuous curve he separated the data into two groups, using a erodibility index value of 0.1 as the seperator. The overall relationship he found is presented in Fig. 6.13. The relationship for earth materials

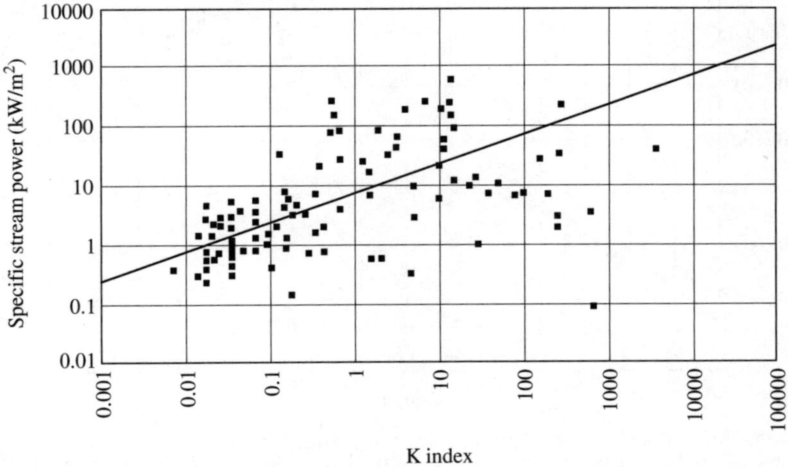

Figure 6.11 Erosion threshold developed by Kirsten (1996) using USDA and South African data.

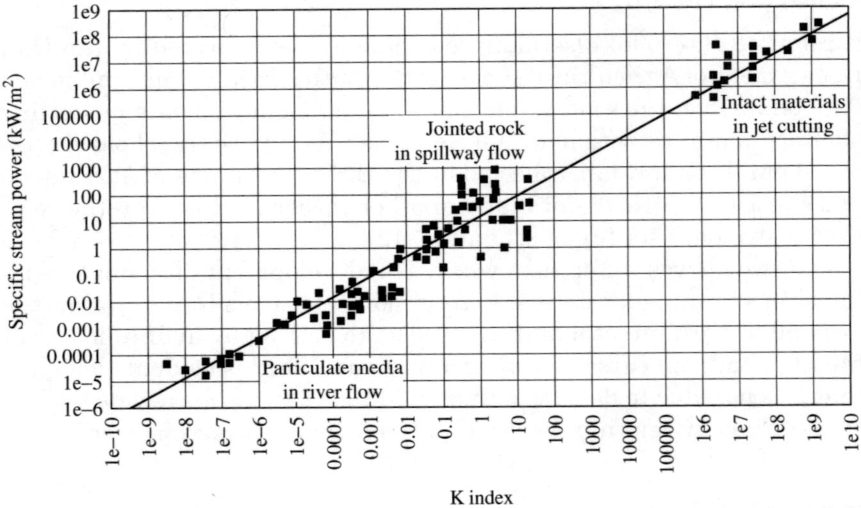

Figure 6.12 Erosion threshold developed by Kirsten (1996) using a comprehensive data set.

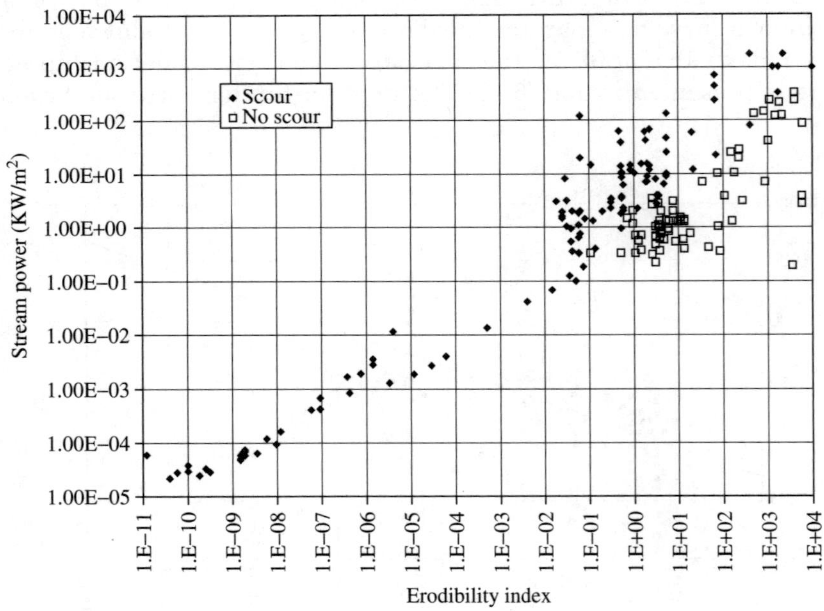

Figure 6.13 Annandale's erosion threshold graph using USDA and South African data, and published data on incipient motion of non-cohesive earth materials (Annandale 1995).

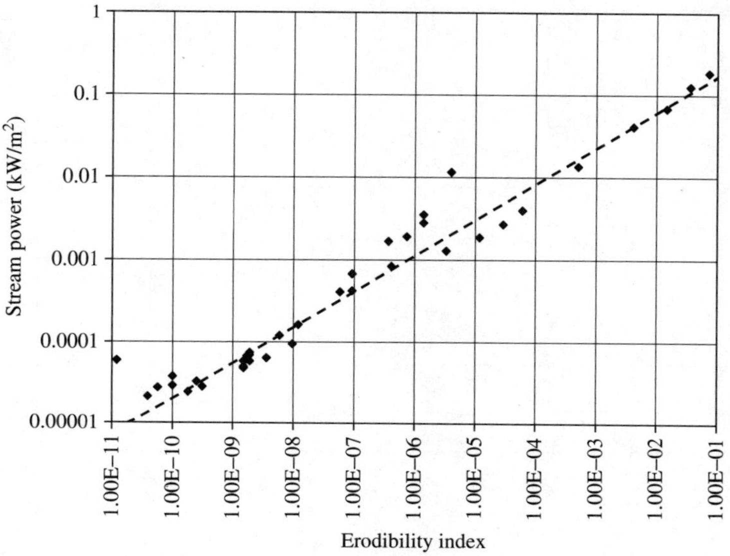

Figure 6.14 Threshold relationship for low erodibility index values (Annandale 1995).

with low erodibility index values is shown in Fig. 6.14, and those with higher erodibility index values in Fig. 6.15.

The equations describing the erosion threshold for lower erodibility index values can be expressed as

$$P_c = 0.48(K)^{0.44} \qquad \text{for } K \leq 0.1 \tag{6.42}$$

where P_c is the critical stream power that will result in incipient motion and K denotes the erodibility index.

The equation describing the threshold relationship for higher values of the erodibility index is

$$P_c = K^{0.75} \qquad \text{for } K > 0.1 \tag{6.43}$$

In addition to analyzing the field and laboratory data the threshold relationship was also validated with near-prototype experiments that were executed at Colorado State University's Engineering Research Center, Fort Collins. These tests were partly funded by the U.S. Bureau of Reclamation. The tests consisted of two series, the one series validated the application of the threshold relationship to predicting scour in non-cohesive granular material and the other set checked the ability of the method to predict the erodibility of a simulated rock formation under near-prototype conditions.

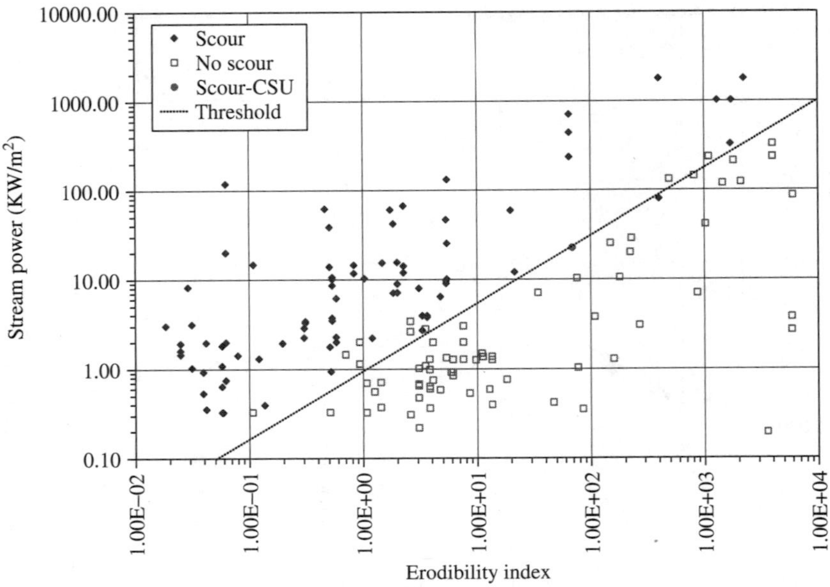

Figure 6.15 Erosion threshold for a variety of earth materials ranging from cohesive and vegetated soils to rock (Annandale 1995). The near-prototype validation of scour using a simulated rock formation executed at Colorado State University, Fort Collins (CSU), plots on the threshold line.

The near-prototype facility consisted of a large basin of approximately 15 m long × 5 m wide × 3 m deep into which an impinging jet with a maximum discharge of approximately 3 m^3/s could be discharged (Fig. 6.16). The foundation material in the basin could vary. Two material types were tested; a non-cohesive road base and a simulated rock formation with a dip of 45 degrees (Fig. 6.17).

Figure 6.16 Near-prototype testing facility at Colorado State University, Fort Collins, for testing scour.

Figure 6.17 Simulated rock foundation with a dip of 45° tested at the facility at Colorado State University, Fort Collins.

The test results from the experiments were very promising. It was found that the threshold of erosion of the simulated rock fell exactly on the threshold line in Fig. 6.15. (Annandale et al., 1998). Additionally, it was found that the calculation of scour depth using Annandale's procedure for the non-cohesive road base correlated very well with measured scour depths (Wittler et al., 1998). The correlation between calculated and predicted scour depth for the non-cohesive granular material tested in the facility is shown in Fig. 6.18.

The validation of the erosion threshold for the simulated rock and the good correlation between measured and calculated scour depths for granular material found in these experiments provide a fair amount of confidence in this erosion threshold relationship.

Comparison

The erosion threshold relationships of Annandale (1995), van Schalkwyk et al. (1995), and Kirsten et al. (1996) can be compared directly as all three use the same units to quantify stream power. By making the assumption that the thickness of the footprint of impinging jets used in the analysis by Temple and Moore (1994) equals 1 m it is possible to add this information to the comparison as well. This is done by multiplying the ordinate parameter in their threshold relationship with the unit weight of water and dividing it by 1 m, which then provides an estimate of stream power per unit area, i.e., kW/m^2.

Comparing the four methods is considered important because the approach is relatively new and if it is concluded that the independent findings of some of the researchers correlate reasonably well it provides

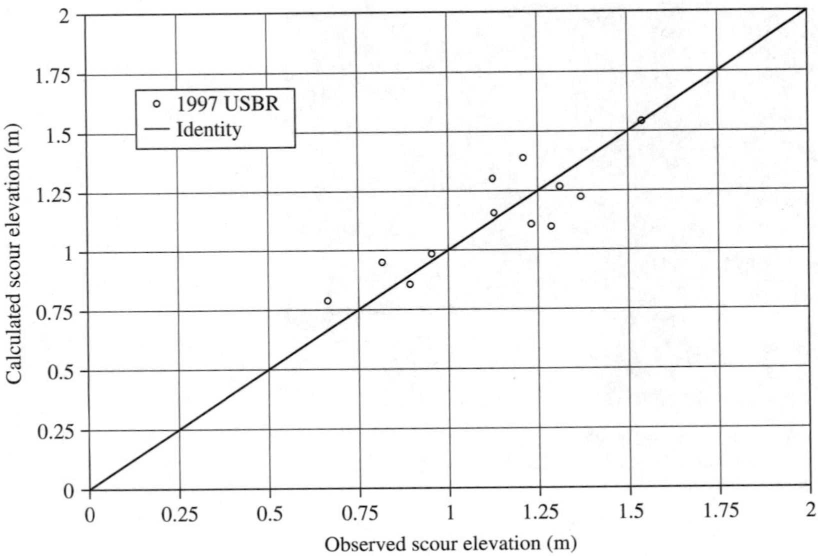

Figure 6.18 Comparison between observed and calculated scour elevations for granular soil in a near-prototype experiment using the erodibility index method (Wittler et al., 1998).

confidence in the approach. This is done by plotting all five of the threshold relationships (one from each of the researchers, except for Kirsten who has two) on Annandale's (1995) erosion threshold relationship (Fig. 6.19). The representative threshold relationship by van Schalkwyk et al. (1995), shown in this graph, is the threshold that signifies significant erosion, i.e., erosion in rock in excess of 2 m. The reason for this selection is that erosion of rock that is less than 2 m is considered relatively inconsequential and is most probably the result of removal of loose blocks of rock on the stratum surface.

The comparison shows good correlation between the erosion threshold relationships by Annandale (1995) and van Schalkwyk et al. (1995), and a somewhat disparate relationship between Kirsten's two threshold relationships and that of Temple and Moore (1994). The comparison also shows that the threshold relationships of Kirsten et al. (1996) and Temple and Moore (1994) are less conservative than those by Annandale (1995) and van Schalkwyk et al. (1995).

The independent verification of the threshold relationships developed by Annandale (1995) and van Schalkwyk et al. (1995) in addition to the near-prototype verification of scour of simulated rock and non-cohesive granular material conducted at Colorado State University provide a level of confidence in these two methods. Additionally, they are more conservative that the erosion threshold lines by Temple and Moore (1994) and by Kirsten et al. (1996) and are therefore preferred.

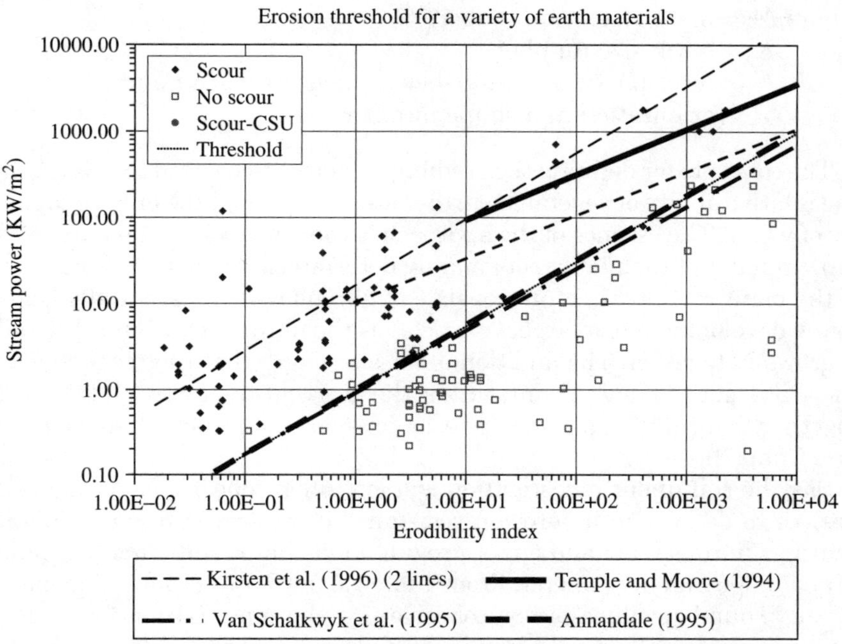

Figure 6.19 Comparison of erosion threshold relationships between Kirsten et al. (1996), Temple and Moore (1994), van Schalkwyk et al. (1995), and Annandale (1995).

Comparison of the results of case studies on Caborra Bassa Dam, Mozambique, by making use of Annandale's (1995) method and Bollaert's (2002) Comprehensive scour method that accounts for brittle fracture, fatigue failure, and dynamic expulsion of rock blocks also provides an additional measure of assurance in the erosion threshold relationships by Annandale (1995) and van Schalkwyk et al. (1995) (see Bollaert 2002).

The preferred erosion threshold relationship based on the erodibility index that is used in this book is the relationship developed by Annandale, i.e., Figs. 6.14 and 6.15.

Vegetated earth material

The erodibility index method developed by Annandale (1995) is also based on a substantial amount of data related to erosion of vegetated earth materials, in addition to erosion data for rock, chohesive, and noncohesive granular materials. An approach has therefore been developed to assess the erodibility of vegetated earth.

As indicated in Chap. 4 the Erodibility Index K is defined as

$$K = M_s K_b K_d J_s$$

where M_s = mass strength number
K_b = block size number
K_d = inter-particle or inter-block shear strength number
J_s = orientation and shape number

The concepts for defining the erodibility index can be used to determine the relative ability of vegetated earth materials to resist the erosive capacity of water. The essence of the approach is that the root architecture and growth habit of the plant roots mainly determines its erosion resistance. If the plant root architecture consists of fibrous roots it essentially leads to the development of a larger "particle" as the fibrous roots binds the soil it grows in together. The additional mass strength that the roots offer to the soil is not that significant; rather the fact that the roots bind the soil together to form a larger unit to resist erosion is of principal importance (see Chap. 3).

Say the soil under investigation is a loose non-cohesive sand and one desires to determine its erosion resistance if it is covered with a plant that has fibrous roots and a root growth habit that results in a root bulb of about 300 mm in diameter. From Table 4.5 in Chap. 4, containing mass strength number values for non-cohesive granular material, it is found that an appropriate value for M_s representing loose sand is

$$M_s = 0.04$$

Based on the assumption that the mass strength of the soil is not really affected by the presence of the plant and its roots, but that the essential value that the plant offers lays in the fact that the soil is bound together by the fibrous root. One can calculate the magnitude of the block size number for the effective particle size. If it is known that the root bulb of the plant has a diameter of 300 mm, this number is used to calculate the value of the block size number K_b. Using the equation for calculating the block size number for non-cohesive material, it follows that,

$$K_b = 1000(D)^3 = 1000(0.3)^3 = 27$$

Furthermore, by assuming that the roots do not change the internal angle of friction of the soil significantly it is possible to calculate the shear strength number. If it is assumed that the internal angle of friction is 30°, then the value of K_d is

$$K_d = \tan(30^\circ) = 0.577$$

The default value of the shape and orientation number is 1.0, i.e., $J_s = 1.0$.

Therefore, the value of the erodibility index is calculated as

$$K = 0.04 \times 27 \times 0.577 \times 1.0 = 0.623$$

This value is greater than 0.1, so one would use Eq. (6.43) to calculate the value of the threshold stream power for the vegetated soil, i.e.,

$$P_c = (K)^{0.75} = (0.623)^{0.75} = 0.7 \text{ KW/m}^2 = 700 \text{ W/m}^2$$

If the soil is planted with vegetation that on average creates an effective particle size of 300 mm throughout the area that might be affected by the erosive capacity of flowing water, the erosion resistance offered by the vegetated soil would be on the order of 700 W/m^2. This is significantly higher than the erosion resistance offered by loose, non-cohesive fine sand, which is on the order of a few watts per square meter.

Engineered earth materials. The erosion resistance of engineered earth materials like concrete or anchored rock can also be determined by making use of the erodibility index method. In the case of concrete the material is treated in exactly the same way that rock would be treated. Its erodibility index is calculated and its threshold stream power determined by making use of Fig. 6.15 or Eq. (6.43).

In the case of anchored rock the principal value of anchoring is obtained with the increase in the effective size of the rock blocks. In essence, the anchors play a role similar to the fibrous roots in vegetated soils. They do not increase the strength of the rock, but they do increase the effective block size of the rock and therefore its erosion resistance. A method that can be used to assess the effectiveness of anchored rock to resist the erosive capacity of water is presented in Chap. 9.

Summary

This chapter presents methods and guidelines to determine erosion thresholds for a variety of earth materials ranging from non-cohesive soil, to cohesive and vegetated soil, rock, and engineered earth materials. The distinction between physical and chemical gels is used throughout. It has been shown that the classification of chemical gels can change to that of physical gels if the chemical gel breaks up into smaller pieces.

The information presented in the chapter illustrates that fluctuating turbulent pressures have a different effect on chemical gels than they have on physical gels. Chemical gels essentially fail in brittle fracture or fatigue, while the elements of physical gels are removed individually by the fluctuating pressures of turbulent flow. It has been shown in this chapter and Chap. 3 that the erosive capacity of the water required to cause incipient motion of non-cohesive granular material depends on flow conditions. When the absolute roughness introduced by the non-cohesive granular material is small the conditions right at the boundary (within the viscous sublayer) is laminar. Such flow conditions attempt to move the non-cohesive particles as an assembly, which requires higher values of erosive capacity than when the flow is in the transition or rough turbulent zones.

When the roughness of non-cohesive particles is large enough to penetrate the buffer layer within the near-bed region of the flow it causes turbulent pressure fluctuations on the bed. In the transition range of flow the areas occupied by the fluctuating pressures are large enough to occupy entire particles. Such particles are then removed from the bed by individually being sucked from the bed. The fact that individual particles are removed by fluctuating pressures requires lower values of the erosive capacity of the water than when the flow right at the bed in the near-boundary region is laminar. Once the flow conditions move to the rough turbulent zone at the boundary the erosive capacity required to remove particles increases again. The reason for this is that the particles are not fully covered by the low-pressure zones associated with the fluctuating pressures. They are therefore removed partly by suction and partly by a movement that develops over such particles due to the presence of the fluctuating turbulent pressures.

Brittle fracture in chemical gels occurs when the stress intensity caused by the fluctuating pressures at the tips of close-ended cracks or other imperfections in the chemical gel exceeds the fracture toughness of the material. When the stress intensity at the tip of a crack is greater than the fracture toughness of a chemical gel it will fail in an explosive manner.

In cases where the stress intensity introduced by fluctuating pressures into a chemical gel is less than the fracture toughness of the material it can still fail. This can occur if the fluctuating pressures are applied long enough. This failure mode is similar to the failure of a paper clip that is continuously bent to and fro until it eventually breaks by fatigue.

Rock, a chemical gel, is considered a brittle material and is known to have failed in brittle fracture and fatigue. The erosion of another chemical gel, clay, is explained in detail in Chap. 4 and can be viewed to act like a brittle material as well. One particular characteristic of clay is that its strength properties change as a function of temperature, pH, and salinity of its interstitial and ambient water. If the erosive capacity of the water is significantly larger than the ability of the clay to resist it, it will fail suddenly. However, if the erosive capacity of the water is within certain bounds the clay erodes at a certain rate. This can be viewed as equivalent to fatigue failure. Additional information on the rate of erosion of clay is presented in Chap. 8.

In order to provide the reader with a qualitative idea of the relationship between the relative magnitude of the erosive capacity of water, expressed in terms of stream power, and the relative ability of different material types to resist erosion, also expressed in terms of stream power, a figure that compares certain familiar flow situations to varying material types was prepared (Fig. 6.20). The relative magnitudes of the erosive capacity of water and the relative ability of earth materials to resist scour are presented in terms of stream power per unit area (W/m^2).

The illustration is intended to provide the reader with a perspective of the range of erosive capacities and scour resistance that might be encountered in practice. The reader is cautioned not to use this figure in analysis and design. Presentation of comprehensive, quantitative methods for determining the relative magnitude of the erosive capacity of water and for quantifying the relative ability of earth materials to resist scour form the basis of this book and those methods should be used in analysis and design.

Figure 6.20 shows that the stream power in lowland rivers, without any manmade features, can range between values that are less than 1 W/m^2 in rivers with very mild slopes to, say, about 10 to 50 W/m^2 in natural rivers with steeper slopes and in bends. The stream power in sharp bends in lowland rivers can reach values of 1 kW/m^2 or greater. The stream power in mountain streams can vary from as low as, say, 10 W/m^2 to as high as about 100 to 200 W/m^2 in very steep reaches. However, it is possible to reach values of up to 1 to 2 kW/m^2 in such rivers.

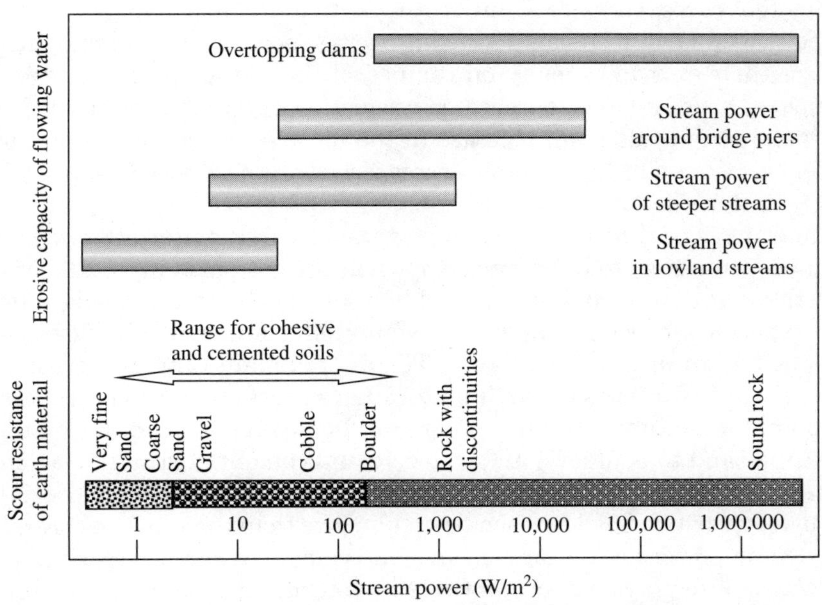

Figure 6.20 Schematic illustration of the relative magnitude of stream power in rivers, around bridge piers and downstream of overtopping dams and the resistance to scour offered by earth materials. This figure is qualitative and should not be used for analysis and design.

In the lowlands the stream power initiating scour around bridge piers in rivers can be on the order of about 50 to 80 W/m^2, but can be as high as 20 kW/m^2 at bridges on bends. The stream power around bridge piers in mountain streams can be much higher, say, on the order of about 800 W/m^2 to as high as 20 kW/m^2. The reason why the stream power around bridge piers is higher than that in the river itself is due to significant turbulence that can develop at bridge piers as water flows around them. Initially these values can be as much as 20 times the erosive capacity of the water in the approach flow.

The range of possible values of the erosive capacity of water discharging over dams depends on both fall height and discharge and can be significant. The heights of large dams can vary from a few meters; say about 15 m, to heights on the order of about 300 m. The magnitudes of floods discharging over such facilities can also vary significantly. It is therefore reasonable to expect quite a large range of magnitudes characterizing the erosive capacity of water discharging over dams. The range indicated in Fig. 6.20, i.e., from about 1 kW/m^2 to several MW/m^2, is reasonably representative of what can be anticipated when overtopping jets impinge downstream of dams.

The estimate of the range of stream powers associated with overtopping dams does not take account of the fact that plunging jets might break up, nor does it take account of the likely presence of plunge pools. When jets plunge through significant heights they can break up and, once broken up completely, can experience significant reduction in their effective erosive capacity. Similarly, if jets discharge into plunge pools it can lead to reduction in their erosive capacity on the bed of the plunge pool. However, plunge pool hydraulics is complicated because pool geometry relative to jet characteristics leads to an increase in the erosive capacity of the water under certain conditions and to a decrease under others. (see Chaps. 5, 7, and 9 for more detail).

The ranges of threshold stream powers for earth materials ranging from very fine sand to intact rock, shown at the bottom of Fig. 6.20, provide the reader with an indication of how well such materials might be able to resist the erosive capacity of water associated with the flow situations shown in the same figure. The approximate ranges of erosion resistance of two types of earth material are presented for powers less than approximately 200 W/m^2. These are non-cohesive granular material (fine sand to boulders) and cohesive and cemented granular earth material. The erosion thresholds for non-cohesive granular material are more clearly defined for each particular size than the range of threshold stream power values for cohesive and cemented earth material.

Particle size plays a dominant role in determining the relative ability of non-cohesive granular earth material to resist the erosive capacity of water, while other factors become more important in the case of cohesive and cemented granular earth material. The possible range of threshold stream power for cohesive granular earth material is quite large. For example, a cemented fine to coarse sand can have threshold stream power values that can be as high as, say, 100 W/m^2, but could also be as low as, say, 1 or 2 W/m^2.

A general method for quantifying the relative ability of any earth material to resist the erosive capacity of water has been presented. This method is known as the erodibility index method. It has been shown that application of the geomechanical index used to quantify the relative ability of earth materials to resist the erosive capacity of water has been applied by a number of researchers. Of the threshold relationships that were developed using this approach it has been found that the relationships developed by Annandale (1995) and van Schalkwyk et al. (1995) correlate well. The other erosion thresholds are much higher than the thresholds developed by these two authors. This means that the threshold relationships developed by Temple and Moore (1994) and Kirsten et al. (1996) potentially significantly over-predict the ability of earth materials to resist the erosive capacity of water.

The erosion threshold proposed by Annandale (1995) has been validated with near-prototype experiments at Colorado State University, Fort Collins. These tests validated Annandale's erosion threshold relationship for a simulated rock formation and for non-cohesive granular earth material. Subsequent practical experience in applying this method in practice and by comparing it with case studies of known scour indicates good correlation between calculated and observed scour (see Chaps. 9 and 10). Annandale's erosion threshold is the preferred method used in this book and is thought to represent a realistic relationship between the erosive capacity of water and the relative ability of earth materials to resist erosion at the point of incipient motion.

Chapter

7

Scour Extent

Introduction

Scour extent is the maximum scour depth resulting from the interaction between flowing water and earth material. If an earth material experiences scour its scour extent increases with increasing erosive capacity of the water. This chapter presents calculation techniques for determining scour extent in both physical and chemical gels. Temporal aspects of scour, i.e., the rate at which scour occurs, are dealt with in Chap. 8.

The approach for calculating scour extent is based on cause and effect. This approach differs from standard approaches that are based on empirical equations relating anticipated scour to a number of parameters. Examples of such equations are Melville's bridge pier scour equation (Melville 2002) and the bridge pier scour equation recommended in HEC-18 (FHWA 2000).

Melville's equation is expressed as

$$d_s = K_{\mathrm{yB}} K_I K_d K_s K_\theta K_G K_t \tag{7.1}$$

where d_s = scour depth
K_{yB} = depth-size factor K_{yb} = (depth-size factor for piers) = K_{yL} (depth-size factor for abutments)
K_I = flow intensity factor
K_d = sediment size factor
K_s = pier or abutment shape factor
K_θ = pier or abutment alignment factor
K_G = channel geometry factor
K_t = time to scour factor

Each of the factors in Melville's equation can be quantified using graphs and tables (see e.g., Melville 2002).

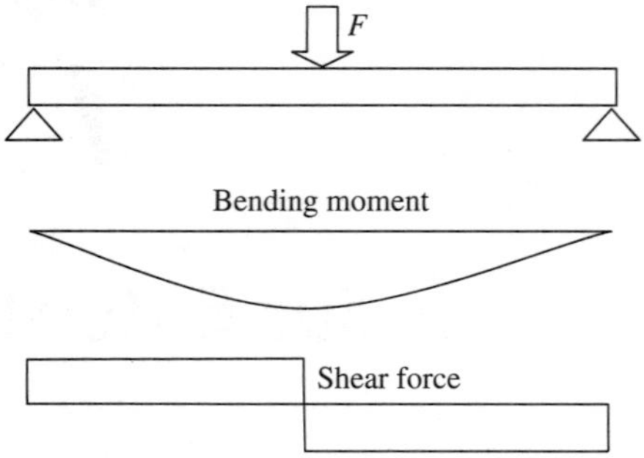

Figure 7.1 Conventional engineering analysis of a structure, like a simply supported beam, entails calculating bending moments and shear forces caused by an imposed load F. Once these are known it is possible to calculate stresses in the beam and determine its performance characteristics. A similar cause-and-effect approach can be followed when conducting scour analyses.

The cause and effect approach followed in this book is more in line with conventional engineering approaches to solving problems. For example, when analyzing a structural component like a beam it is common practice to draw a free body sketch, apply the load to the structural configuration in the sketch, and then use the theory of mechanics to calculate bending moments and shear forces (Fig. 7.1). If the dimensions and material properties of the beam are known it is further possible to calculate the magnitude of the stresses that will develop in the beam as well as other performance characteristics like deflection. In the case of scour analysis the erosive capacity of the water is equivalent to the load on the beam. The ability of the earth material to resist the erosive capacity of the water is equivalent to the material properties of the beam, and scour extent is equivalent to the beam's performance characteristics like deflection.

Conceptual Approach

When calculating scour extent by comparing the erosive capacity of water to the relative ability of earth or engineered earth materials to resist it, it is necessary to determine the spatial variations of the erosive capacity of water and material resistance to scour. The strength of earth materials, and therefore their ability to resist the erosive capacity of water, normally varies as a function of space. For example, a rock formation consisting of various different kinds of rock may vary in strength as a function of elevation below the ground surface. Stronger rock layers will be more resistant to scour.

When quantifying the spatial variation of the erosive capacity of water it is also found that its magnitude varies. For example, when a jet impinges into a plunge pool its erosive capacity gradually decreases as a function of pool depth. Or, if water flows around a bridge pier the erosive capacity of the water at the base of the pier is initially very high, but as the depth of the scour hole around the bridge pier increases the flow becomes more streamlined. This leads to a reduction in the turbulence intensity of the water and a concomitant decrease in the erosive capacity of the water flowing around the bridge pier.

If one were to therefore quantify the relative ability of earth materials to resist the erosive capacity of water as a function of space, and quantify the change in the erosive capacity of water as a function of the same space, it is possible to calculate the extent of scour by comparing these two sets of values. This concept is presented schematically in Fig. 7.2 for scour in, e.g., a plunge pool or around a bridge pier.

The graph on the top left-hand side of the figure shows that the erosive capacity of water on the boundary decreases as a function of elevation as a scour hole increases in depth. On the top right hand side of the figure the variation of the threshold resistance of the earth material is shown, also as a function of elevation. By combining these two relationships, as shown in the bottom of the figure, it is possible to determine the maximum scour depth. This occurs at the elevation where the erosive capacity of the water becomes less than the resistance offered by the earth material.

Application of this concept can be accomplished by making use of all the methods presented in this book, i.e., the erodibility index method, the

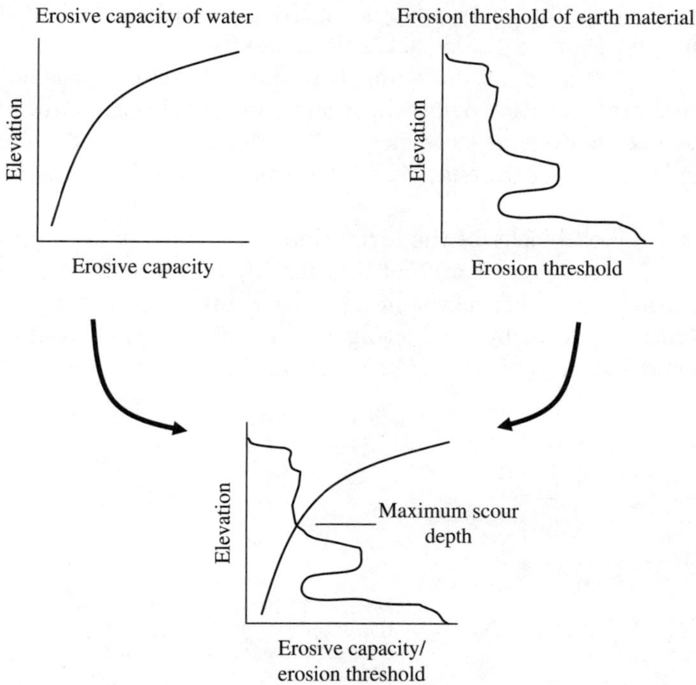

Figure 7.2 Conceptual cause-and-effect approach for calculating scour extent by comparing the magnitudes of the erosive capacity of water and the erosion threshold of earth materials as a function of elevation.

dynamic impulsion method and the comprehensive fracture mechanics method, i.e., scour by brittle fracture and fatigue failure (Annandale 1995 and Bollaert 2002).

Scour Extent of Physical Gels

Physical gels include earth materials like non-cohesive soils and jointed rock formations. Scour calculation of jointed rock formations is explained to illustrate the cause and effect approach to scour calculation for physical gels.

Rock block removal

This subsection demonstrates how the potential for rock block removal and the depth of scour can be determined by making use of the information presented in previous chapters. The demonstration takes the form of an example calculation.

Consider the following problem. A rectangular jet with a velocity V_j = 25 m/s at the point of impingement and a jet thickness D = 4 m will be discharged from a dam into a pool. The objective is to prepare a plunge pool design that will not scour, once constructed. This can be accomplished by pre-excavating the pool to a depth that is equivalent to the maximum scour depth that is expected to occur. Filling the pool with water, deep enough to dissipate the energy of the jet, will prevent scour.

In order to simplify this example assume that the rock quality does not change as a function of elevation below the ground surface. In cases where the rock quality does change, the essence of the calculation procedure is slightly modified to allow for the changes in material quality as a function of elevation.

Assume that the rock blocks in the formation are square in plan and measures 1 m by 1 m (i.e., x_b = 1 m) and that the dip of the rock is vertical (Fig. 7.3). The average height of the rock blocks is unknown. Therefore, assess the potential for scour by conducting a sensitivity analysis assuming optional vertical dimensions of 0.75 m, 1 m, and 2 m.

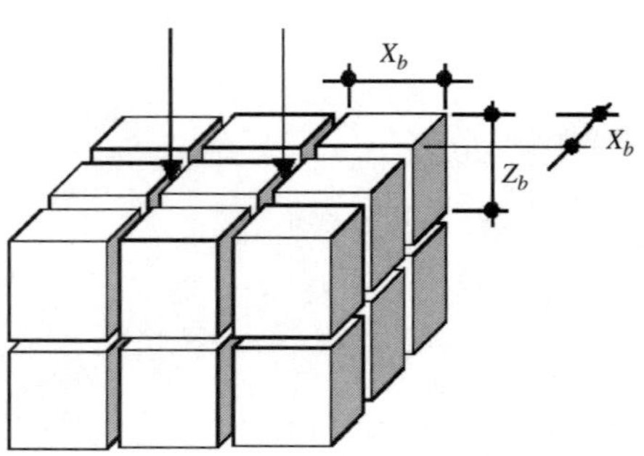

Figure 7.3 Schematic presentation of the rock block configuration used in the example calculation (Bollaert 2002).

The density of the rock is 2650 kg/m^3 and the anticipated air content of the water in the pool can be as high as 1 percent. It is assumed that the shear force on the sides of the rock is zero, i.e., $F_{sh} = 0$ kN. This might not be completely true in practical situations, but considered reasonable for execution of this example.

The equation for calculating the upward movement of a rock block from a rock formation is (see Chap. 6)

$$h_{up} = \left[2\frac{(x_b + 2z_b)}{c}\right]^2 \frac{1}{2g \cdot x_b^4 \cdot z_b^2 \cdot \rho_s^2}\left[C_I \cdot \phi \cdot \gamma \cdot \frac{V_j^2}{2g} \cdot x_b^2 - (\gamma s - \gamma) \cdot x_b^2 \cdot z_b - F_{sh}\right]^2 \tag{7.2}$$

In addition to the information already provided it is necessary to determine the values of the dynamic impulsion coefficient C_I and the pressure wave celerity of the water c. It is assumed that the coefficient $\phi = 1.0$.

For constant pressure and temperature the density of a mixed fluid can be estimated as

$$\rho_{mix} = \rho_{air}\frac{V_{air}}{V} + \rho_{liq}\frac{V_{liq}}{V} \tag{7.3}$$

where V_{air} = volume of air
V_{liq} = volume of the liquid
ρ_{air} = density of air (1.29 kg/m^3)
ρ_{liq} = density of the water
V = total volume

For constant temperature and pressure approximate the wave celerity of the mixture C_{mix} as

$$c_{mix} = \sqrt{\frac{1}{\rho_{mix}} \cdot \frac{1}{\dfrac{V_{liq}/V}{\rho_{liq} \cdot c_{liq}^2} + \dfrac{V_{air}/V}{\rho_{air} \cdot c_{air}^2}}} \tag{7.4}$$

where c_{liq} is the pressure wave celerity of the liquid (approximately 1000 m/s) and c_{air} is the pressure wave celerity of air (340 m/s).

Figure 7.4 is a plot of Eq. (7.4), which shows that the pressure wave celerity in water containing free air changes rapidly as a function of air content. The pressure wave celerity in water changes from 1000 m/s, when it contains no air, to about 100 m/s when it contains about 1 percent of free air by volume. For purposes of the calculation assume that the pressure wave celerity of the water containing 1 percent air is 100 m/s, i.e., $c = 100$ m/s.

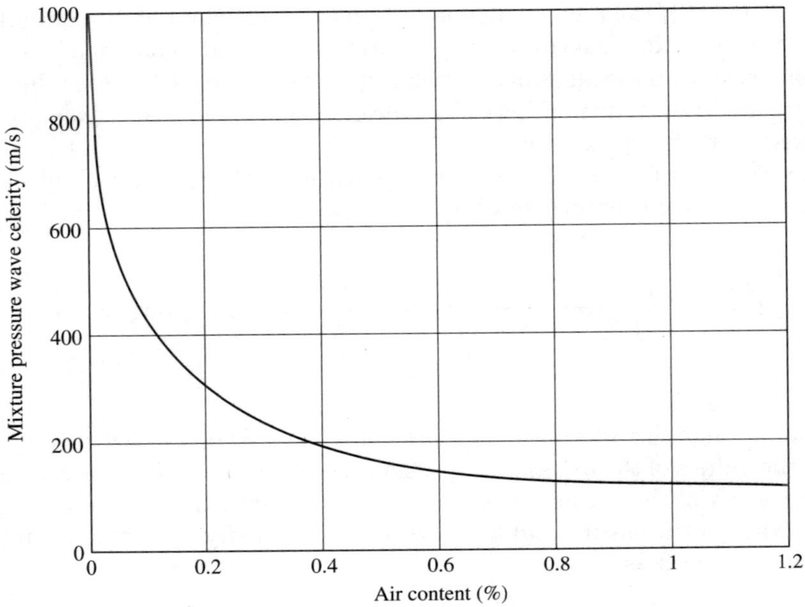

Figure 7.4 Change in pressure wave celerity in water containing air as a function of air content.

The dynamic impulsion coefficient C_I is determined with the following equation (Bollaert 2002; also see Chap. 5):

$$C_I = 0.0035\left(\frac{Y}{D}\right)^2 - 0.119\left(\frac{Y}{D}\right) + 1.22 \qquad \text{for } \frac{Y}{D} < 18 \tag{7.5}$$

The coefficient is a function of the dimensionless depth of the plunge pool Y/D and should therefore form part of the equation when solving for the expulsion distance h_{up} as a function of dimensionless plunge pool depth.

The analysis results can be expressed in dimensionless form by dividing the expulsion height h_{up} by the block height z_b and determining how it varies as a function of the dimensionless pool depth Y/D. The result of this a calculation is shown in Fig. 7.5.

The criteria for block removal, proposed by Bollaert (2002), are shown in Table 7.1 and are illustrated in Fig. 7.5. This figure shows that the vertical dimension of the rock blocks play an important role determining the potential for block removal. If it is assumed that rock blocks are removed from their matrix when the dimensionless uplift is greater than 0.5 it can be concluded that the potential rock scour, depending on the assumed vertical dimension of the rock blocks, can vary between dimensionless plunge pool depths ranging between approximately 5 and 9.

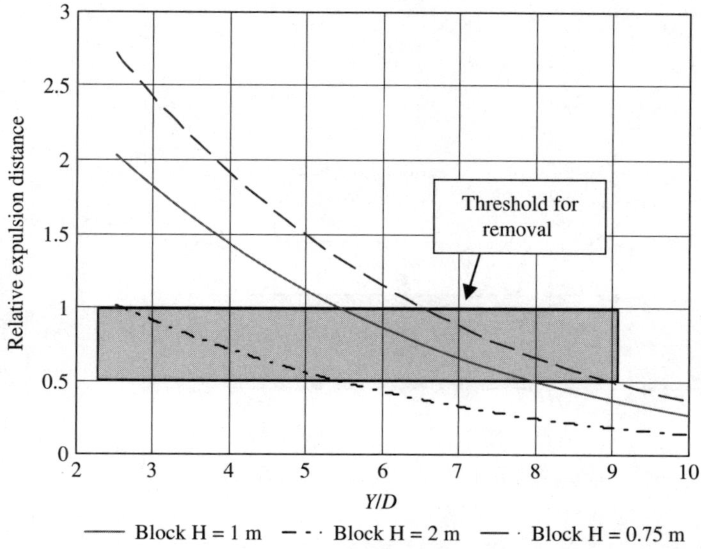

Figure 7.5 Relative expulsion of rock blocks with varying height as a function of dimensionless plunge pool depth *Y/D*.

This means that the anticipated maximum plunge pool scour depth can range between 20 and 36 m. At this point in the investigation it becomes necessary to determine if it is worthwhile spending more money on subsurface investigations to obtain better information on the rock block size, i.e., its vertical dimension. Assuming that it is decided to execute such an investigation and that it is found that the average block height is 1 m, then the predicted scour depth ranges between 22 m and 32 m (i.e., *Y/D* values ranging between 5.5 and 8). Further refinement

TABLE 7.1 Proposed Criteria to Assess Rock Scour Potential by Dynamic Impulsion

$\frac{h_{up}}{z_b} \leq 0.1$	Rock block remains in place.
$0.1 < \frac{h_{up}}{z_b} < 0.5$	Rock block vibrates and most likely remains in place.
$0.5 \leq \frac{h_{up}}{z_b} < 1.0$	Rock block vibrates and is likely to be removed, depending of ambient flow conditions.
$\frac{h_{up}}{z_b} \geq 1.0$	Rock block is definitely removed from its matrix.

SOURCE: From Bollaert 2002.

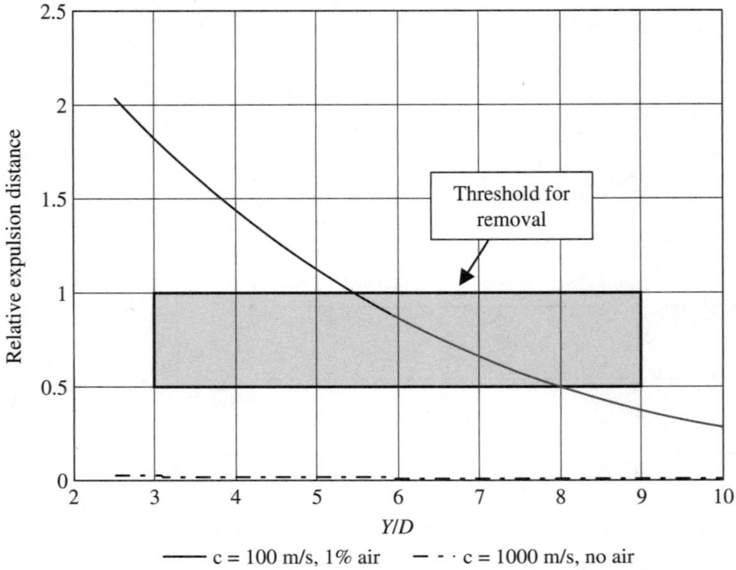

Figure 7.6 Relative expulsion of a rock block of 1 × 1 × 1 m for two different pressure wave celerity in water.

of this calculation can include incorporating estimates of the shear stress on the sides of the rock.

Additionally it is required to conduct a calculation to determine the effect of the air content of the water. The estimate that the air content of the water can reach values as high as 1 percent requires a sensitivity analysis to determine what the scour depth would be if the air content of the water is lower than 1 percent. The results of a calculation for a 1 m block height with and without air are shown in Fig. 7.6.

The result indicates that the potential for expulsion of a rock block if the pressure wave celerity in the water is equal to 1000 m/s, i.e., if there is no free air in the water, is close to zero. This comparison illustrates the sensitivity of rock expulsion to free air in the water. When the water is pure (no free air) the result indicates that there is no need for a plunge pool. However, if the water contains 1 percent air, a plunge pool with a depth of 32 m is required if the criteria in Table 7.1 are to apply.

Scour Extent of Chemical Gels—Brittle Fracture

Brittle materials, like massive rock, will fail in brittle fracture if the stress intensities in close-ended fissures within such materials exceed its fracture toughness. Prediction of brittle fracture in rock requires quantification of the pressure fluctuations due to flowing water at the rock surface and within the close-ended fissures in the rock. Turbulence is the cause of pressure fluctuations at the rock surface, i.e., at the interface with the flowing water. Fluctuating pressures in the close-ended fissures originate from the fluctuating pressures at the rock surface and can experience amplification due to resonance. Conditions conducive to resonance usually develop with the presence of air in the water. The reason is that the pressure wave celerity of the water decreases and leads to conditions that are favorable for resonance to occur.

To illustrate the calculation procedure for determining scour extent resulting from brittle fracture of rock, consider a plunge pool with boundaries consisting of massive, intact rock. The unconfined compressive strength of the rock is 45 MPa and it contains imperfections consisting of close-ended fissures. A rectangular jet with a thickness of 1 m plunges into the pool at a velocity of 25 m/s. The free air content of the water is 10 percent and the issuance turbulence intensity of the jet is 5 percent. It is known that the jet will remain reasonably intact, and that its breakup ratio will definitely not exceed 0.5. The objective of the analysis is to determine the depth of scour that might occur due to brittle fracture of the rock.

Calculate the dynamic pressure at the plunge pool boundary with the equation (Ervine et al., 1997),

$$P_b = C_{\text{pa}}\phi\gamma\frac{V_j^2}{2g}$$

and the fluctuating dynamic pressure in the close-ended fissures with the equation (Bollaert 2002),

$$P_f = \Gamma_{\max}C'_{\text{pa}}\phi\gamma\frac{V_j^2}{2g}$$

Therefore, the total dynamic pressure in a fissure is (also see Chap. 5)

$$P_{\text{dyn}} = (C_{\text{pa}} + \Gamma_{\max}\cdot C'_{\text{pa}})\phi\gamma\frac{V_j^2}{2g}$$

The average dynamic pressure coefficient for a jet with a jet breakup ratio of less than 0.5 is determined as (Ervine et al., 1997; also see Chap. 5)

$$C_{\text{pa}} = 38.4(1-\alpha_i)\left(\frac{D}{Y}\right)^2 \qquad \text{for}\ \frac{Y}{D} > 6.4 \tag{7.6}$$

$$C_{\text{pa}} = 0.85 \text{ otherwise}$$

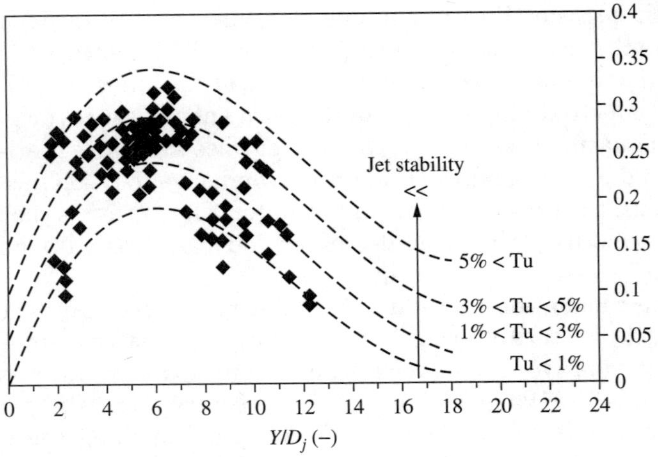

Figure 7.7 Fluctuating dynamic pressure coefficient C'_p (Bollaert 2002).

The air concentration $\alpha_i = \beta/(1+\beta)$, where $\beta = 10\%$ is the air content.

The fluctuating dynamic pressure coefficient C'_p is determined from Fig. 7.7, or calculated from the following equation (Bollaert 2002):

$$C'_{\text{pa}} = a_1 \cdot \left(\frac{Y}{D_j}\right)^3 + a_2 \cdot \left(\frac{Y}{D_j}\right)^2 + a_3 \cdot \left(\frac{Y}{D_j}\right) + a_4 \tag{7.7}$$

for $Y/D_j \leq 20$.

The coefficient values in this equation are dependent on the issuance turbulence intensity of the jet and can be determined from Table 5.5 in Chap. 5. For moderate jet turbulence, i.e., a turbulence intensity not exceeding 5 percent the values of the coefficients are

$$a_1 = 0.00215$$

$$a_2 = -0.0079$$

$$a_3 = 0.0716$$

$$a_4 = 0.100$$

The amplification factor $\Gamma_{\max}$ is determined from Fig. 7.8 or calculated with the following equations:

$\Gamma^+ = 4 + 2Y/D_j$	for $Y/D_j < 8$	
$\Gamma^+ = 20$	for $8 \leq Y/D_j \leq 10$	curve of maximum values
$\Gamma^+ = 40 - 2Y/D_j$	for $10 < Y/D_j$	

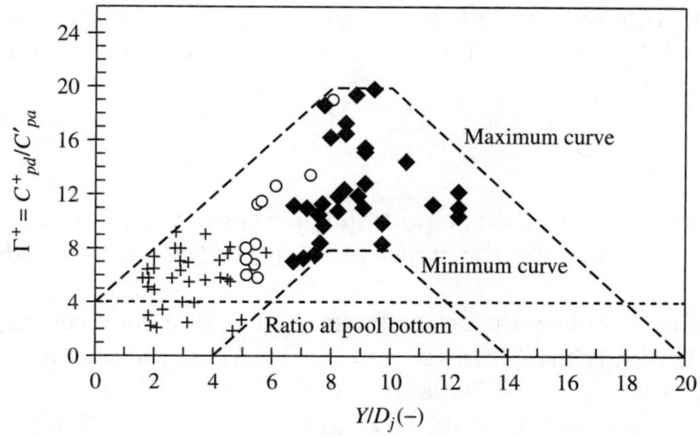

Figure 7.8 Amplification factor Γ^+ (Bollaert 2002).

$\Gamma^+ = -8 + 2Y/D_j$ for $Y/D_j < 8$

$\Gamma^+ = 8$ for $8 \leq Y/D_j \leq 10$ curve of minimum values

$\Gamma^+ = 28 - 2Y/D_j$ for $10 < Y/D_j$

Express the total dynamic pressure in a fissure as a function of dimensionless plunge pool depth as shown in Fig. 7.9. The figure shows that the total dynamic pressure in the fissure reaches a maximum at a dimensionless pool depth of about eight. This means that the pressures in fissure are lower if a pool is shallow; it then increases to reach a maximum when the dimensionless pool depth is about eight, where after it decreases again. This is an important observation when investigating brittle fracture and fatigue failure in brittle materials.

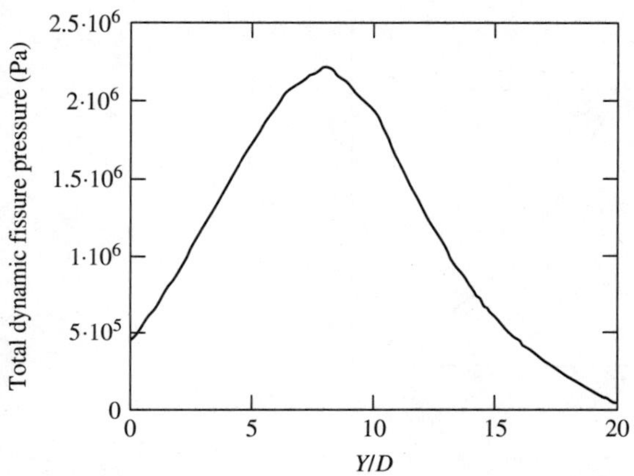

Figure 7.9 Total dynamic pressure in a fissure, including an allowance for resonance.

In order to determine whether the rock will fail in brittle fracture it is required to calculate the stress intensity within the fissures, as follows (see Chap. 4):

$$K_I = \Delta p_c f \sqrt{\pi L_f}$$

(*Note*: the variable f used in this equation represents the boundary correction factor for calculating the stress intensity in the close-ended fissure.)

Assuming that the boundary correction factor $f = 1.0$, that the average fissure length $L_f = 1$m, and that the pressure in the fissure is approximately 0.8 times the maximum pressure (Bollaert and Schleiss 2005) the stress intensity K_I can then aslo be expressed as a function of Y/D (Fig. 7.10).

In order to determine if the rock will fail in brittle fracture, the stress intensity should be compared to the fracture toughness of the rock. Assuming that the internal pressure of the rock can be neglected (i.e., $\sigma_i = 0$), the fracture toughness can be estimated as (see Chap. 4)

$$K_I = 0.08(\text{UCS}) + 0.054\sigma_i + 0.43 = 4 \text{ MPa}\sqrt{\text{m}}$$

By plotting the values of the stress intensity and the fracture toughness on the same graph (Fig. 7.10) it is possible to determine how much scour will occur. For example, if the plunge pool has an initial dimensionless depth of less than about five, it is unlikely that brittle fracture would occur. However, if the dimensionless plunge pool depth is

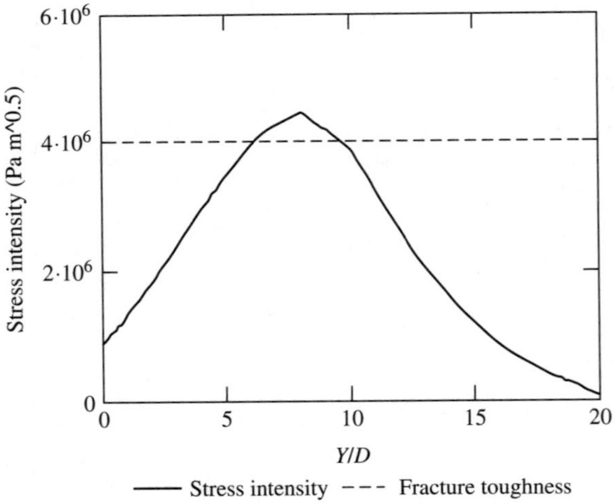

Figure 7.10 Stress intensity and fracture toughness as a function of *Y/D*.

approximately six it is possible for brittle fracture to commence. In such a case scour will continue by means of brittle fracture until a dimensionless pool depth of about 10 is reached. At that point the stress intensity in the fissures will decrease again below the fracture toughness of the rock and scour will cease. The maximum depth of scour that can therefore be expected for this example is about 10 m.

In order to apply this method to other flow scenarios, like e.g., flow around bridge piers, it is necessary to develop techniques to calculate the magnitude of pressure fluctuations resulting from water flowing around a bridge pier. Similar needs exist for determining the magnitude of pressure fluctuations in other flow situations of interest.

Erodibility Index Method

The erodibility index method is based on a scour threshold relationship that can be used to determine the scour potential for both physical and chemical gels. When using the erodibility index method the extent of scour is determined by comparing the stream power that is *available* to cause scour with the stream power that is *required* (i.e., the threshold steam power) to scour the earth material under consideration. The available stream power represents the relative magnitude of the erosive capacity of the water discharging over the earth material, whereas the required or threshold stream power is the stream power that is required by the earth material for scour to commence. If the available stream power is exactly equal to the required stream power, the material is at the threshold of erosion. In cases where the available stream power exceeds the required stream power, the material will scour. Otherwise, it will remain intact.

Figure 7.11 shows how the available and required stream power, both plotted as a function of elevation beneath the riverbed, are compared to determine the extent of scour. Scour will occur when the available stream power exceeds the required stream power. Once the maximum scour elevation is reached, if the available stream power is less than the required stream power, and scour ceases.

The required (threshold) stream power is determined by first indexing a geologic core or borehole data by making use of the erodibility index method (see Chap. 4). The values of the erodibility index thus determined can vary as a function of elevation below the ground surface, dependent on the variation in material properties. Once the index values at various elevations are known, the required stream power is determined as conceptually shown in Fig. 7.12. This figure indicates that the stream power required to scour a particular earth material (i.e., the threshold stream power) is determined by entering the erosion threshold graph on the abscissa, with the erodibility index known, moving vertically to the erosion threshold line, and reading the required (threshold) stream power on

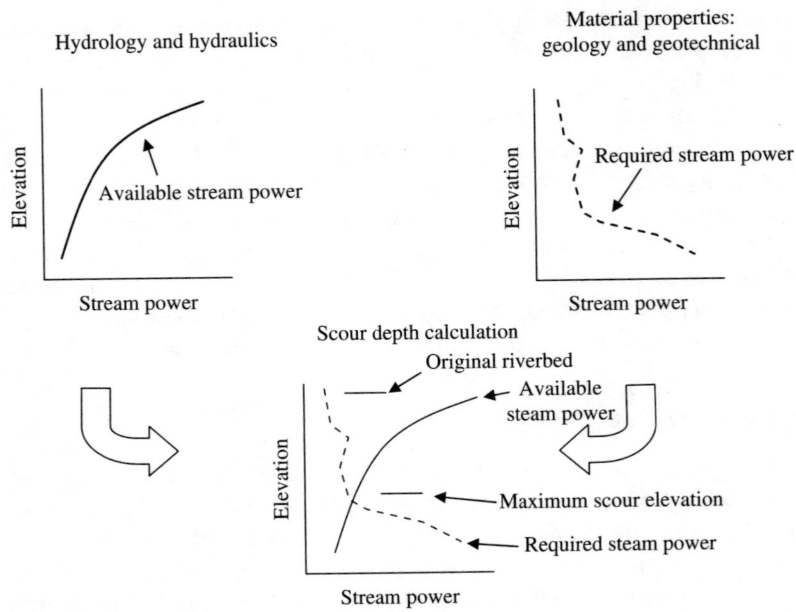

Figure 7.11 Determination of the extent (depth) of scour by comparing available and required (threshold) stream power using the erodibility index method.

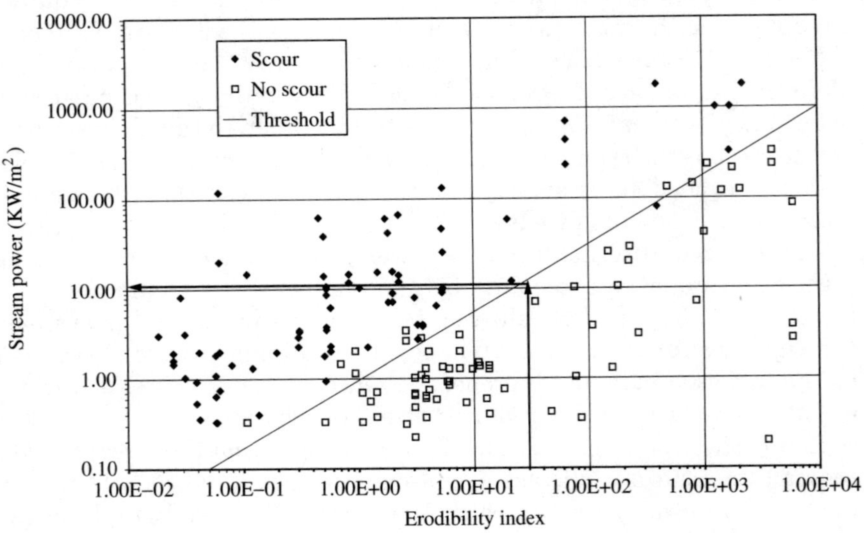

Figure 7.12 Determination of stream power that is required to scour earth material once the value of the erodibility index is known using Annandale's (1995) threshold relationship.

the ordinate. Fig. 7.13 illustrates that the process is repeated as a function of elevation below the riverbed. The values of the required stream power are plotted as a function of elevation, indicating the variation in threshold stream power below the ground surface.

Example

The methodology for implementing the erodibility index method to estimate extent of scour is best illustrated by example. The example illustrates

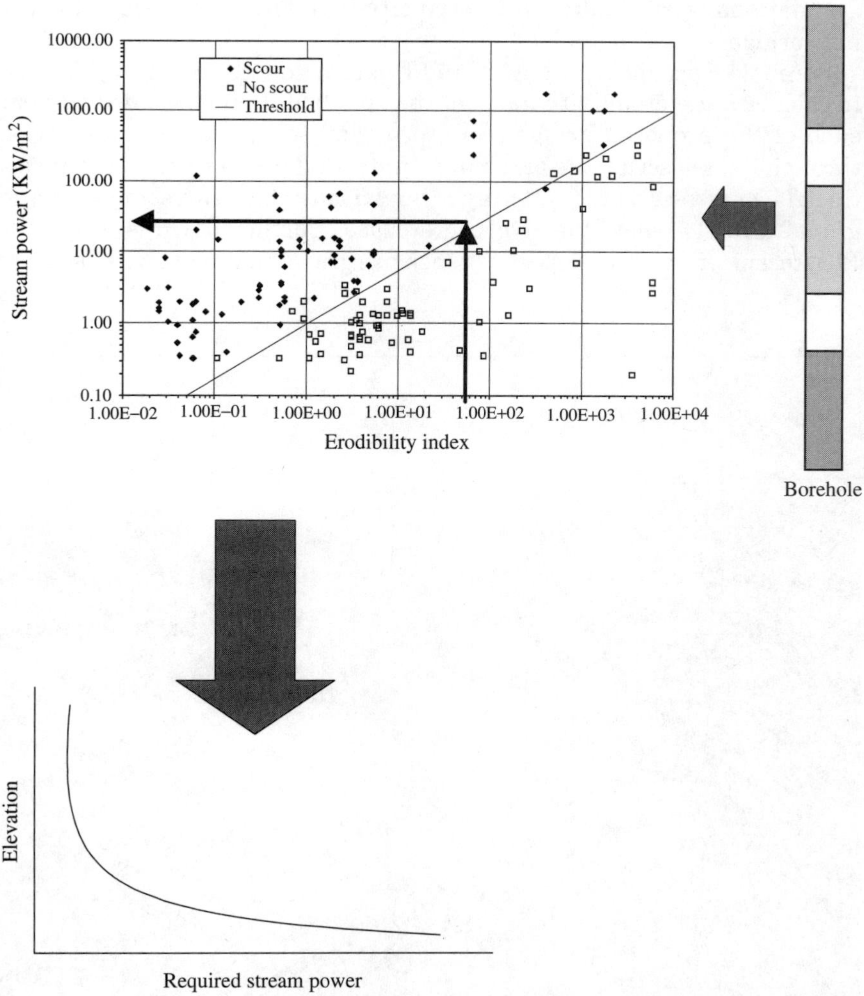

Figure 7.13 Development of a relationship between required (threshold) stream power and elevation below a riverbed by indexing a core or borehole data.

the scour calculations that were performed at the Woodrow Wilson Bridge across the Potomac River, which carries six lanes of Washington, D.C.–area traffic between Alexandria, Virginia, and Oxon Hill, Maryland. The bridge contains a drawbridge that is opened on a regular basis to allow ship traffic to pass. Congestion and the frequency of drawbridge openings for marine traffic cause traffic delay at the bridge. The Woodrow Wilson Bridge is one of a few on the interstate highway system that contains a movable span. Under current coast guard regulations, the 15.2-m-high drawbridge opens approximately 240 times per year to allow for the passage of marine traffic traveling the Potomac River. The traffic congestion at the bridge led to a decision to replace it (Fig. 7.14). Scour calculations were conducted to estimate foundation elevations for the new bridge.

Borehole logs, shear strength, and Dilatometer test results were used to calculate the erodibility index of the riverbed. Boreholes, drilled near each of the proposed bridge piers, provided soil property information through field descriptions and blow counts. Soil profiles generally have a thick layer of very soft to soft gray to brown silty clay, with some sand and gravel directly below the riverbed surface. Further down is a layer of Pleistocene era terrace deposits, which are gray and brown, with dense to

Figure 7.14 Potomac River and the old Woodrow Wilson Bridge that is replaced by a new bridge at the same location (Photo: Federal Highway Administration).

very dense sand with silt, gravel and clay lenses. Finally, the Cretaceous period Potomac group consists of hard gray clay. Dilatometer test results were used to estimate the undrained shear strength of the soil and its residual angle of friction. The shear strength test results were used to confirm the estimates made with the Dilatometer test results.

The analysis entailed assessment of riverbed material properties, hydraulic analysis, and scour analysis. Potential scour depths for the 100- and 500-year floods were calculated for each of the 25 proposed bridge piers. An example of a scour calculation for pier M10 is shown in Table 7.2.

Columns (1) and (2) contain the depth below the surface and elevation respectively. Column (3) represents the estimate of undrained shear strength or SPT blow count. (The whole numbers represent blow counts). Columns (4) to (7) contain the values of the mass strength number, block size number, inter-particle shear strength number, and the orientation and shape number. Their product quantifies the erodibility index, presented in column (8). The required stream power as a function of elevation is contained in column (9). This has been determined from the erodibility index threshold stream power graph using the approach illustrated in Figs. 7.12 and 7.13. Columns (10) to (12) and (15) to (17) contain calculations to quantify the available stream power, as explained further on. This is the stream power around the bridge pier that is available to cause scour. Columns (13) and (14), and (18) and (19) indicate whether scour will occur and provides a ratio between required and available stream power. Scour is considered to occur whenever the available stream power exceeds the threshold (required) stream power.

Intact material strength number

In situ dilatometer test (DMT) results were used to determine a relationship between depth and undrained shear strength of the very soft to soft clay material. A relationship was developed for the soft clay deposits, which begin at the riverbed surface and extend to various depths throughout the bridge cross-section of the riverbed. The soft and very soft alluvial deposits have a cohesive intercept of 3.5 kPa and a residual angle of friction ϕ of 8.1°. A simplified relationship between undrained shear strength and depth below the original ground surface was developed from the field data and was expressed as

$$S_u = 3.5 + 1.42H \tag{7.8}$$

where S_u is the undrained shear strength of soft and very soft alluvial deposits (in kPa) and H is the depth to the point in question from the original ground surface (m).

For borehole depths where the very soft to soft clay was found, Eq. (7.8) was used to estimate the undrained shear strength. The unconfined

TABLE 7.2 Calculation of Scour Depth at Pier M10 of the Woodrow Wilson Bridge

									100-Year Flood					500-Year Flood				
(1)	(2)	(3)	(4)	(5)	(6)	(7)	(8)	(9)	(10)	(11)	(12)	(13)	(14)	(15)	(16)	(17)	(18)	(19)
Depth, H (m)	Elevation (m)	S_u (kPa) OR SPT	M_s	K_b	K_d	J_s	K	Required stream power (KW/m^2)	Dimension-less scour depth	Relative stream power	Available stream power (KW/m^2)	P_R/P_A	Scour (Y/N)?	Dimension-less scour depth	Relative stream power	Available stream power (KW/m^2)	Scour (Y/N)?	P_R/P_A
0	−0.728	3.5	0.003	1	0.142	1	0.000	0.017	0					0				
0.305	−1.033	3.933	0.004	1	0.142	1	0.001	0.018	0.018	8.15	0.0784	0.23	yes	0.016	14.12	0.3381	yes	0.05
0.610	−1.338	4.366	0.004	1	0.142	1	0.001	0.019	0.036	7.88	0.0758	0.25	yes	0.032	13.55	0.3247	yes	0.06
0.914	−1.643	4.798	0.005	1	0.142	1	0.001	0.020	0.054	7.62	0.0733	0.27	yes	0.047	13.02	0.3118	yes	0.06
1.219	−1.948	5.231	0.005	1	0.142	1	0.001	0.020	0.071	7.36	0.0709	0.29	yes	0.063	12.50	0.2994	yes	0.07
1.524	−2.252	5.664	0.006	1	0.142	1	0.001	0.021	0.089	7.12	0.0685	0.31	yes	0.079	12.00	0.2875	yes	0.07
1.829	−2.557	6.097	0.006	1	0.142	1	0.001	0.022	0.107	6.88	0.0662	0.33	yes	0.095	11.52	0.2760	yes	0.08
2.134	−2.862	6.530	0.007	1	0.142	1	0.001	0.023	0.125	6.66	0.0641	0.36	yes	0.111	11.06	0.2650	yes	0.09
2.438	−3.167	6.963	0.007	1	0.142	1	0.001	0.023	0.143	6.44	0.0619	0.38	yes	0.126	10.62	0.2545	yes	0.09
2.743	−3.472	7.395	0.008	1	0.142	1	0.001	0.024	0.161	6.22	0.0599	0.40	yes	0.142	10.20	0.2444	yes	0.10
3.048	−3.777	7.828	0.008	1	0.142	1	0.001	0.025	0.179	6.02	0.0579	0.43	yes	0.158	9.80	0.2347	yes	0.11
3.353	−4.081	8.261	0.009	1	0.142	1	0.001	0.026	0.196	5.82	0.0560	0.46	yes	0.174	9.41	0.2253	yes	0.11
3.658	−4.386	8.694	0.009	1	0.142	1	0.001	0.026	0.214	5.62	0.0541	0.48	yes	0.190	9.03	0.2164	yes	0.12
3.962	−4.691	9.127	0.010	1	0.142	1	0.001	0.027	0.232	5.44	0.0523	0.51	yes	0.205	8.67	0.2077	yes	0.13
4.267	−4.996	9.559	0.010	1	0.142	1	0.001	0.027	0.250	5.26	0.0506	0.54	yes	0.221	8.33	0.1995	yes	0.14
4.572	−5.301	2	0.01	1	0.577	1	0.007	0.053	0.268	5.08	0.0489	1.08	no	0.237	8.00	0.1915	yes	0.28
4.877	−5.605	2	0.01	1	0.577	1	0.007	0.053	0.286	4.92	0.0473	1.12	no	0.253	7.68	0.1839	yes	0.29
5.182	−5.910	2	0.01	1	0.577	1	0.007	0.053	0.304	4.75	0.0457	1.16	no	0.269	7.37	0.1766	yes	0.30
5.486	−6.215	2	0.01	1	0.577	1	0.007	0.053	0.321	4.60	0.0442	1.20	no	0.284	7.08	0.1696	yes	0.31
5.791	−6.520	2	0.01	1	0.577	1	0.007	0.053	0.339	4.44	0.0428	1.24	no	0.300	6.80	0.1628	yes	0.33

6.096	−6.825	1	0.01	1	0.577	1	0.003	0.040	0.357	4.30	0.0414	0.96	yes	0.316	6.53	0.1564	yes	0.25
6.401	−7.129	1	0.01	1	0.577	1	0.003	0.040	0.375	4.15	0.0400	0.99	yes	0.332	6.27	0.1501	yes	0.26
6.706	−7.434	1	0.01	1	0.577	1	0.003	0.040	0.393	4.02	0.0387	1.03	no	0.348	6.02	0.1442	yes	0.28
7.010	−7.739	1	0.01	1	0.577	1	0.003	0.040	0.411	3.88	0.0374	1.06	no	0.363	5.78	0.1384	yes	0.29
7.315	−8.044	1	0.01	1	0.577	1	0.003	0.040	0.429	3.76	0.0361	1.10	no	0.379	5.55	0.1329	yes	0.30
7.620	−8.349	20	0.07	1	0.577	1	0.042	0.119	0.446	3.63	0.0349	3.40	no	0.395	5.33	0.1276	yes	0.93
7.925	−8.653	20	0.07	1	0.577	1	0.042	0.119	0.464	3.51	0.0338	3.52	no	0.411	5.12	0.1226	yes	0.97
8.230	−8.958	20	0.07	1	0.577	1	0.042	0.119	0.482	3.40	0.0327	3.64	no	0.427	4.91	0.1177	no	1.01
8.534	−9.263	20	0.07	1	0.577	1	0.042	0.119	0.500	3.28	0.0316	3.76	no	0.442	4.72	0.1130	no	1.05
8.839	−9.568	20	0.07	1	0.577	1	0.042	0.119	0.518	3.17	0.0305	3.89	no	0.458	4.53	0.1085	no	1.10
9.144	−9.873	14	0.06	1	0.577	1	0.033	0.107	0.536	3.07	0.0295	3.62	no	0.474	4.35	0.1042	no	1.03
9.449	−10.177	14	0.06	1	0.577	1	0.033	0.107	0.554	2.97	0.0286	3.75	no	0.490	4.18	0.1000	no	1.07

compressive strength (UCS) of earth materials can be approximated as twice the value of the undrained shear strength, i.e.,

$$\mathrm{UCS} = 2S_u \tag{7.9}$$

The intact material strength number for cohesive soils can therefore be calculated with the following equation (see Chap. 4):

$$M_s = 0.78(\mathrm{UCS})^{1.05} = 0.78(2S_u)^{1.05} \tag{7.10}$$

where UCS is the unconfined compressive strength (MPa), which must be less than 10 MPa for this equation to be valid (see Chap. 4). The intact material strength number for non-cohesive granular material was based on SPT blow counts and values from the mass strength number tables for non-cohesive granular material in Chap. 4.

For pier M10 column (3) of Table 7.2 shows the value of S_u (kPa) or the SPT blow count, whichever is applicable according to the log, at various depths below the riverbed surface. Note that S_u values appear as decimal numbers; blow counts appear as whole numbers. Column (4) of Table 7.2 shows the estimated values of M_s.

Block/particle size number

The borehole log material descriptions were used to determine the particle/block size number, K_b. K_b was assigned a value of one for all materials except the very hard clay. The reason for this is that all the materials are cohesive, except that the properties of the very hard clay are unique. The very hard clay of the cretaceous period Potomac group was assigned a value of 100. The reason for using $K_b = 100$ for the cretaceous period clay is that the clay is so hard that it can be viewed as soft intact rock with no significant discontinuities. K_b determinations for pier M10 are shown in column (5) of Table 7.2.

Discontinuity or interparticle bond shear strength number

The shear strength number, K_d, was calculated using the following equation:

$$K_d = \tan(\phi) \tag{7.11}$$

where ϕ is 8.1° for the very soft to soft clay material, but is 30° for all other materials. K_d values as a function of elevation for pier M10 are shown in Table 7.2 column (6).

Relative shape and orientation number

A value of one was assigned to the ground structure number, J_s, in all cases [column (7) of Table 7.2].

Erodibility index and required power

The erodibility index (K), i.e., the product of M_s, K_b, K_d, and J_s, is shown in Table 7.2 column (8). The power required to scour the Potomac River's bed material was determined using the following equation (see Chap. 4):

$$P_R = 0.48(K)^{0.44} \tag{7.12}$$

where P_R is the power required to scour earth material with erodibility index values less than 0.1. The units of P_R are in KW/m^2. Required stream power is calculated in Table 7.2 column (9) for pier M10.

Available stream power

The available stream power at each proposed bridge pier was determined with the HEC-RAS computer model developed by the U.S. Army Corps of Engineers. The available power around the bridge piers was expressed as a function of scour-hole depth and quantified through a three-step process using the available data.

First, the available stream power of the Potomac River at a point upstream of the proposed bridge was calculated for each proposed pier using the following equation:[1]

$$P_a = \gamma v d s \tag{7.13}$$

where P_a = available stream power in the river upstream of a bridge pier (kW/m^2)
γ = unit weight of water (kN/m^3)
v = velocity of water (m/s)
d = flow depth (m)
s = energy slope of flow in the river

Data for approach velocity, depth of flow, and energy slope in the Potomac River were obtained from the HEC-RAS model for a river section approximately 1.5 m upstream of the proposed Woodrow Wilson Bridge. The HEC-RAS model was designed to calculate a velocity distribution across the river cross section, thus allowing velocity upstream of each proposed pier to be approximated. The number of HEC-RAS model flow tubes affected the velocity calculated at the piers; thus the

[1]This analysis was completed prior to the discovery that the applied stream power at the bed is best expressed as $P = 7.853\rho(\tau_w/\rho)^{1.5}$.

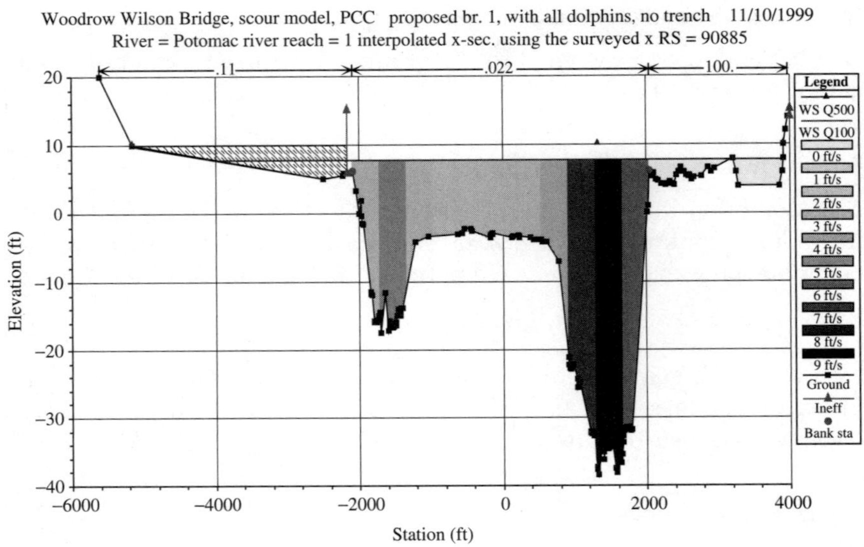

Figure 7.15 Estimated velocity distribution from the HEC-RAS model of the Potomac River at a cross section just upstream of the proposed bridge.

number of tubes was varied to achieve maximum velocities at the bridge piers. A schematic of velocity across the upstream section is shown in Fig. 7.15. Available stream power upstream of the bridge was calculated for the 100- and 500-year floods. Hydraulic data and resulting available stream power are presented in Table 7.3 for pier M10.

In the second step, a relationship between dimensionless stream power at the base of a scour hole and dimensionless scour hole depth is determined using Fig. 7.16, which were obtained from experimental studies at the FHWA laboratories in Reston, Virginia. (The reader may prefer to rather use the relationships presented in Chap. 5 on other projects.) The stream power is expressed in dimensionless form on the ordinate of the graph as the ratio P/P_a and scour depth as the ratio y_s/y_{max}. P_a is the magnitude of

TABLE 7.3 Hydraulic Data for Pier M10 of the Woodrow Wilson Bridge

Hydraulic variable	100-year flood	500-year flood
Upstream velocity	1.21 m/s	1.67 m/s
Water surface elev.	3.31 m	4.48 m
Ground surface elev.	−0.73 m	−0.73 m
Flow depth	4.04 m	5.21 m
γ	9.82 kN/m^3	9.82 kN/m^3
Energy slope	0.0002	0.00028
Upstream stream power	0.010 kW/m^2	0.024 kW/m^2
HEC-18 scour depth	17.1 m	19.3 m

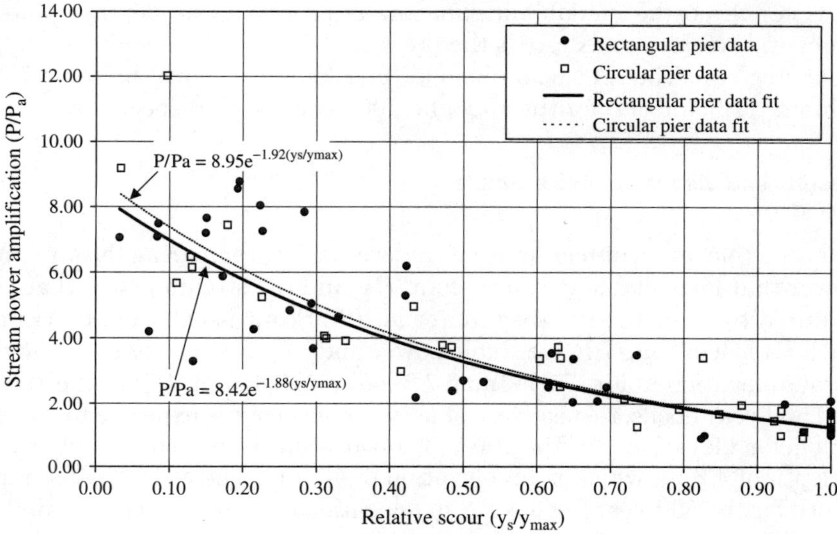

Figure 7.16 Dimensionless stream power around bridge piers as a function of relative scour depth used in the scour calculations for Woodrow Wilson Bridge.

the stream power in the river upstream of the pier as determined by Eq. (7.13), and P is the magnitude of the stream power at the base of the scour hole as it increases in depth. The variable Y_{max} represents the maximum scour depth that can develop around a bridge pier under given flow conditions, whereas y_s represents variable scour depth ($y_s < y_{\text{max}}$). The maximum depth was assumed to be that calculated by HEC-18. The reason for this is that the pier scour equation recommended in HEC-18 (Federal Highway Administration, 2001a) is an envelope curve embracing known scour depth data, and as such presents the maximum possible scour depth that can be attained around a bridge pier.

The equations of the two relationships in Fig. 7.16 are

Rectangular Piers:

$$P/P_a = 8.42e^{-1.88(y_s/y_{\text{max}})} \tag{7.14}$$

Circular Piers:

$$P/P_a = 8.95e^{-1.92(y_s/y_{\text{max}})} \tag{7.15}$$

The dimensionless scour depths for pier M10 are shown in Table 7.2 columns (10) and (15) for the 100- and 500-year floods, respectively. Columns (11) and (16) show the 100- and 500-year flood relative stream power calculations using Eq. (7.14) for pier M10.

In step three, the available stream power at a given scour depth, P (subsequently referred to as *PA*), is the product of P_a from step one and P/P_a from step two. The *PA* calculations for pier M10 are shown in Table 7.2 columns (12) and (17) for the 100- and 500-year floods, respectively.

Results and discussion for example pier M10

The scour elevation at pier M10 was determined by comparing the stream power that is available to cause scour, *PA*, and the stream power that is required to scour the riverbed material, *PR*. Scour is expected to occur until *PA* is less than *PR*. Available power and required power are shown versus elevation in Fig. 7.17. Table 7.2 columns (14) and (18) for the 100- and 500-year floods, respectively, show whether scour is expected to occur at a given elevation. For the 100-year flood, scour is expected to occur to a depth of 4.6 m, which is an elevation of –5.3 m. The calculated scour depth for the 500-year flood is 8.2 m (elevation of –8.96 m). The erodibility index method predicted scour elevations at pier M10 are 12.5 m and 11 m shallower than the HEC 18 predictions for the 100- and 500-year floods, respectively.

The factor of safety quantifies the ability of the earth material to withstand the erosive power of the river at potential scour depths. The factor of safety was calculated as the required stream power divided by the available stream power in columns (13) and (19). The factor of safety at the base of the expected 100-year flood scour hole is approximately 1.

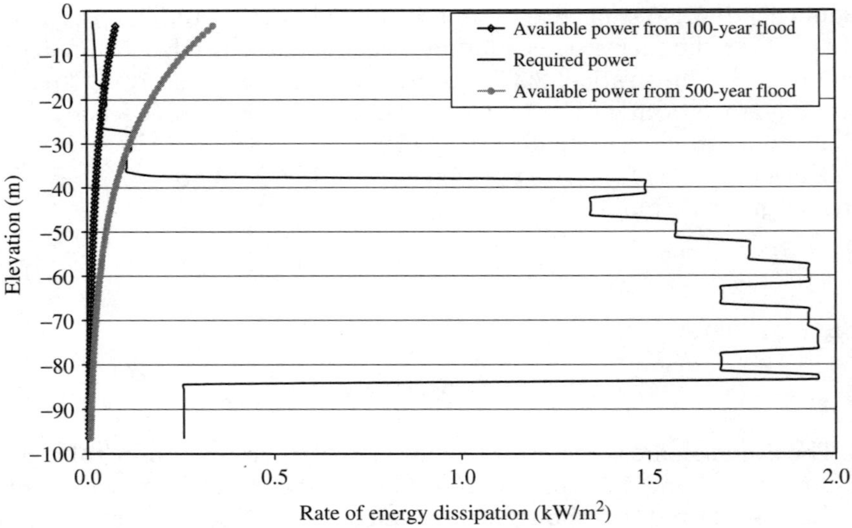

Figure 7.17 Available and required stream power for pier M10.

The variation in the factors of safety with depth, as seen in Table 7.2 is dictated by the relationship between the variation in material properties as a function of elevation below the riverbed and changes in the available stream power. To be conservative in the bridge pier design, Maryland State Highway Department designed pier M10's foundation at the elevation where the 500-year flood factor of safety is greater than 5.0; this elevation is 11.7 m with a factor of safety of 18.3.

Summary

The methods for calculating scour extent presented in this chapter are based on a cause-and-effect approach, which differs from conventional methods to calculate scour. Conventional approaches for calculating scour consist of empirical equations that relate scour depth to various other variables. Such methods can only be used for the conditions for which they were developed. The cause-and-effect approach adopted in this book allows engineers to analyze scour for varying conditions. The reason for this is that the threshold relationships presented in Chap. 6 can be applied to analyze scour for any flow condition, once the erosive capacity of the water has been quantified. Quantification of the erosive capacity of water can be accomplished with the methods presented in Chap. 5, by making use of other computational techniques or by measuring it in physical hydraulic model studies. The preferred indicator parameter for quantifying the erosive capacity of water is stream power, unless it is possible to calculate the magnitude of pressure fluctuations (as illustrated for plunging jets).

This chapter provides examples for calculating scour extent in physical and chemical gels. This was done by presenting examples to calculate rock scour resulting from fluctuating pressures using Bollaert's dynamic impulsion and comprehensive fracture mechanics approaches. Additionally, a discussion and application example using the erodibility index method (Annandale 1995) is provided. The latter method provides engineers with the ability to analyze scour in both physical and chemical gels.

The methods discussed in this chapter quantify the extent of scour. Chapter 8 presents methods for quantifying the temporal aspects of scour, i.e., the rate of scour.

Chapter

8

Temporal Aspects of Scour

Introduction

Physical gels generally scour as soon as the erosive capacity of the water exceeds the ability of the material to resist removal. Temporal aspects are of very little interest in such cases. Therefore, when considering scour of physical gels, such as non-cohesive granular earth materials, the maximum depth of scour is the main topic of interest.

Scour characteristics of chemical gels differ from those of physical gels. Chemical gels can scour in either brittle fracture or fatigue failure. When the stress intensities generated in close-ended fissures by fluctuating turbulent water pressure in chemical gels exceed their fracture toughness they fail immediately in brittle fracture. Such failure is explosive and the scour occurs instantaneously.

If the stress intensity caused by fluctuating pressures within close-ended fissures in chemical gels does not exceed their fracture toughness, failure can occur in subcritical mode (i.e., fatigue failure). This happens when the fluctuating pressures are applied long enough to eventually result in failure. The break-up of the material is not instantaneous but occurs after a certain period of time; which means that the material apparently withstands the applied pressures for some time and then suddenly fails. When investigating scour by subcritical failure in chemical gels both the maximum extent of scour and the time it takes to reach the maximum scour depth are of interest. Scour by subcritical failure can occur in chemical gels like rock and clay, and in engineered earth materials like concrete. The subcritical failure of clay occurs at a different rate, i.e., it scours at a rate that is proportional to the erosive capacity of the water. It can still be viewed as subcritical failure, but with a differing rate to that experienced in other chemical gels like rock.

Estimation of the rate of scour is important. For example, if a concrete arch dam is overtopped by a flood with a duration of, say, 20 h and the rock foundation downstream of the dam requires 100 h of continued exposure to fluctuating turbulent pressure before it fails in fatigue the dam will not experience scour during that particular flood event. Continued exposure to fluctuating water pressures, say during subsequent floods, will add up the hours for the rock to fail. Once the total sum of the durations of subsequent floods of given magnitude equals 100 h the rock will suddenly fail. It may therefore appear as if the rock withstands the erosive capacity of the water during the earlier flood events, but then suddenly fails during the last flood event. It is therefore important to know how rock will scour, either in brittle fracture or fatigue failure. If it will fail in fatigue it is important to understand its characteristics to appropriately design mitigation measures.

In what follows the calculation of fatigue failure of brittle materials is first presented. This is followed by a discussion on the erosion of cohesive material.

Subcritical Failure (Fatigue)

The equations and material properties that are required for determining the characteristics of subcritical failure of brittle materials are derived from information presented in Chaps. 4 to 6. Development of an equation that can be used to calculate the duration required for a brittle material to fail in fatigue can be accomplished by solving the equation proposed by Paris et al. (1961), i.e.,

$$\frac{dL}{dN} = C(\Delta K)^m \tag{8.1}$$

where N = number of cycles of the fluctuating pressure that will lead to fatigue failure

C, m = material properties

ΔK = range of stress intensities introduced to the material by the fluctuating pressures

L = distance of crack growth required for the material to fail

The stress intensity range can be calculated by subtracting the minimum stress intensity that is predicted to occur in the close-ended fissures from the maximum stress intensity. The stress intensity can be calculated with the following equation (see Chap. 4):

$$K = \sigma_{\text{water}}\sqrt{\pi a}f$$

The maximum stress intensity is calculated by using the estimated maximum value of the water pressure in the close-ended fissures σ_{water_max} and the minimum water pressure σ_{water_min} is used to calculate the minimum K value. The equation for ΔK can therefore be written as

$$\Delta K = \Delta\sigma_{\text{water}}\sqrt{\pi a}\,f \tag{8.2}$$

where $\Delta\sigma_{\text{water}} = \sigma_{\text{water\_max}} - \sigma_{\text{water\_min}}$.

Equation (8.1) can then be expressed as

$$\frac{dL}{dN} = C\left(\Delta\sigma_{\text{water}}\sqrt{\pi a}\,f\right)^m$$

from which follows that

$$\int_0^N dN = \int_a^{L_f} \frac{dL}{C\left(\Delta\sigma_{\text{water}}\sqrt{\pi L}\,f\right)^m} \tag{8.3}$$

where a = initial length of the close-ended fissure

L = variable length of the close-ended fissure (it changes as the crack grows during the process of fatigue failure)

L_f = the thickness of the material layer (i.e., $L_f - a$ = distance through which the crack must grow in order for fatigue failure to occur)

Solving the integral in Eq. (8.3) leads to the following:

$$N = \frac{1}{C\left(f\Delta\sigma_{\text{water}}\sqrt{\pi}\right)^m}\cdot\frac{1}{\left(1-\dfrac{m}{2}\right)}\cdot\left(a^{1-\frac{m}{2}} - L_f^{\,1-\frac{m}{2}}\right) \quad \text{for } a < L_f \tag{8.4}$$

The pressure in a close-ended fissure can be calculated with the following equation (see Chap. 4):

$$\sigma_{\text{water}} = 0.36p_0 + 0.64p_{\max}$$

where p_0 is the pressure at the entrance to the close-ended fissure, i.e., at the surface of the rock and p_{max} is the maximum pressure at the closed end of the fissure.

The pressure at the surface of the rock is equal to the average dynamic pressure in the plunge pool (at the interface with the rock), and the maximum pressure at the closed end of the fissure is equal to the pressure developing in the fissures due to pressure fluctuations and resonance. These values are respectively calculated as (see Chap. 5)

$$p_0 = C_{pa}\gamma\phi\frac{V_j^2}{2g}$$

and

$$p_{max} = (C_{pa} + \Gamma_{max}^{+}C'_{pa})\gamma\phi\frac{V_j^2}{2g}$$

The maximum water pressure in a close-ended fissure is therefore quantified as

$$\sigma_{water\_max} = (0.36C_{pa} + 0.64(C_{pa} + \Gamma_{max}^{+}C'_{pa}))\phi\gamma\frac{V_j^2}{2g}$$

and the minimum pressure can be calculated as

$$\sigma_{water\_min} = [0.36C_{pa} + 0.64(C_{pa} + \Gamma_{min}^{+}C'_{pa})]\phi\gamma\frac{V_j^2}{2g}$$

The decision for selecting the minimum value depends on the engineer's interpretation of conditions when analyzing a scour problem. At times the minimum water pressure can be assumed to equal zero. It is also reasonable to consider using the extreme dynamic pressure coefficient values (i.e., the maximum and minimum values) presented in Chap. 5.

Figure 8.1 presents the results of a calculation to determine the depth of scour in a rock-lined plunge pool that is subject to a plunging jet of varying discharge. The rock properties used are $C = 1.8 \times 10^{-8}$, $m = 8.2$, and the values of $a = 0.5$ m and $L_f = 2$ m. The drop height of the jet is approximately 100 m.

The figure indicates that a discharge of 100 m^3/s will require about 550 days of cumulative discharge to result in fatigue failure, 200 m^3/s require roughly about 200 days, 400 m^3/s require about 50 days, and flows of 800 to 950 m^3/s will require about 6 to 10 days.

If a discharge of 950 m^3/s were to pound the plunge pool for about 7 days the rock would suddenly fail in fatigue at the end of that period and the plunge pool would rapidly increase in depth to about 25 m before scour temporarily ceases. After that, the pool depth will increase very slowly, should

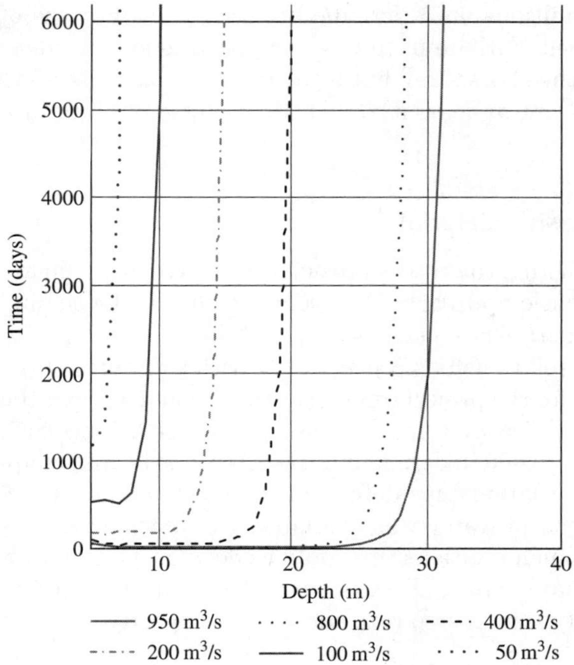

Figure 8.1 Example of the results of a fatigue failure calculation for rock scour subject to a plunging jet with varying discharge.

this high discharge occur for long enough periods of time. The rapid rise of the curve at about 25 m scour depth leads to this conclusion.

As such high flows would most probably only occur very infrequently, and not for long periods of time, whenever they occur, it is unlikely that the rock would scour to any significant depth by fatigue failure in practice. It can also be seen that it would take on the order of about 1200 days (i.e., about 3 years) of cumulative time for a discharge of 50 m^3/s to cause fatigue failure of the rock. Should this occur, it is unlikely that the scour would be more than a few meters deep.

Based on the information presented in Fig. 8.1 it is evident that this methodology can be used to determine the likelihood for rock and other brittle materials subject to plunging jets to scour as a result of fatigue failure. The methodology can therefore be used to design plunge pool linings when using concrete, or to determine if it is necessary to pre-excavate plunge pools to prevent scour in the future.

The methodology can also be applied to other scour conditions, e.g., scour at bridges, but requires additional investigation to quantify the pressure fluctuations around bridge piers. Such investigations might be important in locations where bridges are founded on rock and are subject to intense hydraulic conditions on a periodic basis, e.g., in the case of bridges in hurricane zones. Fatigue in the rock supporting such bridges might not be obvious in the short term, but if the rock foundation suffers fatigue each time a hurricane appears it could lead to catastrophic failure at some point in time.

Rate of Erosion of Cohesive Material

Current practice for predicting the rate of erosion of cohesive material follows a somewhat pragmatic approach. In essence it entails developing relationships between the rate of erosion of clay and the erosive capacity of water, and using such relationships concurrently with estimates of the erosive capacity of water for the project under consideration to predict the rate and maximum extent of scour. Chapter 5 presented such an approach by simulating the rate of erosion in dam foundation fissures. In that chapter, it was assumed that relationships defining rate of erosion as a function of the erosive capacity of water was available. No information was provided indicating how such relationships could be developed. The principle objective of this chapter is to discuss current techniques used to develop relationships between erosion rate and erosive capacity for cohesive soil samples.

The characteristics of clay erosion presented in Chap. 4 indicate complex relationships between erosion rate and shear stress, and the salinity, pH, and temperature of the interstitial and ambient water. It has been shown that the erosion rate for a particular shear stress can vary significantly depending on changes in temperature, salinity, and pH of the interstitial and ambient water.

Practical methods for developing equations to predict the rate of erosion of clays using the understanding developed in Chap. 4 are presented here. The complexity of these tests differs, as does the amount of useful information obtained from each. For example the test for successfully determining the impact of pH, temperature, and salinity on the rate of erosion of clays and developing equations that capture the essential character of the material is more detailed than the others. Execution of this test is warranted when the consequences of failure are significant. For example, predicting scour around pylons supporting oil rigs in the ocean may become important if they are founded in cohesive material. Changes in the salinity of the water, its pH, and its temperature may have greater impacts on the erosion of the cohesive material than the shear stress itself. What this means is that the erosion rate for a particular shear stress might be much greater for prototype conditions than for test conditions if the influence of pH, salinity, and temperature are not investigated and not taken into account in the scour calculations for the pylons. This will happen if the pore and ambient water used in the tests, as well as the water temperature differ from conditions on site.

The methods that are presented include the Couette flow device (CFD) (see e.g., Croad 1981 and Tan 1983), the vertical jet tester (VJT) (Hanson and Cook 2004), the hole erosion test (HET) (Wan and Fell. 2002) and the erosion function apparatus (EFA) (Briaud et al., 2002). Of the four devices the CFD is the only device that is considered appropriate and effective for investigating the effects of pH, salinity, and temperature on the scour rate of clay while concurrently obtaining enough information to develop fundamentally-based equations incorporating their effect. The HET can be used to empirically assess the effects of changes in these parameters and develop the rate of erosion and threshold relationships for implementation in practice. When using the VJT it is more difficult to conduct such sensitivity analyses. The VJT is usually performed in situ and uses local water and in situ materials to conduct tests. The EFA as it is implemented at writing of this book is not considered sensitive enough to determine the effects of such changes. The reasons for this statement are presented in what follows.

Couette flow device (CFD)

The CFD is based on Couette flow conditions that have been studied in detail and is well understood (see e.g., Schlichting and Gersten 2000). Couette flow occurs between two plates when the one plate moves relative to the other (Fig. 8.2). The flow therefore occurs due to the movement of the one plate relative to the other and pressure differences do not play a role.

The CFD (Fig. 8.3), originally developed by Moore and Masch (1962), consists of a circular soil sample mounted on a support that is attached to a torsion wire. An outer drum, coaxial with the sample contains the eroding water. This outer drum is rotated using a variable speed motor and a shear stress is consequently transmitted to the surface of the soil sample. The shear stress is measured directly by knowing the torsion acting on the soil sample through the angular displacement of the torque wire. The influence of end effects on the soil sample is minimized by independently mounted coaxial end pieces at both ends of the soil sample. They act to give a uniform shear stress over the surface of the sample.

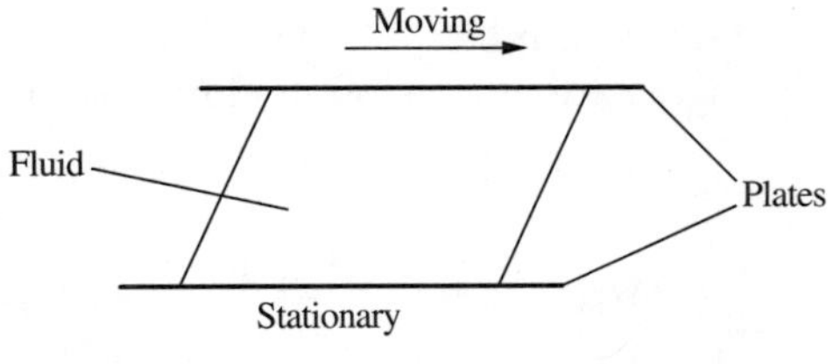

Figure 8.2 Couette flow between two plates.

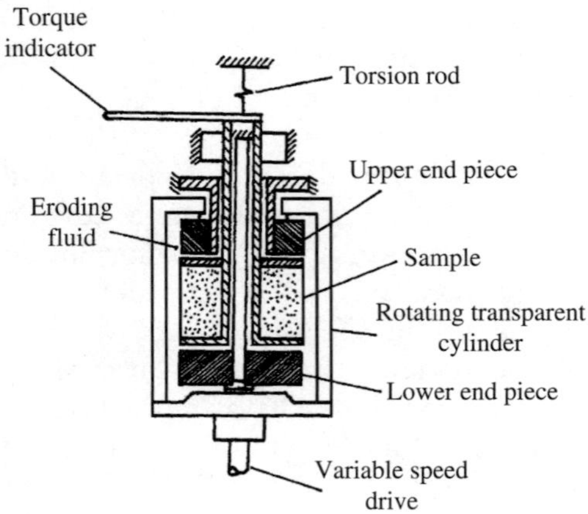

Figure 8.3 Couette flow device for erosion testing.

The advantage of the device is that the average shear stress is measured directly from the torque wire and is not inferred or calculated indirectly. The role of end-effects is also minimal and it is relatively easy to control the temperature and water chemistry in the device. The erosion rate of the material is proportional to the weight loss of the sample.

The device has only a few disadvantages. For example is has been found that the turbidity levels can become high but it has not been found to be a serious disadvantage (Croad 1981). It is also required that the soil samples should not distort under their own weight. One of the disadvantages of the device is that although it is ideal for research applications, it is most probably too complex to operate for routine laboratory application. However, if the consequences of scour in clay are significant it might merit application of this testing procedure.

Research at the University of Auckland, New Zealand, by Hutchison (1972), Croad (1981), and Tan (1983) made it possible to use the CFD to quantify the activation energy and the number of fixed bonds required to calculate the rate of scour with the Arhenius equation (see Chap. 4). The magnitude of the activation energy and the number of fixed bonds are material and site specific. This means that it has not been possible to develop generic values of activation energy and the number of bonds for clays. Clays differ from site to site, and tests are required to determine the magnitude of the activation energy and the number of fixed bonds.

In Chap. 4 it has been shown that the rate of erosion of clays can be expressed in the form

$$\dot{e} = k[S][F] = \vartheta^* \frac{0.35\rho_s}{(1+V')} \cdot \frac{\nu}{u_*} \cdot P \exp\left(\frac{-E_a}{RT} + \frac{k'\lambda L\tau_t}{2n_B RT}\right) \tag{8.5}$$

The values of all the variables, except for the activation energy E_a and the number of fixed bonds n_B, are either known or can be estimated (see Chap. 4).

Determination of the number of fixed bonds n_B. The term $\vartheta^* \frac{0.35\rho_s}{(1+V')} \cdot \frac{\nu}{u_*}$ in Eq. (8.5) has units of mass per unit area per second. (The term ϑ^* represents the frequency of turbulent bursts, which are associated with the frequency of pressure fluctuations at the boundary.) The rest of the expression in Eq. (8.5) on the right-hand side of the equal sign is dimensionless. Therefore, Eq. (8.5) can be written in dimensionless form, as follows:

$$\dot{E} = \frac{\dot{e}}{A_1} = P\exp\left(\frac{-E_a}{RT} + \frac{k'\lambda L}{2n_B RT}\tau_t\right) \tag{8.6}$$

where

$$A_1 = \vartheta^* \frac{0.35\rho_s}{(1+V')} \cdot \frac{\nu}{u_*}$$

When the temperature is held constant Eq. (8.6) can be rewritten as

$$\ln(\dot{E}) = A_2 + \frac{\beta}{n_B}\tau_t \tag{8.7}$$

where

$$A_2 = \ln(P) - \frac{E_a}{RT}$$

and

$$\beta = \frac{k'\lambda L}{2RT}$$

By plotting $\ln(\dot{E})$ as a function of shear stress τ for erosion tests executed under constant temperature, it is possible to calculate values of A_2 and n_B (Fig. 8.4). Express $\vartheta^* = 1/T_B$ as frequency of turbulent bursts. The value of T_B is estimated by making use of an equation developed by Kim et al. (1971), i.e.,

$$\frac{\bar{u}\bar{T}_B}{\delta_b} = 5 \tag{8.8}$$

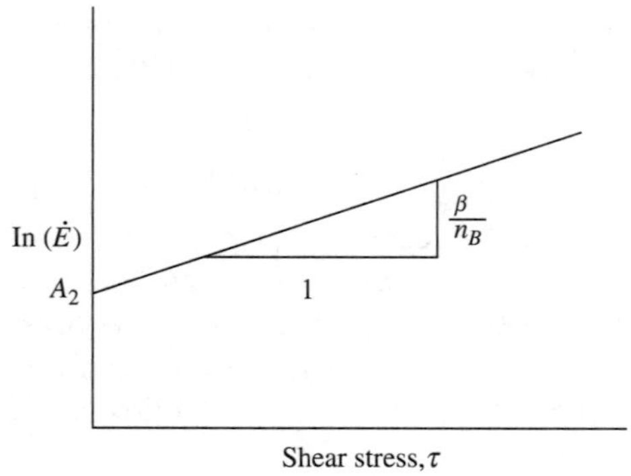

Figure 8.4 Plot for determination of the number of fixed bonds between clay particles.

where δ_b is the thickness of the boundary layer. In fully developed rough turbulent flow in the CFD is assumed, Eq. (8.8) can be expressed as

$$\frac{\overline{u}\,\overline{T}_B}{(r_2 - r_1)} = 5 \tag{8.9}$$

where r_1, r_2 are the inner and outer radii of the space between the sample and the outer edge of the section in the CFD containing the fluid.

The void ratio V' and unit weight ρ_s of the cohesive material is determined using standard geotechnical techniques. The shear velocity is calculated as $u_* = \sqrt{\tau_t/\rho}$ and the kinematic viscosity for water, which is temperature dependent, can be calculated with equations provided in Chap. 4. A representative value of the kinematic viscosity is $\nu = 1.13 \cdot 10^{-6}$ m^2/s at approximately 15°C (288 K). With these values and associated rates of erosion $\dot{e}$ known, the variable $\ln(\dot{E})$ can be calculated from test data for varying values of τ_t.

The value of A_2 is determined directly from the data plot, it is the intercept of the curve where $\tau_t = 0$. The value of β/n_B is equal to the slope of the plot between $\ln(\dot{E})$ and τ_t.

Tan (1983) and Croad (1981) determined the value of β/n_B by making use of data from a Couette flow device. Both assumed a value of the fluctuating pressure coefficient $k' = 18$, which leads to the value of

$$\beta = \frac{9\lambda L}{RT} \tag{8.10}$$

The value of the Avogadro number is $L = 6.022 \times 10^{23}$ 1/mol and the universal gas constant is $R = 8.315$ J/K·mol. The value λ represents the distance of separation at which the maximum interaction force between two clay particles is experienced. This distance has been estimated by Croad (1981) as

$$\lambda = 3 \cdot 10^{-10}\,\text{m} \tag{8.11}$$

With the values of these variables and the value of the slope of the curve between $\ln(\dot{E})$ and τ_t known, it is possible to calculate the number of fixed bonds between clay particles. Typically the value of n_B is on the order of about 10^{12} per m^2. Experimental results for determining the number of bonds by Croad (1981) are shown in Fig. 8.5 and results in Table 8.1.

Additionally Croad (1981) also analyzed experimental data for San Francisco Bay mud, Kaolinite, Grundite and Halloysite published by others (Partheniades 1965; Christensen and Das 1973; Hutchison 1972; Raudkivi and Hutchison 1974; and Rao 1971) and determined that the average number of bonds per square meter for their data ranged between 9.6×10^{10} and 8.2×10^{11}.

Determination of the activation energy E_a. Determination of the value of the activation energy for clay requires testing of the erosion rate as a

TABLE 8.1 Number of Bonds n_B from Experimental Data

No.	Soil	Fluid	pH	Temp (°C)	Correlation coefficient	n_B (bonds/m^2)
1a	Rheogel	Deionized water	10.1	21.1	0.86	3.5×10^{12}
1b	Rheogel	Deionized water	10.1	24.4	0.90	4.2×10^{12}
2	Rheogel	NaOH solution	11.8	23.8	0.98	4.6×10^{12}
3	Rheogel	HCl solution	1.8	25.8	0.98	6.3×10^{12}
4	Rheogel	0.010 M NaCl		23.0	0.72	6.9×10^{12}
5	Rheogel	0.0010 M NaCl		22.8	0.80	2.9×10^{12}
6	Wyoming clay	HCl solution	2.8	24.5	0.98	2.9×10^{12}
7	Wyoming clay	Deionized water	8.2	24.5	0.95	5.2×10^{12}
8	Wyoming clay	Deionized water	9.4	24.0	0.95	3.8×10^{12}
9	Panther creek	Deionized water	9.1	23.8	0.97	1.9×10^{12}
10	Panther creek	NaOH solution	11.6	24.0	0.91	8.2×10^{12}
11	Kaolinite	HCl solution	1.6	23.0	—	1.7×10^{12}
12a	Kaolinite	HCl solution	2.5	29.7	0.97	8.0×10^{11}
12b	Kaolinite	Deionized water	2.4	44.9	0.89	1.3×10^{12}
13	Kaolinite	Deionized water	4.4	24.0	0.96	2.8×10^{12}
14	Kaolinite	NaOH solution	12.0	23.1	1.00	4.4×10^{12}
15a	Kaolinite	Deionized water	5.6	20.4	0.87	1.5×10^{12}
15b	Kaolinite	Deionized water	5.7	46.8	0.77	1.7×10^{12}
17a	Kaolinite	NaOH solution	11.1	8.8	0.93	3.8×10^{12}
17b	Kaolinite	NaOH solution	11.1	17.4	0.87	2.8×10^{12}
17c	Kaolinite	NaOH solution	11.1	24.8	0.91	9.0×10^{11}
18	Kaolinite	0.100 M NaCl	5.7	22.5	0.99	1.1×10^{12}
19	Kaolinite	0.100 M NaCl	5.7	24.0	0.80	1.5×10^{12}

SOURCE: From Croad 1981.

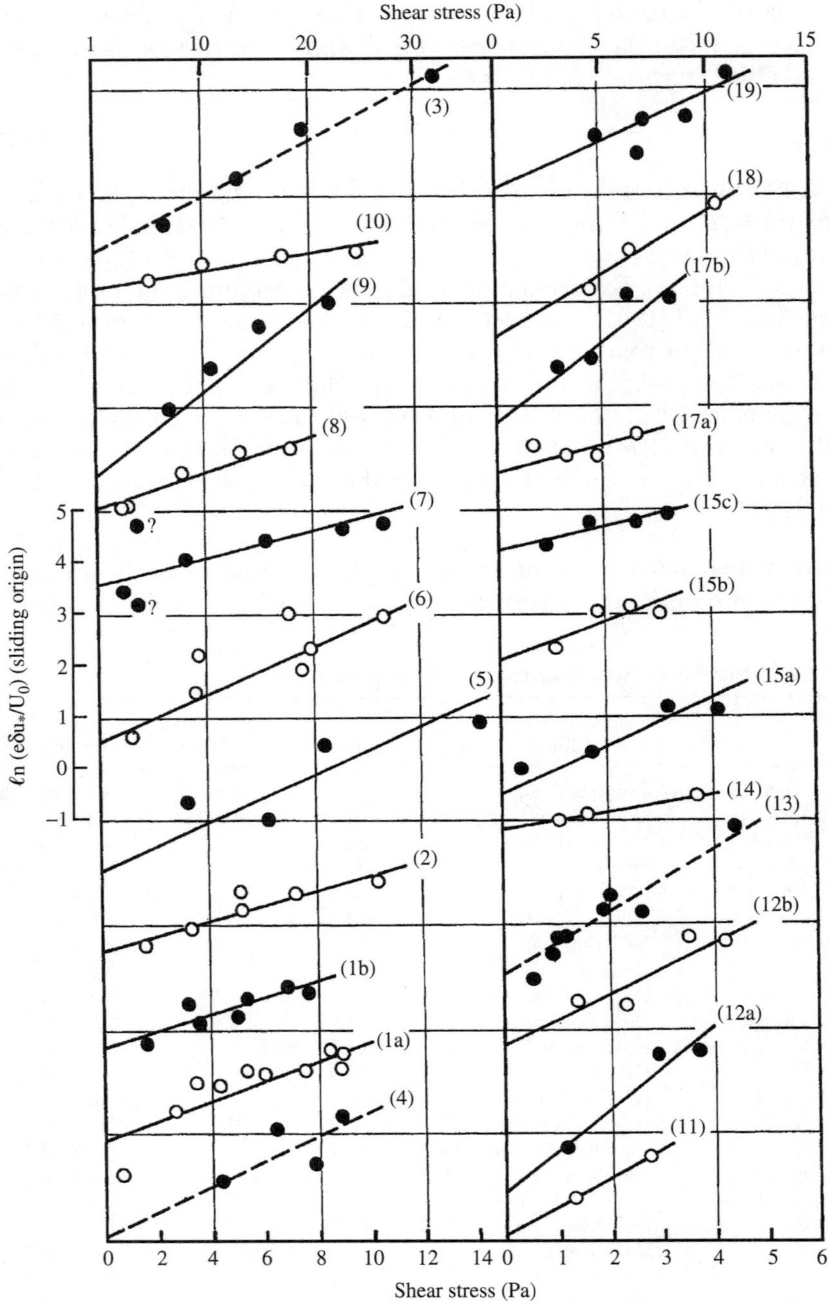

Figure 8.5 Determination of the number of bonds n_B associated with clays. The upper τ-axis is associated with the broken lines and the lower τ-axis with the solid lines (Croad 1981).

function of temperature. In order to accomplish this write the rate of erosion equation for clay in a different format, i.e.,

$$\ln(\dot{E}) - \frac{k'\lambda L}{2n_B RT}\cdot \tau_t = \ln(P) - \frac{E_a}{R}\cdot\frac{1}{T} \tag{8.12}$$

The value of the activation energy can be determined by plotting the left hand side of the equation as a function of the inverse of the absolute temperature, expressed in Kelvin (Fig. 8.6). The intercept of the curve is equal to ln(P) and the slope is equal to E_a/R. Once the slope of the curve is known, the value of E_a can be calculated. Typical values of E_a are seldom less than about 10 kJ/mol, and are rarely above 140 kJ/mol.

Experimental data published by Croad (1981) are plotted in Fig. 8.7 and estimated values of the activation energy, using this data, are presented in Table 8.2.

With the number of bonds and the activation energy for clay under investigation known it is possible to quantify the erosion rate equation that was developed in Chap. 4. Improved understanding of the erosion characteristics of the clay as a function of temperature, pH, and salinity is available at completion of these tests.

Vertical jet tester (VJT)

Hanson and Cook (2004) describe the vertical jet tester (VJT), test procedures, and analytical methods for determining the in situ rate of erosion of cohesive earth materials. The VJT approach is based on the assumption that the relationship between the rate of erosion of a cohesive earth material and shear stress is linear and can be expressed by the equation,

$$\varepsilon_r = k_d(\tau_e - \tau_c) \tag{8.13}$$

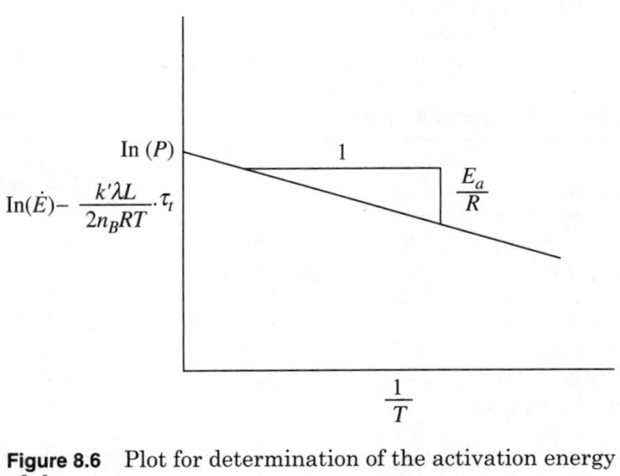

Figure 8.6 Plot for determination of the activation energy of clay.

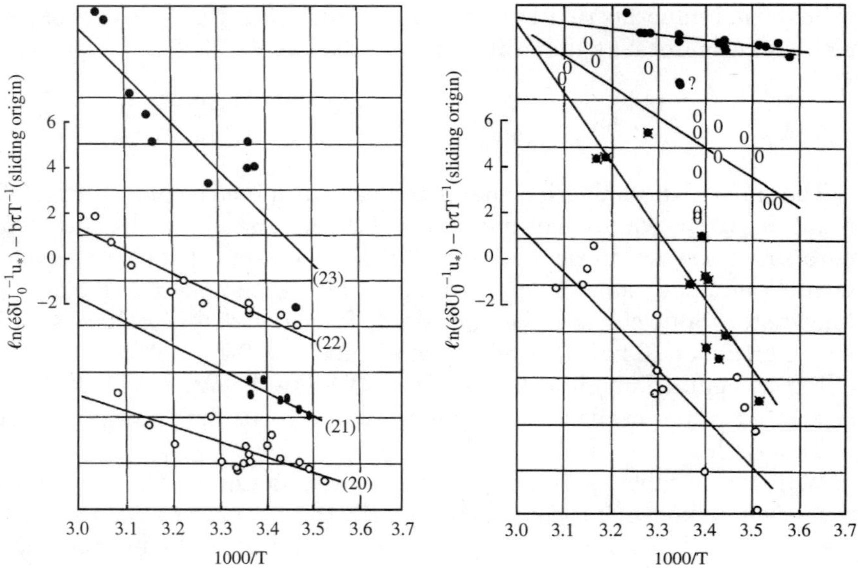

Figure 8.7 Determination of the activation energy using data by Croad (1981).

where ε_r = rate of erosion (m/s)

k_d = erodibility coefficient (m^3/Ns)

τ_e = effective stress (Pa)

τ_c = critical shear stress (Pa)

Although Hanson and Cook's justification for using this equation is based on the premise that it has been used by numerous investigators, it obviously has limitations when considered in the light of the clay scour characteristics presented in Chap. 4. However, it is possible that the

TABLE 8.2 Activation Energy for Seven Clay Samples

No.	Soil	Fluid	pH	Correlation coefficient	$E_a \pm$ std. error (kJ/mol)
20	Rheogel	Deionized water	10.1	0.80	27.8 ± 5.5
21	Rheogel	NaOH solution	11.8	0.85	41.5 ± 11.4
22	Wyoming clay	Deionized water	8.2	0.95	41.7 ± 4.2
23	Panther cr.	Deionized water	9.1	0.88	90.6 ± 17.4
24	Kaolinite	HCl solution	2.5	0.80	90.7 ± 19.9
25	Kaolinite	Deionized water	4.4	0.90	132.3 ± 21.7
26	Kaolinite	Deionized water	5.7	0.75	52.6 ± 12.1
27	Kaolinite	NaOH solution	11.7	0.75	11.5 ± 2.9

SOURCE: From Croad (1981).

linear relationship can hold over particular ranges of interest and it is most probably a reasonable pragmatic assumption for a number of practical situations.

The units of k_d is expressed in volume per force-second by Hanson and Cook because the purpose of this equipment is to determine *spatial* changes resulting from erosion. It is also possible to express the equation in terms of *mass* eroded per second. The mass rate of erosion can be estimated by multiplying the volume estimate and the bulk density of the material.

The VJT apparatus is fairly simple, consisting of a jet tube, nozzle, point gauge, an adjustable head tank, and a jet submergence tank (Fig. 8.8). A schematic showing the different components and dimensions is presented in Fig. 8.9. The jet tube is 0.92 m long, is made of 50 mm internal diameter acrylic tubing with 6.4 mm wall thickness. Clear tubing is used to allow visual observation of air accumulation in the jet tube. The jet tube has an 89-mm-diameter orifice plate 12.7 mm thick with a 6.4 mm diameter nozzle opening in the center of the plate. Water is delivered to the tube from an opening 0.41 m above the orifice plate via a 32-mm overall diameter hose. An air relief valve and a point gauge are attached to the top of the jet tube. The air relief valve is used to remove air that has accumulated in the jet tube during initial filling. Once a test is started, scour readings are taken with the point gauge. The point gauge is axially aligned with the jet nozzle so that the point gauge can pass through the nozzle to the

Figure 8.8 Vertical jet tester.

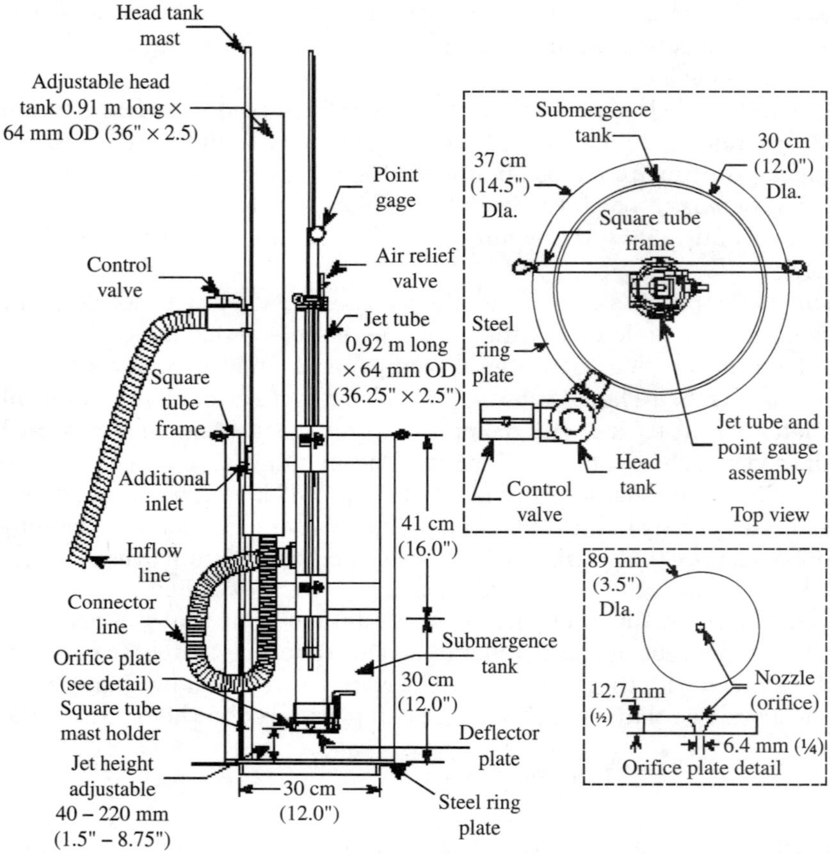

Figure 8.9 Schematic of vertical jet tester showing dimensions (Hanson and Cook 2004).

bed to read the depth of scour. The point gauge diameter is nominally equivalent to the nozzle diameter so that when the point gauge rod passes through the nozzle opening, flow is effectively shut off. A deflector plate is attached to the jet tube and is used to deflect the jet, protecting the soil surface during initial filing of the submergence tank. At test initiation the deflector plate can be moved out of the way of the jet, allowing the jet to impinge directly onto the soil surface.

The adjustable head tank, 0.91-m-long, is made of 50 mm internal diameter acrylic tubing with 6.4 mm wall thickness. Clear tubing is used to allow visual observation of the water level in the head tank. The height of the head tank can be adjusted by sliding it up and down on the mast.

The submergence tank is 0.3 m in diameter, 0.3-m-high, and is made of 16-gauge steel. The tank is open on both ends and has a 25-mm-square tube frame attached to hold the jet tube in the center of the tank. The frame allows the jet tube and nozzle height to be conveniently set prior to initiating a jet test. The tank also has a 32-mm-square tube attached to the outside perimeter to hold the head tank mast during testing. A steel ring plate is attached to the outside perimeter of the tank, 25 mm from the bottom end. The tank is driven into the soil to a depth of 25 mm until the steel ring plate makes contact with the soil surface. The intent for driving the tank into the soil is to create a seal that will allow the tank to be filled with water, submerging the jet orifice. During testing excess water overflows the top rim of the tank.

The testing procedure commences with site selection and preparation. The soil surface to be tested should be horizontal, and usually requires preparation with a shovel or spade. The bottom 25 mm of the tank is inserted into the soil to create a seal. Creating an adequate seal at the base of the tank is at times a problem in practice. What has been found to work fairly well is to pre-create a round slot in the ground into which the tank can fit. This can be done by means of a 300 mm overall diameter PVC pipe cerated on one end. The cerated edge is then placed on the ground and the pipe is turned until a round slot that is slightly deeper than 25 mm has been created in the soil. This slot is filled with bentonite that has a consistency that will allow it to be rolled by hand for insertion into the slot. By placing the tank into the bentonite rim, a seal is created.

Once the nozzle tank has been inserted into the soil and leveled, the other components are assembled to this base and a water supply line attached to the head tank. When working in river environments the water can be supplied by making use of a small pump. When water is not readily available, it might be necessary to use a temporary storage tank or a water truck.

The erosive capacity provided by the submerged jet is a function of the distance between the nozzle and the ground. The distance between the end of the nozzle and the ground should be somewhere between 6 and 35 nozzle diameters, i.e., between 40 and 220 mm. The recommended initial distance is about 12 nozzle diameters, i.e., about 80 mm.

The test commences by taking zero readings, i.e., use the point gauge to measure the distance between the end of the nozzle pipe and the surface of the ground prior to setting the jet in motion. Once these measurements have been recorded the test can commence. Set the deflector plate below the nozzle hole and allow water to discharge into the tank until it overflows. The head on the jet can be adjusted by increasing/decreasing the height of the head tank. The air release valve at the top of the jet tube is used to release air from this tube.

Once the system is filled with water, the point gauge is lifted about 12 nozzle diameters above the level of the nozzle, i.e., approximately 80 mm. This is done to prevent interference with flow from the jet tube through the jet orifice. The deflector plate is then turned so that the jet discharges onto the ground. The test commences at this point in time and should be continuously timed. Check the head and the amount of erosion that occurred every 5 to 10 minutes, depending on how erodible the soil is. From a practical point of view it is advisable to feel the surface conditions at the bottom of the scour hole created by the jet when measurements are made. The reason for this is that small pieces of gravel might armor the bottom of the hole, which can provide lower erosion rates than those characteristic of the material. If pieces of gravel are found, they should be removed. This should be noted in the record of measurements.

The setting of the head tank is determined by estimating the anticipated magnitude of the erosive capacity of the water that is expected to occur in the prototype. A calculation can then be made to determine the head that is required by the VJT to produce the same range of magnitudes of erosive capacity.

The maximum shear stress between the interface of the jet and the ground surface is calculated with the equation shown next (Hanson and Cook 2004):

$$\tau_i = \tau_o \left(\frac{J_p}{J_i} \right)^2 \qquad \text{for } J_i \geq J_p \tag{8.14}$$

where τ_i = maximum shear stress at the interface between the jet and the ground surface

$\tau_o = C'_f \rho U_o^2$

C'_f = friction coefficient defined by Hanson and Cook to equal 0.00416

ρ = density of water

$U_o = \sqrt{2gh}$ = flow velocity at the exit from the nozzle

h = differential head measurement

g = acceleration due to gravity

J_p = potential core length of the jet = $C_d d_o$

C_d = 6.3 = diffusion constant

d_o = nozzle diameter

J_i = distance between the orifice and the ground surface

J_e = distance between the orifice and the ground surface when equilibrium scour conditions are reached, i.e., the scour hole does not deepen as a function of time (see Fig. 8.10)

The potential jet core length J_p is the distance over which the velocity of the jet remains constant and equal to the velocity right at the orifice. This distance is roughly equal to about six orifice diameters.

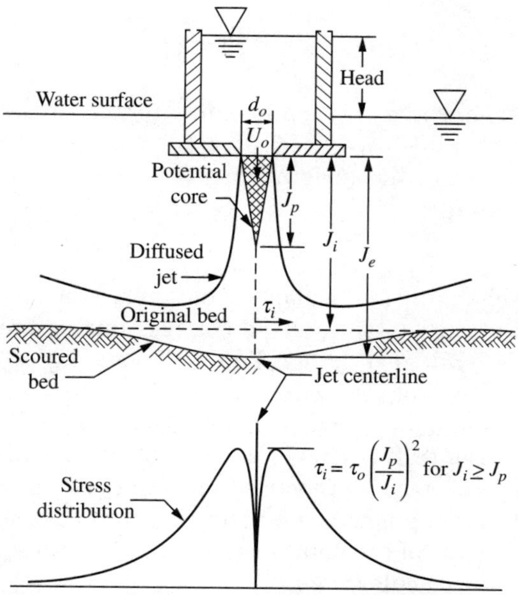

Figure 8.10 Schematic showing shear stress distribution of the round jet associated with the VJT and relevant variables (Hanson and Cook 2004).

Although it is possible to vary the head during the test by changing the elevation of the adjustable head tank, it is generally not advised. If it does change, it is necessary to record the change in head. The data collected during the field test, i.e., the change in erosion depth as a function of time and the pressure head are analyzed to determine the values of the erosion rate coefficient k_d and the critical shear stress τ_c. These are required for using Eq. (8.13) in design projects.

The basic assumption in calculating the values of the erosion rate coefficient and the critical shear stress is that the shear stress associated with the equilibrium erosion depth J_e is the critical shear stress. This obviously presents a practical problem because it is often difficult to identify equilibrium conditions in the field. The reason for this is that the time required to reach this condition can be very long.

In order to overcome this problem Hanson and Cook (2004) used an approach that was developed by Blaisdell et al. (1981) who used a hyperbolic function representing the relationship between scour depth and time to estimate the equilibrium scour depth. The equation proposed by them is

$$x = \left[(f - f_o)^2 - A^2\right]^{0.5} \qquad (8.15)$$

where $x = \log[(U_o t)/d_o]$
$f = \log[J/d_o] - \log[U_o t/d_o]$
$f_o = \log[J_e/d_o]$
A = value of the semi-transverse and semi-conjugate axis of the hyperbola
U_o = velocity of the jet at the orifice (origin)
t = time of the data reading
d_o = orifice diameter

The two principal variables changing in the above equation are the erosion depth J and the measured time t. (If the head changes during the course of the test the value of U_o will also change, but is a known input.) The unknowns in Eq. (8.15) are therefore the values of f_o and A, which can be determined by minimizing the difference between the measured values and the value calculated with Eq. (8.15). This is done by starting with assumed values of $f_o = 1$ and $A = 1$, which are varied until the best fit is obtained. Once this has been accomplished, the value of f_o thus determined is used to calculate the value of the equilibrium erosion depth J_e. The critical shear stress τ_c is then calculated as

$$\tau_c = \tau_o \left(\frac{J_p}{J_e} \right)^2 \tag{8.16}$$

A dimensionless time function, shown below, is used to calculate the value of the erosion rate coefficient k_d, i.e.,

$$T^* = -J^* + 0.5 \ln\left(\frac{1+J^*}{1-J^*} \right) \Bigg|_{J_i^*}^{J^*} \tag{8.17}$$

where $T^* = t_m / T_r$
t_m = measured time
T_r = a reference time = $J_e/(k_d \tau_c)$
J^* = dimensionless scour term, J/J_e
J_i^* = dimensionless scour term at J_i/J_e
J = distance from the orifice to the centerline depth of scour
J_i = initial distance from orifice to soil surface

The calculations are done by rewriting Eq. (8.17) in terms of the measured time, i.e.,

$$t_m = T_r \left[0.5 \ln\left(\frac{1+J^*}{1-J^*} \right) - J - 0.5 \ln\left(\frac{1+J_i^*}{1-J_i^*} \right) + J_i^* \right] \tag{8.18}$$

TABLE 8.3 Example of Values of k_d and τ_c That Were Calculated from Testing Pleistocene Soils Varying in Clay Content and Degree of Cementation

Date	Test number	k_d (m^3/Ns)	τ_c (Pa)
6/23/2004	1	7.70E-08	3.108
	2	1.49E-06	0.286
	3	3.91E-06	0
	4	2.44E-06	0.052
	5	1.62E-06	0
7/14/2004	1	2.94E-06	0
	2	1.63E-05	0.002
	3	4.88E-08	1.949
	4	2.69E-06	0.001
	5	5.34E-06	0
7/19/2004	6	2.96E-07	0.008
	7	2.43E-06	0.362
	8	1.13E-07	0.006
	9	1.29E-06	0
	10	2.23E-07	0

The value of k_d is then determined by minimizing the difference between the measured values of time and the value of t_m determined from Eq. (8.18) by varying the value of k_d.

Hanson and Cook (2004) developed a spreadsheet that is freely available in the public domain that can be used to perform these calculations. Table 8.3 provides the reader with an indication of the range of k_d and τ_c values that were obtained by testing Pleistocene soils with varying degrees of cementation and clay content in Maricopa County, Arizona.

Discussion

Practical experience has been that the VJT apparatus provides reasonably consistent relationships between the rate of erosion and shear stress for cohesive soils. In this regard it is important to state the background against which this statement is made. General experience in sediment transport technology indicates that the range of results when conducting tests on the interaction between flowing water and earth materials can be very large, even for the same conditions. Therefore, the statement that the results obtained from the VJT apparatus are reasonably consistent does not mean that they are exactly replicated, but that one can make reasonable conclusions for application in practice when investigating the outcome.

A significant benefit of the test is that it is conducted in situ, which is helpful if the materials that are tested are close to the surface.

Sample disturbance is minimal when conducting in situ testing. However, if the erosion characteristics of cohesive material in deep subsurface layers are sought, it is not possible to use this equipment unless an excavation is made, which might not always be possible. Such testing could e.g., be required when investigating the erosion resistance of foundation materials of existing dams that might be subject to internal erosion.

It is appropriate to highlight some of the limitations of this testing method. The assumption by Hanson and Cook (2004) that the relationship between the erosive capacity of water and the rate of erosion of cohesive earth materials like clay is always linear is not necessarily supported by the known erosion behavior of these materials. It is known that the relationship between shear stress and rate of erosion could be convex or concave (see e.g., Croad 1981 and Tan 1983 and the discussions in Chap. 4).

Additionally, as argued throughout this book, it is preferable to quantify the erosive capacity of water in terms of stream power rather than shear stress. The erosion process in turbulent flow is the result of fluctuating pressures that act on the surface of earth materials at its interface with the flowing water, and is not the result of shear action. Stream power has been shown to be a reasonably good indicator of the relative magnitude of pressure fluctuations resulting from the turbulent action of water (see e.g., Annandale 1995).

Should one decide to use the VJT to quantify the erosion behavior of cohesive earth materials and additionally desire to use stream power as an indicator parameter (rather than shear stress) it is possible to derive relationships that are based on Hanson and Cook's (2004) assumptions for accomplishing this goal. This can be done by writing equations relating erosion rate to shear stress and to stream power respectively, and then equating them, i.e.,

$$\varepsilon_r = k_d(\tau_e - \tau_c) \tag{8.19}$$

and

$$\varepsilon_r = k'_d(P_e - P_c) \tag{8.20}$$

where k'_d = erosion rate coefficient associated with stream power
P_e = effective stream power
P_c = threshold stream power

For a given erosion rate ε_r the two equations are equal, i.e.,

$$k'_d(P_e - P_c) = k_d(\tau_e - \tau_c)$$

from which follows:

$$k'_d = k_d \frac{(\tau_e - \tau_c)}{(P_e - P_c)} \tag{8.21}$$

Using Hanson and Cook's (2004) definition of shear stress, i.e., $t = C'_f \rho U_0^2$, it follows that

$$U_o = \sqrt{\frac{\tau}{C'_f \rho}} \tag{8.22}$$

By writing the expression for stream power as

$$P = \tau U_o = \tau \sqrt{\frac{\tau}{C'_f \rho}}$$

it can be shown that

$$P = \frac{1}{\sqrt{C'_f}} \rho \left(\frac{\tau}{\rho} \right)^{3/2} \tag{8.23}$$

This equation for stream power is equivalent to the equation for applied stream power derived from boundary layer theory in Chap. 5 [i.e., $P_{\text{applied}} = 7.853\rho(\tau/\rho)^{3/2}$], from which follows that

$$\frac{1}{\sqrt{C'_f}} = 7.853 \tag{8.24}$$

When using this relationship it is found that

$$C'_f = 0.016 \tag{8.25}$$

This value is somewhat higher than the value of 0.00416 proposed by Hanson and Cook. The origin of their assumed value is not stated in their paper. If it is accepted that the boundary layer process of turbulence production is universal, which is commonly believed to be the case (e.g., Schlichting and Gersten 2000), then the assumed value by Hanson and Cook appears to be on the low side. This means that the shear stresses using their standard analysis technique is most probably underestimated. However, this statement should not be considered conclusive and needs to be checked experimentally.

However, returning to the derivation of an expression for the rate of erosion coefficient to relate erosion rate to stream power by using parameters estimated from Hanson and Cook's standard calculation procedure it follows that

$$k'_d = k_d \sqrt{C'_f \rho}\left(\frac{\tau_e - \tau_c}{\tau_e^{3/2} - \tau_c^{3/2}}\right) \tag{8.26}$$

The equation for the rate of erosion as a function of stream power is therefore written as

$$\varepsilon_r = k_d \sqrt{C'_f \rho}\left(\frac{\tau_e - \tau_c}{\tau_e^{3/2} - \tau_c^{3/2}}\right)(P_e - P_c) \tag{8.27}$$

This means that if the assumption of a linear relationship between shear stress and rate of erosion is valid, then the relationship between erosion rate and stream power is not linear. This can for example be seen in Fig. 8.11, which shows erosion rate in mm/h as a function of shear stress and stream power for values of $\tau_c = 2.119$ Pa, $k_d = 2.648 \times 10^{-5}$ m^3/N · s, and $C'_f = 0.00416$ (i.e., using Hanson and Cook's assumptions).

Erosion function apparatus (EFA)

Briaud et al. (2001) developed the erosion function apparatus (EFA) to execute site specific erosion studies that minimizes sample disturbance.

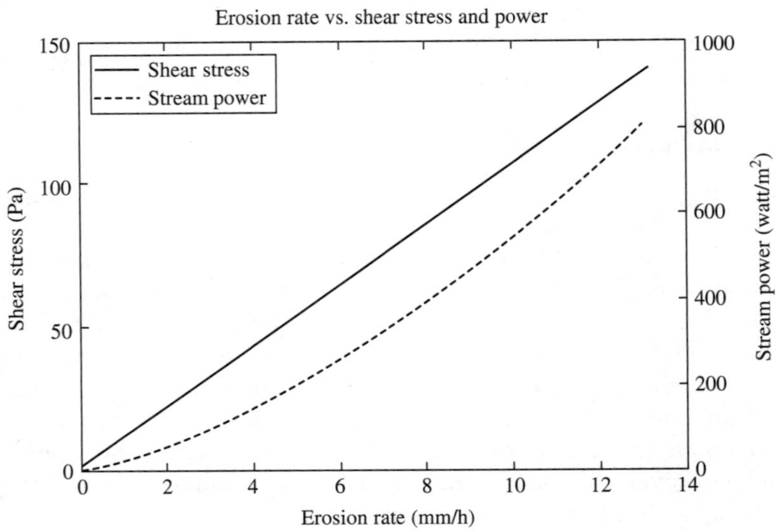

Figure 8.11 Erosion rate as a function of shear stress and stream power; determined by making use of the VJT apparatus.

This is done by measuring the rate of erosion as a function of shear stress for the soil tested and determining the critical (threshold) shear stress of the soil. The apparatus accommodates fine grained soil samples that were collected by pushing a Shelby tube with a 76.2 mm external diameter into the soil (ASTM 1999).

One end of the Shelby tube filled with the sample is placed in an opening at the bottom of the EFA, which consists of a rectangular pipe with internal dimensions of 101.6 × 50.8 mm. The square pipe is 1.25 m long and has flow straighteners at the upstream end. Water is circulated through the pipe and can reach velocities ranging between 0.1 to 6 m/s. The end of the Shelby tube is held flush with the bottom of the rectangular pipe and a piston at the bottom of the Shelby tube pushes the soil into the rectangular pipe at the upper end of the Shelby tube. The protrusion of the soil sample into the rectangular pipe should be no more than 1 mm, which is eroded by the water flowing over it (Fig. 8.12).

The accuracy of the test is very much dependent on operator skill. The operator monitors the erosion by eye and moves the sample up as erosion occurs ensuring that a 1 mm protrusion of the sample continuously exists within the apparatus. The upward movement of the sample and the flow conditions (i.e., flow velocity) is continuously monitored. The test starts off with a low velocity, say about 0.3 m/s and it is determined how long it takes for 1 mm of the soil to erode. Then, once the protruded material has eroded, or after one hour whichever comes sooner, the flow velocity is increased to, say, 0.6 m/s and the sample is pushed up to protrude 1 mm again. The time it takes for the 1 mm of soil to erode at this new velocity is measured once again and the process continued by gradually increasing the velocity every time. The erosion rate and the flow velocity data is then used to determine a relationship between erosion rate and shear stress. Briaud

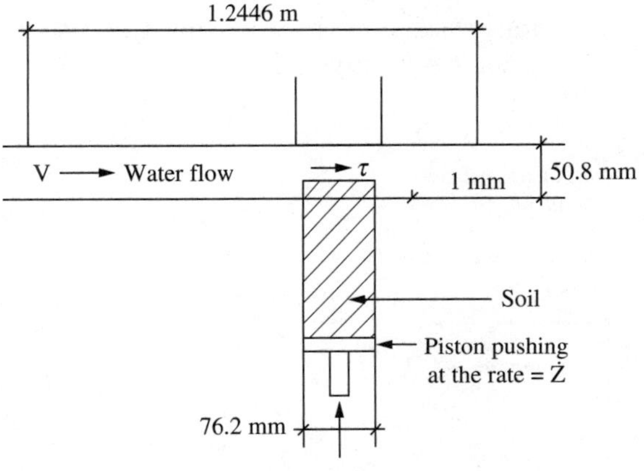

Figure 8.12 Schematic showing principle of operation of EFA (Briaud et al., 2001).

et al. (2001) assume that the erosion of soils is a shear process, a mechanism that is questioned in this book for turbulent flow.

During development Briaud et al. (2001) first experimented with estimating the shear stress over the sample by conducting pressure measurements up and downstream of the sample. They abandoned this approach by incorrectly reasoning that the pressure difference over the sample did not accurately represent the roughness contributed to the flow by the individual particles in the samples. They argued that if samples with sandy particles of the order of 1 mm were used then the pressure difference measured over the sample appropriately represented the roughness caused by the particles, but not when using very fine grained soils, like clay. They reasoned that the clay is very fine and that the 1 mm protrusion of the sample provided an incorrect representation of the grain roughness of the soil. This argument is obviously incorrect as the absolute roughness of individual soil particles over such a short distance (i.e., the diameter of the soil sample) makes only a small contribution the total magnitude of the turbulence, and thus the erosive capacity of the water, that develops over the sample.

The flow in the upstream portion of the conduit, i.e., upstream of the testing section, is most probably smooth turbulent flow. As the flow passes over the test section, which is assumed to be raised, the flow "trips" over the protrusion at the upstream end of the sample and by doing so introduces turbulence over the sample (Fig. 8.13). The presence of the "trip" is therefore a critical component determining the nature of the flow over the sample, the generation of turbulence and, consequently, the magnitude of the erosive capacity of the water acting on the sample. The roughness on the top surface of the sample can only contribute to additional turbulence (and therefore greater pressure fluctuations) if it is "rougher" than the absolute roughness introduced by the protrusion (see e.g., Fig. 8.14).

The erosive capacity of the water that develops over the sample is much more dependent on a 1 mm protrusion than it would be on the roughness contributed by fine particles in a clay sample. However, if the

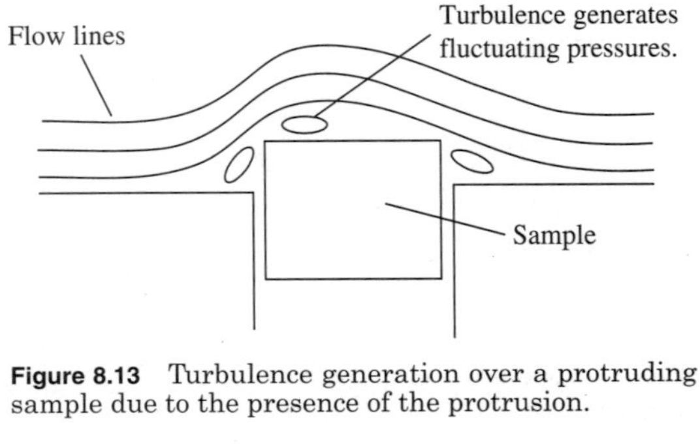

Figure 8.13 Turbulence generation over a protruding sample due to the presence of the protrusion.

Figure 8.14 An extremely rough sample surface in the EFA apparatus could have a greater impact on determining the magnitude of the erosive capacity of the water over the test section than the "average" protrusion of the sample.

roughness of the top surface of the sample increases significantly (Fig. 8.14) it will also contribute to turbulence generation.

During development of the equipment Briaud et al. (2001) assessed the role that a protrusion plays in generating erosive capacity. They used an aluminum tube instead of a soil sample and measured the pressure difference up- and downstream of the test section for conditions when the aluminum tube was flush with the bottom of the test section and for conditions when it protruded 1.2 mm. They found that the estimated average shear stress calculated using pressure measurements differed by approximately 50 percent at high velocities (Fig. 8.15). The estimated average shear stress over the test section was higher in the presence of the protrusion.

This finding by Briaud et al. (2001) is insightful and it is not immediately clear why the approach to measure pressure differences over the test section for calculating average shear stress was abandoned. Briaud (2004) indicates that the estimated shear stress over the test section using this approach resulted in critical (threshold) shear stress values that were higher than published data he could find.

Briaud et al. (2001) incorrectly abandoned the use of pressure measurements and instead adopted the Moody diagram to estimate the

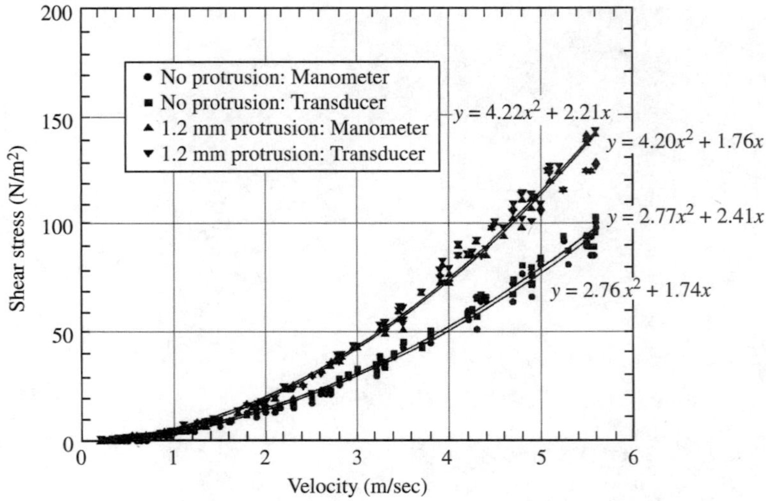

Figure 8.15 Difference in estimated shear stress due to a 1.2 mm protrusion in the EFA test section, determined by pressure measurements (Briaud et al., 2001).

hydraulic roughness, and subsequently the shear stress acting on the sample. They use the equation,

$$\tau = \frac{f}{8}\rho U_o^2$$

to estimate shear stress, with f the friction factor obtained from the Moody chart (Fig. 8.16).

Briaud et al. (2001) assumed that the roughness coefficient f obtained for pipe flow from the Moody diagram can be used to estimate the shear stress over the sample in the test section. They assumed that the absolute roughness ε for determining the dimensionless ratio ε/D, (where D is the hydraulic radius) is equal to $0.5D_{50}$ (where D_{50} is the median grain diameter of the soil in the sample). This recommendation is based on the argument that only half of the grains are exposed to the flow. They then use the appropriate curve represented by the calculated value of ε/D and the calculated value of the Reynolds number to estimate f (see e.g., Fig. 8.16; the discussion on hydraulic roughness in Chap. 5 can also be used for guidance).

The hydraulic radius of the pipe is calculated as

$$D = \frac{4A}{P} = \frac{2ab}{(a+b)} \tag{8.28}$$

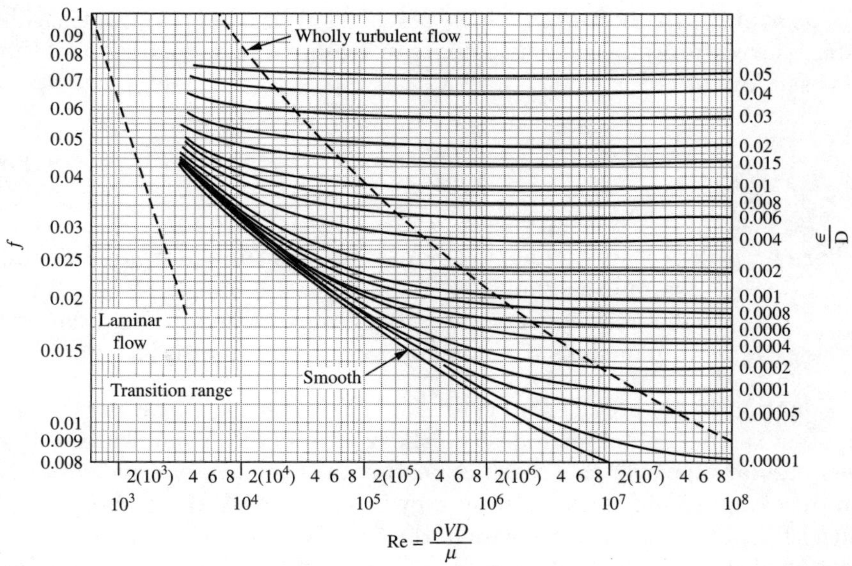

Figure 8.16 Moody chart for determining the friction factor f.

where A = cross-sectional area of the tube
P = wetted perimeter of the tube
a, b = the side dimensions of the rectangular tube

Discussion

The concept used by the EFA has merit but the equipment should be improved by minimizing operator error and by providing a better means to estimate the shear stress acting on the sample. The former can be accomplished by designing a sensor system that continuously monitors the elevation of the upper surface of the sample and automatically moves the sample upwards so that it remains, on average, 1 mm above the invert of the test section. This will eliminate operator error and ensure more consistency in the test procedure.

Use of the Moody diagram to calculate shear stress when using the EFA apparatus is not defensible. The Moody diagram is based on pipe flow, assuming a reasonably uniform absolute roughness throughout the whole of the pipe. Using the grain roughness over the test section as an estimate of the absolute roughness is incorrect and not representative of actual conditions over the test section.

The average shear stress over the test section should preferably be calculated from pressure measurements upstream and downstream of the test section, as

$$\tau = \frac{ab(p_1 - p_2)}{2l(a+b)} \tag{8.29}$$

where p_1, p_2 = pressure measurements up-and downstream of the sample section
l = distance between the two pressure measurement locations
a, b = dimensions of the rectangular conduit cross-section

Equation (8.29) provides an estimate of the average shear stress over the test section, which includes the shear stress on the side walls of the rectangular pipe over this reach. An estimate of the average shear stress on the sample itself can be improved by installing a third pressure transducer in the apparatus upstream of the test section. This will allow estimation of the average shear stress on the smooth wall of the rectangular pipe as well. By subtracting this shear stress (i.e., the shear stress calculated between transducers 1 and 2 in Fig. 8.17) from the average shear stress measured over the test section (i.e., between transducers 2 and 3 in Fig. 8.17) an improved estimate of the shear stress on the sample itself can be obtained.

Should one wish to express the erosion rate as a function of stream power, it can be calculated as

$$P = 7.853\ \rho(\tau/\rho)_{3/2} \tag{8.30}$$

Hole erosion test (HET)

Wan and Fell (2002) invented the hole erosion test (HET) that can be used to develop relationships between the erosive capacity of water and the rate of erosion. They elected to use shear stress to represent the erosive capacity of the water.

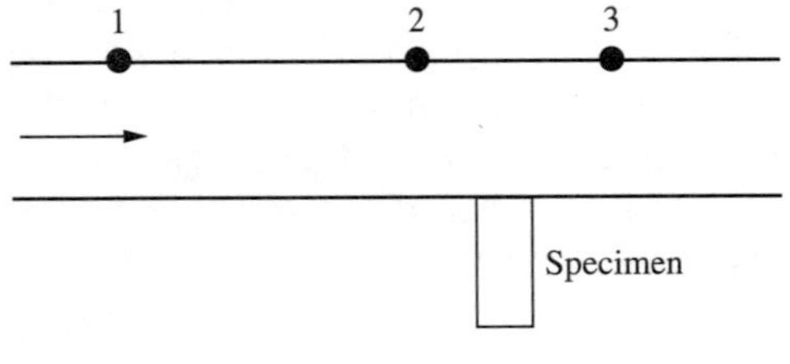

Figure 8.17 Locations of pressure transducers that could improve the estimate of average shear stress on sample in the EFA apparatus.

Wan and Fell's intent was to develop an index known as the erosion rate index (I_{HET}) than can be used to classify the degree of erodibility of earth materials, principally cohesive earth materials. From a modeling point of view this test could also be used to develop rate of erosion relationships for use in simulation of, say, dam failure or dam foundation erosion (see e.g., Annandale 2004).

The overall assembly of the equipment is shown in Fig. 8.18. The test section itself consist of three components, i.e., the upstream flow equalizing section, the middle section where the test specimen is placed, and the downstream section, through which the water flows prior to discharge (Fig. 8.19). The equipment is designed in a manner that allows the use of a standard Shelby tube. The test specimen is prepared by cutting the Shelby tube containing the sample to the required length of 115 mm. The standard procedure is then to drill a 6 mm hole into the specimen along its longitudinal axis (Fig. 8.20). However, the diameter of the hole can be changed. For example, if the material tested is highly erosion resistant the hole diameter can be enlarged to increase the magnitude of the erosive capacity of the water on the sample.

The remainder of the assembly consists of a head tank connected to the upstream end of the test section, two air release valves and two

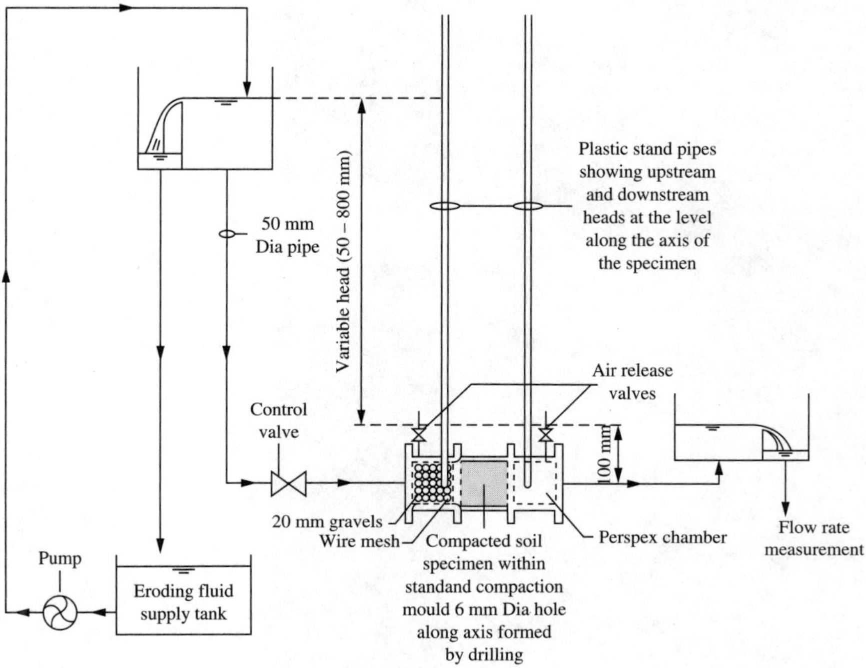

Figure 8.18 Hole erosion test equipment assembly (Wan and Fell 2002).

Figure 8.19 HET test section.

Figure 8.20 Drilling a hole along the axis of the soil specimen.

manometers connected respectively to the centers of the pipes up- and downstream of the test section. A tank to maintain a minimum head on the downstream end of the test section is also provided. The pressure head on the equipment is regulated by varying the elevation of the head tank on the upstream side. The up- and downstream manometers measure the pressure differential over the test section. Additionally, the discharge from the equipment is measured at the downstream end.

The essential theory used to analyze the test results entails the following:

The shear stress is expressed as

$$\tau_{\text{time}} \cdot P_t \cdot L = \Delta p_t \cdot A_t$$

where τ_{time} = shear stress in the sample at time t
P_t = wetted perimeter at time t
L = length of the test section
A_t = cross section area of flow in the test section (it is assumed that the flow through the specimen surrounding the drilled hole is negligible)

The pressure difference Δp_t at any point in time t is calculated as

$$\Delta p_t = \rho g \Delta h_t = \rho g s_t L$$

where Δh_t is the pressure head difference over the test section determined from the manometers at time t, and s_t is the hydraulic gradient over the section at time t.

The shear stress is expressed as

$$\tau_{\text{time}} = \rho g \frac{A_t}{P_t} \cdot \frac{\Delta h_t}{L} \tag{8.31}$$

The wetted perimeter is calculated as $\vartheta_t = \pi\phi_t$ and the cross-sectional area as $A_t = (\pi\phi_t^2)/4$, where ϕ_t is the diameter of the eroding hole at time t.

Equation (8.31) can therefore be rewritten as

$$\tau_{\text{time}} = \rho g \frac{\Delta h_t}{L} \cdot \frac{\phi_t}{4} \tag{8.32}$$

Assuming that the radius of the hole at any point in time is R_t, it can be shown that the incremental change in the cross-sectional area of the hole can be expressed as (Fig. 8.21)

$$dA = 2\pi R_t dR_t$$

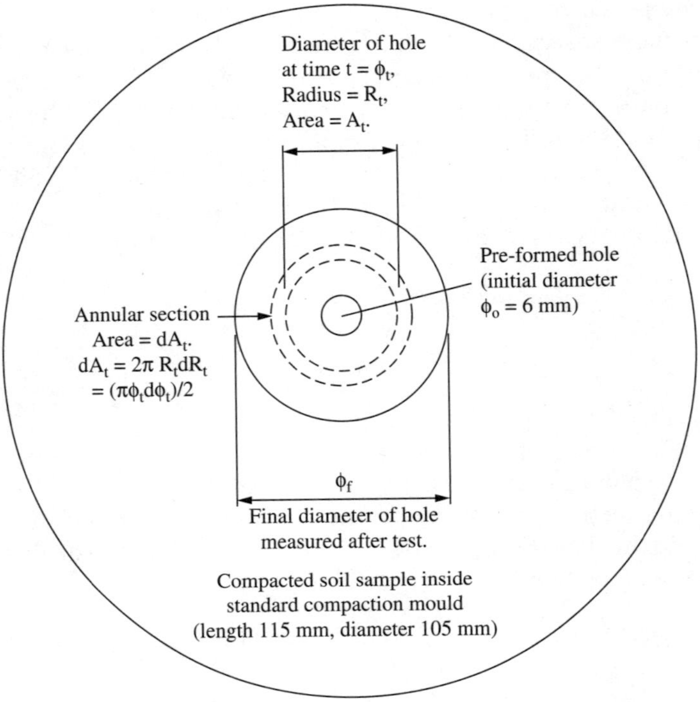

Figure 8.21 Cross section of sample in the HET test showing hole and its assumed variation with time (Wan and Fell 2002).

where dR_t is the incremental change in the radius of the hole as it erodes. This equation can also be written in terms of the diameter at a particular point in time, i.e.,

$$dA_t = \pi\phi_t \cdot \frac{d\phi_t}{2} \tag{8.33}$$

where $d\phi_t$ is the change in hole diameter at time t.

The rate of erosion of the material, in terms of mass eroded, can be expressed as

$$\dot{\varepsilon}_t = \rho_d \frac{d\phi_t}{2dt} = \frac{\rho_d}{2}\frac{d\phi_t}{dt} \tag{8.34}$$

where ρ_d denotes the dry density of the soil sample.

In order to facilitate analysis of the data Wan and Fell (2002) adopted the following procedure: for laminar flow at any particular point in time the shear stress is assumed to be directly proportional to the average flow velocity, i.e.,

$$\tau = f_L' U_o \tag{8.35}$$

where f'_L is a friction factor for laminar flow and U_o is the average flow velocity through the hole.

In the case of turbulent flow, which Wan and Fell (2002) assume occurs when the Reynolds number exceeds 5000, the relationship between shear stress and average flow velocity is assumed to be

$$\tau = f_T' U_o^2$$

where f'_T is a friction factor for turbulent flow.

By calculating the average flow velocity as

$$U_o = \frac{4Q}{\pi\phi^2}$$

They then define the two friction factors as

$$f_L' = \frac{\rho g \pi \phi^3}{16Q} \cdot \frac{\Delta h}{L} \tag{8.36}$$

and

$$f_T' = \frac{\rho g \pi^2 \phi^5}{64Q^2} \cdot \frac{\Delta h}{L} \tag{8.37}$$

Therefore, at any point in time t with the values of Δh_t, Q_t, f'_{Lt} and f'_{Tt} known it is possible to calculate the hole diameter ϕ_t at those same points in time with the following equations, depending on whether the flow is respectively laminar or turbulent:

$$\phi_t = \left(\frac{16 Q_t f_{Lt}' L}{\pi \rho g \Delta h_t} \right)^{1/2} \tag{8.38}$$

and

$$\phi_t = \left(\frac{64 Q_t^2 f_{Tt}' L}{\pi^2 \rho g \Delta h_t} \right)^{1/5} \tag{8.39}$$

Solving these two equations obviously requires estimates of the two friction factors defined by Wan and Fell (2002). This is done by making the assumption that they vary linearly as a function of the initial and final conditions. The initial value of the friction factor is calculated with a known diameter of the hole in the sample. At the end of the test the final hole diameter is measured once more and the final value of the friction factor calculated. This assumption obviously additionally implies that the cross section of the hole is uniform right throughout the sample. An example of a sample hole at conclusion of a test is shown in Fig. 8.22. This photograph shows that the assumption can, at times, be questionable. However, the engineer responsible for the testing is required to determine whether the assumptions are true for the samples tested and make the required adjustments, as necessary.

The analysis procedure, using data published by Wan and Fell (2002) is as follows:

- Calculate the Reynolds number for the test and decide which friction factor(s) should be estimated.
- Estimate the initial and final friction factors using the initial hole diameter and the final diameter and Eqs. (8.36) or (8.37) depending on whether the flow is laminar or turbulent.
- Estimate the hole diameter at different points in time ϕ_t with the known values of Q_t and Δh_t, and the estimated values of f'_{Lt} or f'_{Tt} as a function of time t using Eqs. (8.38) or (8.39) depending on the governing flow conditions.
- Plot a curve relating the estimated diameters as a function of time. This curve can be used to estimate the value of $d\phi_t/dt$ at any point

Figure 8.22 Example of the condition of a hole in a specimen at conclusion of a HET test.

in time by determining the slope of the curve at different times (Fig. 8.23).

- Use Eq. (8.39) and the values of $d\phi_t/dt$ calculated in the previous step to determine how erosion rate $\dot{\varepsilon}_t$ varies as a function of time.
- Plot the erosion rate and the shear stress as a function of time (Fig. 8.24). This graph shows the progression of shear stress and erosion rate as a function of time, i.e., in this case shear stress increases continuously but the erosion rate first decreases and then increases.
- Plot the erosion rate as a function of shear stress (Fig. 8.25). This graph can be used to develop relationships between erosion rate and shear stress and to determine the threshold shear stress.

Wan and Fell (2002) are of the opinion that the initial downward trend in the tests are associated with loose material that is removed from the hole during the initial stages of the test. This convex shape appears to be a common feature of these tests. Wan and Fell (2002) propose to determine the magnitude of the threshold shear stress by linearly extrapolating the portion of the curve with the positive slope backwards to intersect the shear stress axis where the erosion rate is zero (Fig. 8.25).

In order to determine the relationship between erosion rate and stream power the analysis procedure is exactly the same, except that

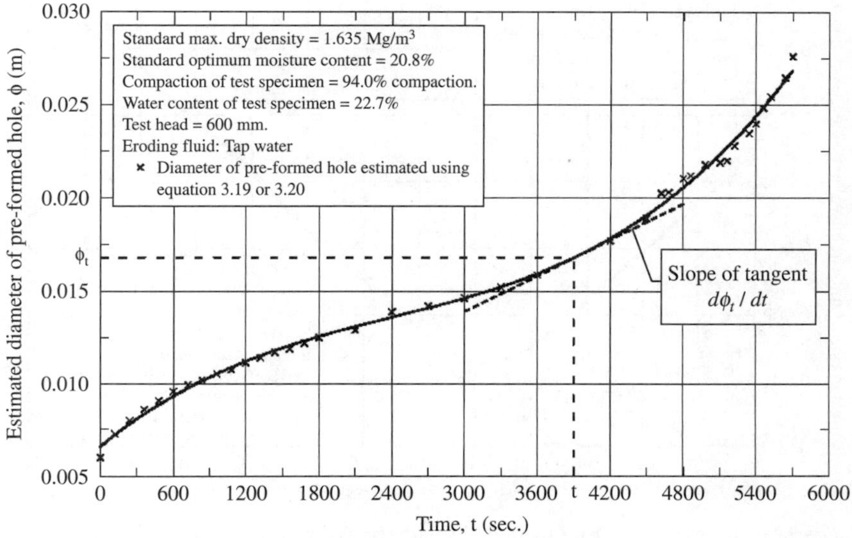

Figure 8.23 Estimated change in hole diameter in HET as a function of time (Wan and Fell 2002).

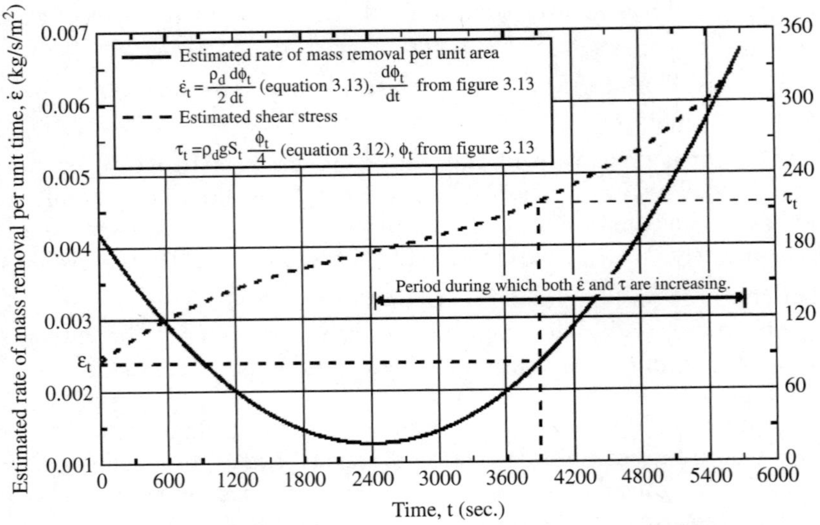

Figure 8.24 Erosion rate and shear stress plotted as a function of time (Wan and Fell 2002).

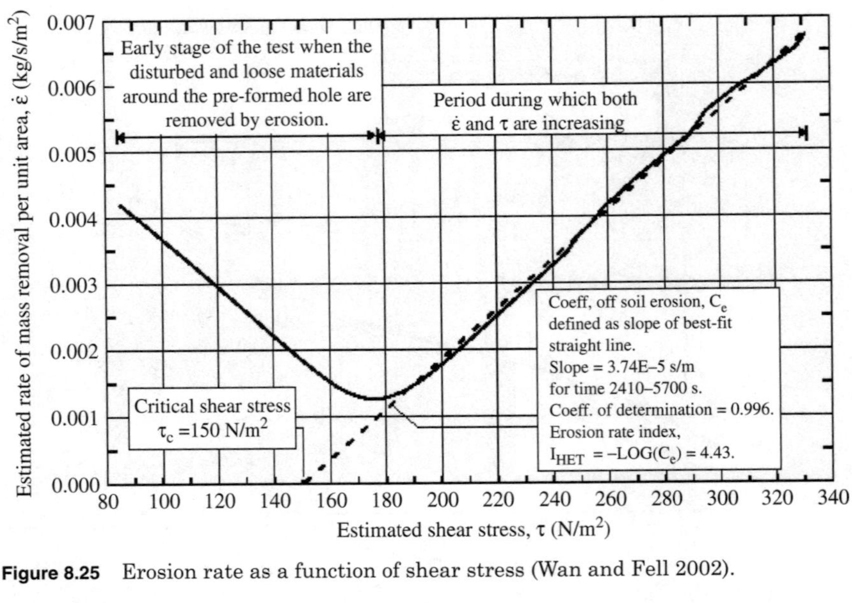

Figure 8.25 Erosion rate as a function of shear stress (Wan and Fell 2002).

the erosion rate is plotted against stream power instead of shear stress. The stream power for the HET is calculated as

$$P_t = \frac{\rho g Q_t \Delta h_t}{\pi \phi_t L} \tag{8.40}$$

where P_t is the stream power per unit area at time t.

Discussion

The limited experience with this apparatus indicates that it provides fairly consistent erosion rate and threshold results. A weak point of the method is that the measurement of the hole diameter at the end of the test could be difficult. If the hole is irregular, assignment of an "average" diameter might be challenging. As the estimate of the progression of erosion is fully dependent on the measurement and assignment of an average hole diameter at the end of the test it could adversely affect the accuracy of the estimates of the erosion rate.

It may also make sense to reinvestigate the reasons for the convex shape of the relationship between the erosion rate and shear stress. Wan and Fell's (2002) explanation that it is due to loose material being discharged from the sample may well be correct. However, using the erosion rate equation developed in Chap. 4 it is possible to demonstrate that convex relationships between erosion rate and shear stress can exist depending on the combination of activation energy, number of bonds, and temperature (see Figs. 8.26 and 8.27). Experimental results resembling this shape were also found by Croad (1981).

Summary

Methods for determining the rate of erosion of chemical gels are presented in this chapter. Such erosion occurs when a chemical gel is subject to subcritical failure. In the case of some brittle materials like rock or concrete the process of subcritical failure results in the material apparently resisting the erosive capacity of water for a lengthy period of time and then suddenly failing. The rate of subcritical failure in the case of cohesive soils like clay occurs at a higher rate and is characterized by a relationship between the rate of erosion of the clay and the erosive capacity of the water.

Determination of material characteristics for quantifying subcritical failure conditions for materials like rock are currently principally determined from published research. Representative parameters for rock are summarized in Chap. 4. Once the subcritical failure parameters have been estimated for the brittle material under consideration they can be used in a calculation procedure presented in this chapter to calculate the time to failure for subcritical conditions.

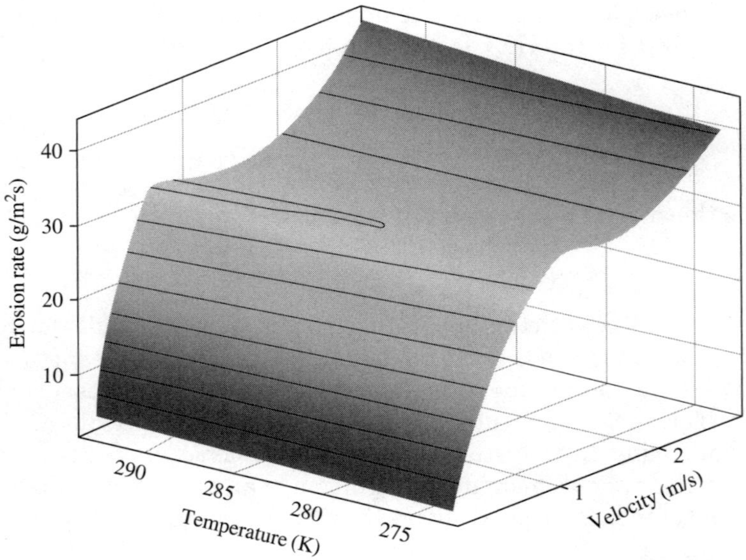

Figure 8.26 Theoretical relationship between erosion rate, temperature, and average flow velocity for E_a = 21 KJ/mol; δ = 20 mm; β = 60 K/Pa; γ = 0.05×10^8 g/m^2 using the erosion rate equation for clay developed in Chap. 4.

Calculation of the rate of erosion of clay under prototype project conditions require establishment of relationships between the erosion rate of the clay and the erosive capacity of the water. Once established, the relationship is used with calculated values of the erosive capacity of water for anticipated project conditions to predict erosion rate and extent.

The relationship between rate of erosion and erosive capacity of water is site-specific. Methods for determining such relationships have been

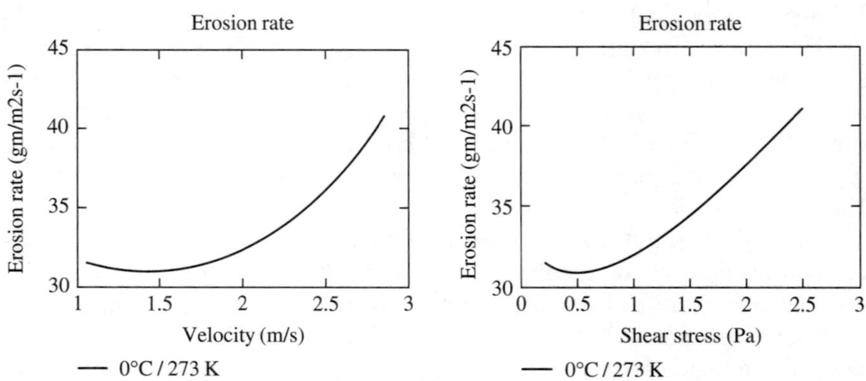

Figure 8.27 Theoretical erosion rate for clay as a fucntion of velocity and shear stress for a temperature of 273 K (taken from Fig. 8.26).

presented. These include the CFD, VJT, HET and EFA test procedures. Of the four test procedures the shear stress measured with the CFD device is the most accurate because it is measured directly. The shear stress in the other three devices is measured indirectly. The current approach for measuring shear stress in the EFA device (at the time of writing) is considered the least accurate of the four devices.

The rate of erosion for the CFD device is measured by weighing and is accurate. In the VJT device it is measured directly with a point gauge, while it is estimated in the HET by taking average measurements of the hole through the sample at the beginning and end of the test. The measurement of rate of erosion in the EFA device is very much dependent on operator skill.

Quantification of the shear stress or stream power when applying the rate of erosion relationships obtained from these devices in practice requires careful consideration. For example, the shear stress measured in the CFD device is the actual shear stress acting on the boundary of the sample. As the characteristics of the Couette flow in the device have been studied in significant detail over many years (see e.g., Schlichting and Gersten 2000) it is reasonable to assume that the measured shear stress is equal to the turbulent shear at the boundary (see Chap. 5).

In the case of the EFA and the HET devices the drag on the sample (wall shear stress) is calculated, which is not equal to the turbulent boundary shear stress. The relationship between the shear due to the drag and the turbulent boundary shear stress has been shown to be (see Chap. 5)

$$\tau_t = 7.853\tau \frac{u_*}{\overline{u}}$$

where τ_t is the turbulent shear at the boundary and τ is the shear due to drag calculated in the EFA and the HET devices.

The shear stress in the VJT device is calculated by assuming

$$\tau_i = \tau_o \left(\frac{J_p}{J_i} \right)^2 \qquad \text{for } J_i \geq J_p$$

where $\tau_o = C'_f \rho U_o^2$.

Hanson and Cook (2004) assume $C'_f = 0.00416$. It has been argued that this value may be too low. Based on the boundary layer theory it appears as if this value should be $C'_f = 0.016$. However, this estimate needs confirmation by testing. The current opinion is that a reasonable estimate of the magnitude of the turbulent boundary shear stress is obtained when using the latter coefficient value.

Should estimates of the applied stream power at the boundary be required for the EFA and the HET devices it can be estimated with the following equation (see Chap. 5):

$$\tau_t \overline{u} = 7.853\rho\left(\frac{\tau}{\rho}\right)^{3/2} = 7.853\frac{\tau^{3/2}}{\sqrt{\rho}}$$

Assuming that the shear stress measured in the CFD device equals the turbulent boundary shear stress, the stream power in the CFD can simply be calculated as

$$P = \tau_t \overline{u}$$

where τ_t is the turbulent boundary shear stress measured directly with the CFD device and $\overline{u}$ is the flow velocity in the CFD device.

A similar calculation, as shown before, can be performed for the VJT device, assuming that using the value of $C_f' = 0.016$ provides an estimate of the turbulent boundary shear stress. However, it needs to be pointed out that the linear relationship between shear stress and erosion rate assumed by Hanson and Cook (2004) is not necessarily true when relating stream power and erosion rate.

Chapter

9

Engineering Management of Scour

Introduction

This chapter presents optional approaches for managing the effects of scour. Design standards do not form part of this discourse, as these can be found in numerous manuals, design guidelines, and regulations. Rather, the information presented in what follows provides a conceptual basis for approaching design and for developing and engineering solutions to scour problems.

The unique conditions associated with each scour problem preclude the use of rule-based engineering design. Rather, solutions to scour problems should be developed by making use of the knowledge and understanding gained in the other chapters of this book, and by making use of the decision-making process outlined in Chap. 2.

In what follows the approach to developing, engineering, implementing, and maintaining solutions to scour problems is first presented. This is followed by a discussion of optional solutions to protect the public, property, and infrastructure against the effects of scour. Each of these options is then discussed in more detail, illustrating their application by example.

Approach

The approach towards protecting the public, infrastructure, and property against the effects of scour consists of six elements; i.e., scour analysis; protection analysis; costing and alternative selection; engineering and preparation of drawings and specifications; construction; and maintenance.

Scour analysis

The first step in the design process, i.e., scour analysis, entails

- determining the magnitude of the design flood,
- establishing the geometric properties of the element under investigation (i.e., a river reach, plunge pool, and the like),
- conducting a hydraulic analysis and estimating the relative magnitude of the erosive capacity of the water,
- characterizing the earth material,
- determining whether the earth material has the ability to resist the erosive capacity of the water, and
- estimate the extent of scour.

If it is found that the earth material can resist the erosive capacity of the water, no additional protection against scour is necessary. However, if the erosive capacity of the water exceeds the ability of the earth material to resist it, protection is required.

If it is determined that scour is possible, the next step is to estimate scour extent. Large scour extents could require protection (depending on the design approach and selection of protection option—see further on), while minimal extents may require little or no protection.

Protection analysis

Protection analysis entails identifying technically feasible approaches for managing scour. This is done by analyzing the scour problem in more detail, identifying optional approaches to limit or control scour, analyzing the potentially viable approaches, and identifying those that can be implemented in practice. Issues considered during protection analysis selection include space limitations, material availability, the ability of the various protection options to resist the erosive capacity of water, and sizing of the protection measures. Adequate data are often not available to accurately analyze the behavior of the protection measures. Decisions regarding technical feasibility are based on subjective and objective decision-making processes discussed in Chap. 2.

Costing and selection

Once the technically feasible protection measures have been identified, the next step is to identify the most economical means of providing such protection. This decision is made by determining the material, construction, and maintenance cost of the optional protection measures that are considered technically feasible. In this regard, it is necessary to calculate life cycle costs of each protection measure in order to gain full understanding of economic implications. The option of providing no protection should also be assessed, and the cost of losing infrastructure or property, or exposing human life to danger due to scour if no or sub-optimal protection is provided, should be considered.

Engineering and preparation of drawings and specifications

The next step in the design process, after selecting the most economical and technically feasible design solution, is to engineer the selected protection measure, and to prepare design drawings, details, and specifications. The drawings should be prepared in a manner clearly communicating the design intent, while the specifications should clearly state the required material properties, placement and maintenance requirements.

Construction

The design engineer should observe the construction and gain assurance that the protection measures are constructed in a manner that will protect the public, infrastructure and property as intended. This is done by ensuring adherence to the construction drawings and specifications, and by making on-site modifications that will satisfy the design intent. The latter is the responsibility of the design engineer and should be conceived and supervised by him or her.

Maintenance

Scour protection, once constructed, requires maintenance on a regular basis to ensure its effectiveness and longevity. The design engineer should clearly communicate maintenance requirements to the owner.

Scour Protection Options

The essential goal of the scour protection options presented herewith entails safeguarding the public and protecting infrastructure and property against unplanned scour events. A scour event is considered unplanned if flow conditions that can reasonably be anticipated adversely impact surrounding property and infrastructure due to the effects of scour, and unnecessarily imperil public safety. Unplanned scour can be prevented by manipulating the three variables affecting scour, i.e., the erosive capacity of the flowing water, the ability of earth or engineered earth materials to resist the erosive capacity of the water, and the geometry of the flow boundary.

This implies that scour can be minimized by reducing the erosive capacity of water, strengthening or protecting the earth material under attack, or by changing the geometric properties of the flow boundary in a way that will reduce scour potential. Six remedies against unplanned scour are flow boundary pre-forming, accommodating protection, earth material enhancement, hard protection, flow modification, and combining these approaches.

Flow boundary pre-forming is aimed at anticipating the effects of scour and, by design, providing the appropriate boundary geometry that will prevent the occurrence of unplanned scour. This approach entails predicting the extent and shape of the scour hole that are compatible with the boundary material and anticipated erosive capacity of the water, and, in advance, constructing the predicted shape of the scour hole. By accommodating the design flood in the preformed scour hole very little to no additional scour occurs. This approach is based on the premise that the erosive capacity of the water at the preformed flow boundary reduces to values that are below the threshold conditions of the boundary earth material once the predicted, "stable" scour geometry has been reached.

Accommodating protection entails allowing scour to occur unimpeded and arranging adjacent infrastructure, infrastructure components, and surrounding property in a manner that will prevent failure and imperilment of public safety when severe flow occurs. This design approach requires prediction of the extent and shape of the scour hole that will form. Once known, infrastructure and property are located in a manner that will not expose them to the effects of scour. A common feature of this approach and the pre-forming approach is that both require prediction of the shape and extent of anticipated scour. In the case of pre-forming the anticipated shape and extent of the scour hole is artificially created prior to the occurrence of the design flood. In the case of accommodating design, no pre-forming occurs but infrastructure and property are placed far enough away from the anticipated boundaries of a scour hole to protect them against the effects of scour.

Earth material enhancement entails improving the ability of naturally occurring earth material to resist the erosive capacity of the water. For example, the ability of naturally occurring soils on a channel boundary to resist the erosive capacity of water can be enhanced by making use of vegetation. The vegetation, once established, increases the erosion resistance offered by the naturally occurring soil and prevents erosion. It is the task of the engineer to develop engineering specifications of the required vegetation. The restoration ecologist uses these specifications to select plant material. Anchoring of rock formations is another example of how the erosion resistance of naturally occurring earth material can be improved. For example, the ability of rock in a dam abutment to resist the erosive capacity of a plunging jet can be enhanced by installing post-tensioned rock anchors.

Hard protection aims at resisting the effects of the erosive capacity of water by hardening the flow boundary. When analysis indicates that naturally occurring earth material will scour when subjected to anticipated flow conditions and that earth material enhancement will not adequately increase its capacity to resist scour, hardening can be used as a protection option. By selecting hardening to protect against the effects of scour, naturally occurring earth is covered with a lining. Hardening includes using concrete, riprap, or other custom or commercial lining systems to protect the underlying earth against scour.

Flow Modification entails changing the flow characteristics in an area of concern in a manner that will reduce its erosive capacity. An example of flow modification entails inclusion of splitters at the end of spillways or spillway chutes to break up plunging jets. A broken jet has less erosive capacity than a coherent jet with the same discharge. Another example, in open channel flow, is to streamline the flow boundary in a manner that will significantly reduce the turbulence intensity of the flowing water. A reduction in turbulence intensity leads to a reduction in the erosive capacity of the water.

Combining the optional protection measures can result in design optimization. For example, pre-forming and hard design approaches can be combined by pre-forming a scour hole to reduce the erosive capacity of the water enough to be resisted by an available hard protection approach. This approach can be used if the erosive capacity of the water prior to pre-forming of a scour hole can lead to the destruction of the available hard protection. By changing the geometry of the scour-hole, it is possible to reduce the erosive capacity of the water at the boundary enough to be resisted by the available hard protection option. This approach can lead to less excavation, which, when combined with the selected lining, results in a minimum cost solution. Another example includes making use of a hydraulic jump placed on, say, a concrete apron. The hydraulic jump provides a means of reducing the erosive capacity of the water through flow modification, while the apron offers hard protection over the section in the channel where the flow modification occurs. The hydraulic jump modifies the flow by changing it from supercritical to sub-critical flow. The erosive capacity of the water downstream of the hydraulic jump is lower than on its upstream side.

Earth material enhancement can, in a similar manner, be combined with pre-forming. For example, when preparing river restoration designs creation of meandering pathways with accompanying pools, glides, runs, and riffles can reduce the erosive capacity of the water to such an extent that live vegetation, instead of hard protection measures, can be used to protect river banks against erosion. A meandering flow path increases the length of the stream and reduces its effective slope; resulting in a reduction of the erosive capacity of the flowing water. By preparing a meandering design that adequately reduces the erosive capacity of the water, sole use of vegetation as a means of protection against erosion can be accomplished. The reader can conceive other examples of combining scour protection options.

Scour Analysis

In order to assess whether special action is required to protect infrastructure against the effects of scour it is necessary to conduct a scour assessment. The objective of scour assessment is to determine whether scour is a concern. Scour assessment entails estimating the relative magnitude of the erosive capacity of the water and the relative ability of the earth material under consideration to resist it. If the erosive capacity of the water exceeds the ability of the earth material to resist scour, protection against the effects of scour is required. In cases where the resistance of the earth material exceeds the erosive capacity of the water, no scour will occur. This means that additional protection against scour is not required.

Project examples presented in what follows illustrate scour assessments executed at Moochalabra and Harding Dams in Western Australia, and Gibson Dam in Montana (United States of America). Essential project information illustrates the applicability and validity of the scour assessment technique.

Moochalabra and Harding Dams both have unlined spillways cut in natural rock. These two dams experienced floods in 2001 and 2004, respectively, with no resulting scour of any significance at either. Gibson Dam experienced an overtopping flood in 1964, which caused minor scour damage on the one abutment and very little to no scour on the other.

Moochalabra Dam[1]

The spillway at Moochalabra Dam in Western Australia consists of an unlined rock cut, with a stepped outlet on its downstream end. Figure 9.1 is a close-up view of the water discharging over the end of the spillway chute, cascading down the sand stone beds that form steps up to 2 m high. Figure 9.2 is another view of the spillway, from downstream. The approach channel is 100 m wide. The rock consists of quartz-rich sandstone with thin interbeds of siltstone and shale, horizontally bedded. Where the beds of sandstone have subsequently been exposed, they have been found to be relatively extensive with a wide joint spacing.

---

[1]Information provided by Bob Wark, Technical Director: Dams, GHD Pty (Ltd), Perth, Australia.

Figure 9.1 Moochalabra Dam spillway in action during the 2001 flood. Peak discharge was 270 m^3/s. (Photo: Bob Wark)

The adopted approach for the design and construction was to excavate a bench cut at full supply level to provide the rock fill for the dam and to remove most of the weathered materials from the downstream face of the chute. Subsequent flows have washed the remaining loose and

Figure 9.2 Downstream view of Moochalabra Dam spillway towards end of 2001 flood. (Photo: Bob Wark)

weathered materials from the face. The 2 m high steps represent the maximum thickness of the more massive beds of sandstone, although in many cases multiple beds, exposed on the face, make up the individual steps. The spillway experienced no significant scour in the approach channel during the 2001 flood, nor did the face of the outlet.

The rock is classified as hard to very hard, with an unconfined compressive strength (UCS) of about 35 MPa. The *RQD* is high (90 percent) and the rock matrix is defined by a system of three plus random joint sets. The rock joints are smooth planar, slightly altered, and contains non-softening, non-cohesive rock filling. Some interbeds of siltstone and shale exist. The rock blocks are equi-sided and the dip of the beds is almost horizontal, dipping upstream at about 10° to 20°. The near vertical joints defining the blocks are widely spaced.

The estimated peak discharge during the flood event was 270 m^3/s and the estimated flow depth about 0.98 m. The flow velocity in the approach channel was set at about 2.7 m/s.

The estimated threshold stream power of the rock at Moochalabra Dam spillway is shown in Table 9.1. The values of the variables in the table are determined from the information provided in Chap. 4 and the scour threshold is determined using Annandale's (1995) threshold relationship presented in Chap. 6. The relationship between the erodibility index and threshold stream power, following the latter approach, is

$$P_{\text{thresh}} = EI^{0.75}$$

The stream power of the flowing water in both the approach channel and at the steps was determined. The calculation of the stream power in the approach channel is shown in Table 9.2. The input stream power is calculated with the equation,

$$P_{\text{input}} = \rho g q s$$

The energy slope s is calculated by accounting for the friction in the channel. This is done by making use of the absolute roughness k by

TABLE 9.1 Determination of Threshold Stream Power of Rock at Moochalabra Dam Spillway

Rock properties		
Mass strength of rock	$M_s =$	35
Rock quality designation	$RQD =$	90
Joint set number	$J_n =$	3.34
Joint roughness number	$J_r =$	1
Joint alteration number	$J_a =$	2
Joint structure (orientation) number	$J_s =$	0.56
Erodibility index	$K =$	264
Threshold stream power	$P_{\text{thresh}} =$	66 kW/m^2

TABLE 9.2 Calculation of Input and Applied Stream Power at Moochalabra Dam

Input stream power		
Mass density of water	$\rho =$	1000 kg/m^3
Acceleration due to gravity	$g =$	9.81 m/s^2
Discharge/m	$q =$	2.7 m^2/s
Depth of flow	$d =$	0.98 m
Velocity of flow	$v =$	2.7 m/s
Channel absolute roughness	$k =$	0.3 m
Chezy coefficient	$C =$	28.3
Energy slope	$s =$	0.0096
Input stream power	$P_{\text{input}} =$	0.254 kW/m^2
Applied stream power		
Wall shear stress	$\tau =$	92 Pa
Applied stream power	$P_a =$	0.221 kW/m^2

first calculating the value of the Chezy roughness coefficient, i.e.,

$$C = \sqrt{8g}\, 2 \log_{10}\left(\frac{12y}{k}\right)$$

(It is assumed that the hydraulic depth is equal to the actual flow depth due to the width of the channel.)

Once C is known, the value of s is calculated using the Chezy velocity equation, i.e.,

$$s = \frac{v^2}{C^2 y}$$

Following this approach, it is estimated that the input stream power during this flood event was about 0.254 kW/m^2. As indicated previously, the input stream power is not equal to the applied stream power at the bed. As the latter is of principal importance to assess scour potential, it is necessary to estimate the applied stream power on the bed. This is done by making use of the equation,

$$P_a = 7.853\rho\left(\frac{\tau_w}{\rho}\right)^{3/2}$$

and the wall shear stress τ_w is calculated as

$$\tau_w = \rho g y s$$

TABLE 9.3 Calculation of Stream Power of Jet Impinging on 2 m Steps

Unit stream power		
Mass density of water	$\rho =$	1000 kg/m^3
Acceleration due to gravity	$g =$	9.81 m/s^2
Discharge	$Q =$	270 m^3/s
Step height	$H =$	2 m
Total power	$P_t =$	5297 kW
Overflow width	$b =$	100 m
Overflow depth	$I =$	0.98 m
Footprint area	$A =$	98 m^2
Unit stream power	$P =$	54 kW/m^2

The wall shear stress is estimated as 92 Pa, and the applied stream power as 0.221 kW/m^2. Comparison of the input and applied stream power indicates that the applied stream power is very close to the value of the input stream power. This is mainly due to the relatively low flow depth and high roughness of the bed.

With the water cascading down the outlet over 2 m high steps, the stream power for jets impinging onto each of the steps was estimated. The stream power of the impinging jet was calculated assuming a drop height of 2 m and a jet thickness equaling the depth of flow in the approach channel. These are reasonable assumptions as the drop height is not significant. The calculation of stream power for the cascading jets is shown in Table 9.3.

Comparing the erosive capacity of the water, for both the approach channel flow and the impinging jets, to the threshold stream power of the rock indicates whether scour can be expected or not. The estimated threshold stream power of the rock is 66 kW/m^2, while the approach channel stream power (applied) is only 0.221 kW/m^2 and the stream power of the impinging jet is 54 kW/m^2. The rock has the ability to resist the erosive capacity of the water for this flood in both the approach channel and the cascading outlet. Observations during and after the 2001 flood indicate that this assessment of the ability of the rock to resist scour is reasonable. The rock in the approach channel and outlet remained intact during the flood event, except for removal of some loose rock on the steps.

Harding Dam[2]

The spillway at Harding Dam, also located in Western Australia, consists of a 70 m wide approach channel, cut in the rock and left unlined,

[2]Information provided by Bob Wark, Technical Director: Dams, GHD Pty (Ltd), Perth, Australia.

and an 18 m vertical drop at its outlet. Figure 9.3 shows water discharging through the Harding spillway in 2004 at a flow rate of 952 m^3/s. Figure 9.4 provides a close-up view of the spillway approach channel and drop that is located in the left upper corner of Fig. 9.3.

The rock in the unlined spillway is a very good quality dolerite sill structure. It is extremely hard, with an UCS of about 280 MPa and an *RQD* of 100 percent. Site assessment characterized the rock as having two joint sets, which are tightly healed, rough, irregular, and undulating. The rocks are equi-sided and the joints are vertically dipping. Determination of the threshold stream power of the rock, using Annandale's erodibility index method, is shown in Table 9.4. This table indicates that the threshold stream power of the rock is estimated at 4293 kW/m^2.

In order to assess whether the rock will scour under flow conditions that occurred in 2004, it is necessary to estimate the magnitude of the stream power in the approach channel and at the drop. The stream power in the approach channel is determined using the same methodology explained in the case of Moochalabra Dam (see Table 9.5). The relationship between the absolute roughness and Manning's n has been determined using the equation $n = 0.038 \cdot k^{1/6}$ (see Henderson 1971, p. 98).

An interesting aspect of the stream power in the approach channel at Harding Dam is that the applied stream power is much lower than the

Figure 9.3 Harding Dam, Western Australia experiencing a flood in 2004. The rock-cut spillway is located towards the top left hand corner of the photograph and is illustrated by the close-up photograph in Fig. 9.4. Peak discharge was 952 m^3/s. (Photo: Bob Wark)

Figure 9.4 Rock cut spillway at Harding Dam, Western Australia in action 2004. (Photo: Bob Wark)

TABLE 9.4 Determination of Threshold Stream Power of Rock at Harding Dam

Rock properties		
Mass strength of rock	$M_s =$	280
Rock quality designation	$RQD =$	100
Joint set number	$J_n =$	1.83
Joint roughness number	$J_r =$	3
Joint alteration number	$J_a =$	0.75
Joint structure (orientation)	$J_s =$	1.14
Erodibility index	$K =$	69770
Threshold stream power	$P_{\text{thresh}} =$	4293 kW/m^2

TABLE 9.5 Input and Applied Stream Power in the Approach Channel at Harding Dam During the 2004 Flood

Input stream power		
Density of water	$\rho =$	1000 kg/m^3
Gravitational acceleration	$g =$	9.81 m/s^2
Discharge/m	$q =$	13.6 m^2/s
Depth of flow	$d =$	2.78 m
Velocity of flow	$v =$	4.9 m/s
Canal roughness	$k =$	0.178 m
Manning's n	n	0.0285
Chezy coefficient	$C =$	40
Energy slope	$s =$	0.005325
Input stream power	$P =$	0.710 kW/m^2
Applied stream power		
Wall shear stress	$\tau =$	145 Pa
Applied stream power	$P_a =$	0.434 kW/m^2

TABLE 9.6 Calculation of Jet Stream Power at Harding Dam

Unit stream power		
Mass density of water	$\rho =$	1000 kg/m^3
Acceleration due to gravity	$g =$	9.81 m/s^2
Discharge	$Q =$	952 m^3/s
Step height	$H =$	18 m
Total power	$P_t =$	168104 kW
Overflow width	$b =$	70 m
Impact jet thickness	$I =$	1.26 m
Footprint area	$A =$	88 m^2
Unit stream power	$P =$	1902 kW/m^2

input stream power. The input stream power in the channel is on the order of 0.710 kW/m^2, while the applied stream power at the bed is only 0.434 kW/m^2. The principal reason for this difference is the much lower roughness and deeper channel flow depth at Harding Dam, as compared to Moochalabra Dam.

The stream power of the plunging jet has been estimated by making the same assumptions as in the case of Moochalabra Dam (see Table 9.6). The characteristics of the plunge pool downstream of the drop at Harding Dam during the flood are not fully known and dissipation of the energy of the jet has been ignored. This can be considered a conservative approach, and it is therefore reasonable to assume that the stream power of the impinging jet, at 1902 kW/m^2, is most probably overestimated.

Comparison between the threshold stream power of the rock, estimated at 4293 kW/m^2, and the applied stream power in the approach channel (0.434 kW/m^2), and the stream power of the impinging jet (1902 kW/m^2) indicates that the rock is unlikely to scour under these conditions. The observations made during and after the 2004 flood confirm this conclusion.

Pre-Forming

Conforming to the basic premise of this book, i.e., that scour in turbulent flow is caused by pressure fluctuations, the basis of pre-forming entails creating flow boundaries that will reduce the turbulence intensity of the flowing water. Reduction of turbulence intensity leads to a reduction in the magnitude and frequency of pressure fluctuations, which in turn results in less scour.

The extent and final shape of a preformed flow boundary depend on the magnitude of the erosive capacity of the water relative to the ability of the boundary material to resist scour. The extent of a scour hole formed by a discharge of specified magnitude in, say, non-cohesive gravel will be larger than the extent of a scour hole formed by the same discharge in competent, fractured rock. The extent of a preformed hole in non-cohesive gravel will therefore be greater than the extent of a scour hole in competent, fractured rock.

In what follows a few examples of applying the pre-forming approach to reduce scour potential are provided. These examples are not exhaustive and the reader is free to conceive of other applications where pre-forming can be used to manage scour.

River restoration

An example of pre-forming to reduce the effects of scour can be found in some river restoration approaches. The geometry of a river can be manipulated by amending its cross-sectional shape and its longitudinal slope. Change of the longitudinal slope can be accomplished by increasing the sinuosity of the stream. Sinuosity is the ratio between the length of the flow path in a river and the valley length. The valley length is the straight distance between two locations along a stream. The distance of the valley length is shorter than the distance along the flow path (Fig. 9.5). Therefore, the slope along the flow path will be milder than the slope along the valley. This means that the stream power, representing the relative magnitude of pressure fluctuations on the flow boundary, will be lower for water flowing along the meandering flow path than for water flowing in a channel that directly connects A and B.

Changing the cross-sectional shape of a river for purposes of stabilization entails changing two components, i.e., the relation between the active flow channel and surrounding floodplain (Fig. 9.6), and pre-forming the concave and convex sides of river bends to minimize the potential occurrence of unplanned scour. The active channel is the portion of the river cross section accommodating the majority of the flowing water over time. As such, its size generally accommodates flows with recurrence intervals of approximately 1.5 to two years.

In the case of naturally stable rivers the active channel is accompanied by floodplains located on one or both sides. When floods in excess of the 2-year recurrence interval occurs the water flows over the banks of the active channel onto the adjacent flood plain. If the floodplains are large,

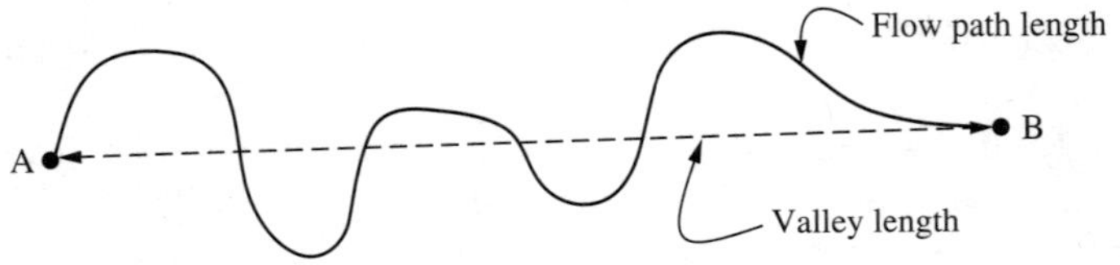

Figure 9.5 Valley length and flow path length of a river.

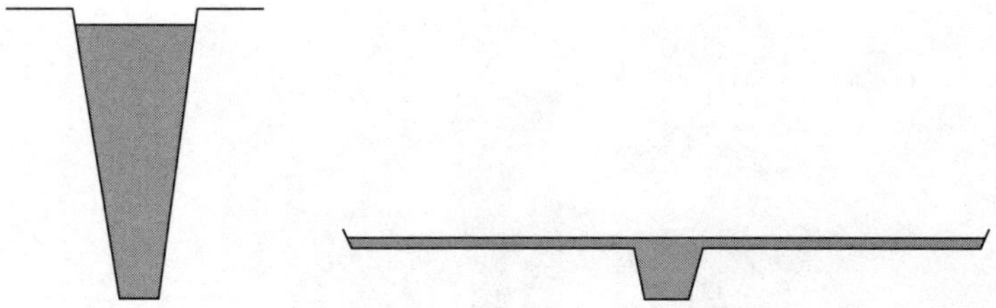

Figure 9.6 Unstable, incised channel (left) and stable, active channel (right) with accompanying floodplain.

the increase in flow depth near the active channel during floods is relatively minor, resulting in only minor increases in the erosive capacity of the water in the active channel. This minor increase in erosive capacity often does not result in additional scour and retains the stability of the active channel (Fig. 9.6).

An example where a pre-forming approach restored a creek can be found in the West Branch of Alamo Creek, Contra Costa County, California. In this particular case the cross section was changed from an incised stream to an active channel with an adjacent floodplain (Fig. 9.7).

Additionally, the cross-sectional shape of a river channel around bends requires pre-forming as well. The flow depth on the concave side of a river bend is usually deeper than on the convex side. The reason for this is that the erosive capacity on the concave side of a bend is greater than that on the convex side. An indication of how to determine the pre-formed cross-sectional shape of a river bend can be estimated by applying theory developed by Odgaard (1986). An example of the results of a calculation using Odgaard's theory to determine bend geometry is shown in Fig. 9.8.

A pre-forming approach was applied to develop the restoration design for the Blue River in Colorado, directly downstream of Dillon Dam. Comparison of pre- and post-restoration conditions (Fig. 9.9) demonstrates the value of providing pools, glides, runs, and riffles. In addition to establishing good trout fishing habitat the design is stable from scour and hydraulic points of view.

In the case of the Blue River, the cobble and boulder sizes used as bed material are somewhat larger than the sizes characteristic of incipient motion. The reason for this is that smaller stone sizes would jeopardize long-term river stability. Stone sizes resulting in removal under design flow

Figure 9.7 Alamo Creek, California. These photographs illustrate the use of pre-forming to change the cross-sectional shape of a river from an incised stream subject to erosion (top left photo - 2001) to an active channel with a floodplain (top right photo - 2001). Photo at bottom shows channel after establishment of vegetation (2005).

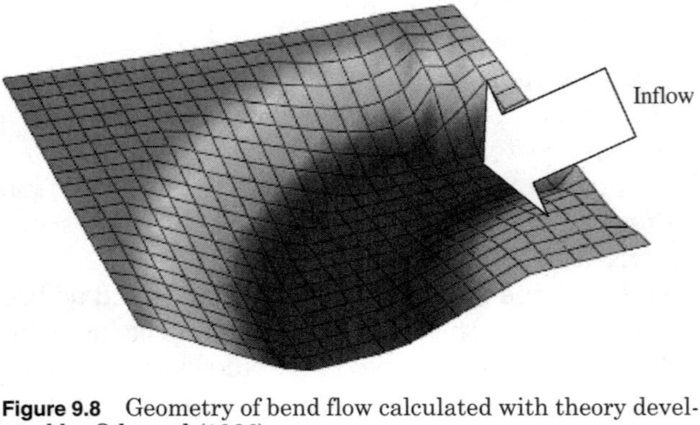

Figure 9.8 Geometry of bend flow calculated with theory developed by Odgaard (1986).

Figure 9.9 Restoration of the Blue River, Colorado, by making use of a pre-forming approach. The photo on the left shows conditions prior to restoration. The original stream is straight and featureless. The photo on the right shows the stream after restoration. The restored stream meanders, flowing through pre-formed pools, glides, runs, and riffles. It is a popular trout fishing spot.

conditions will not be replaced over the long term because the sediment inflow from upstream is jeopardized by the presence of Dillon Dam. By sizing the bed material slightly larger than associated with incipient motion prevents long-term degradation. In a way, this design combines hardening and pre-forming approaches.

Plunge pool scour

Chapters 7 and 8 outline approaches to determine the maximum scour depth in plunge pools subject to impinging jets. For example, the maximum scour depth of a plunge pool subject to a plunging jet can be determined by making use of the erodibility index method (Annandale 1995), or the dynamic impulsion or comprehensive fracture mechanics approaches (Bollaert 2002). Once the maximum depth of scour for a particular material type and design flow is determined the plunge pool can be pre-excavated to this level, which is likely to prevent additional future deepening of the pool for floods equal to or less than the design flood.

Annandale's method is based on a scour threshold that relates the erosive capacity of water to the relative ability of rock to resist scour. This threshold relationship quantifies the threshold stream power of the earth material making up the boundary of the plunge pool. After characterizing the earth material, it is possible to use this relationship to determine the earth material's threshold conditions as a function of elevation below the ground surface.

The erosive capacity of the water is also determined as a function of elevation below the water surface for increasing plunge pool depths. The erosive capacity of water in a plunge pool first increases and then decreases (see Chap. 5 for the theory and an application example in

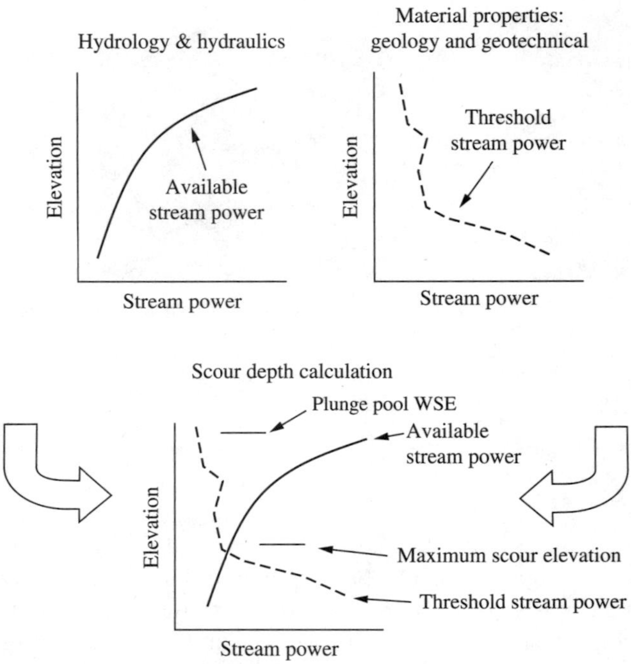

Figure 9.10 Determination of the pre-forming depth for a plunge pool subject to a plunging jet using Annandale's erodibility index method.

Chap. 10). This means that the erosive capacity of the water, expressed in terms of stream power, eventually reduces as the scour hole in the plunge pool increases in depth.

At the elevation where the erosive capacity of the water is lower than the threshold stream power of the earth material the scour ceases (Fig. 9.10). The maximum scour elevation provides an indication of the maximum depth of excavation required for pre-forming of the plunge pool. By making the plunge pool as deep as the anticipated scour depth (i.e., by pre-forming) the erosive capacity of the water is reduced enough to prevent scour of the underlying earth material. (The latter approach can also be viewed as modification of the flow, i.e., reducing the erosive capacity of the water at the boundary of the plunge pool.)

Earth Material Enhancement

Examples of enhancing the properties of naturally occurring earth materials include using vegetation or rock anchors. Vegetation can provide the necessary strength to protect channel beds or banks against erosion.

Rock anchors e.g., can increase the scour resistance of jointed rock and protect dam abutments against the effects of scour.

Vegetation

When using live vegetation to enhance the strength of naturally occurring earth materials it is necessary to commence the design process by developing engineering specifications for the vegetation. Restoration ecologists use the engineering specifications, developed by using the erodibility index method, to select plant material.

When using the presented technique the principal focus is on determining the required root architecture of plants that will enhance the natural soil's ability to resist the erosive capacity of the water. The preferred root architecture consists of fine, fibrous roots that can bind soil to form larger "pseudo" particles. The methodology assumes that the fibrous roots do not add additional strength (either mass strength or shear strength) to the native soil, nor, it is assumed, do the roots provide any enhancement in terms of the shape and orientation factor. The assumption that the roots do not add strength is a conservative assumption.

Therefore, when using the erodibility index method the selection of the mass strength number (M_s) and shear strength number (K_d) essentially reflects the characteristics of the naturally occurring soils. By convention, a value of one is assigned to the orientation and shape number J_s. What remains is to determine the required value of the block/particle size number (K_b) that indirectly represents the size of the "pseudo" particle that will resist the erosive capacity of the water.

The required size of the "pseudo" particle is dependent on the erosive capacity P of the water. Once quantified the desired value of K_b is determined by solving the following equations (see Chaps. 4 and 6) for conditions when the applied stream power $P > 0.1$ kW/m^2 and when $P < 0.1$ kW/m^2 respectively, i.e.,

$$K_b = \frac{(F'P)^{4/3}}{M_s K_d J_s} \tag{9.1}$$

for cases when the value the stream power of the flowing water $P > 0.1$ kW/m^2.

In cases when the value of $P \leq 0.1$ kW/m^2 the required value of K_b is determined with the equation,

$$K_b = \frac{\left(\dfrac{F'P}{0.48}\right)^{1/0.44}}{M_s K_d J_s} \tag{9.2}$$

The symbol F' represents the factor of safety, which is selected by the engineer conducting the study in order to express the degree of uncertainty associated with the analysis. The expressions $P^{4/3}$ and $(P/0.48)^{1/0.44}$ convert the stream power of the flowing water at the threshold of erosion to the equivalent value of the erodibility index (see Chap. 6).

The required root bulb diameter D of the fibrous root system in meters is then calculated with the equation (see Chap. 6),

$$D = \sqrt[3]{\frac{K_b}{1000}} \tag{9.3}$$

Table 9.7 illustrates a calculation for determining the root architecture requirements for vegetative erosion protection along a channel. Column 1 identifies the location where the calculation is performed, i.e., the river station. The stream power in column 4 is the sum of those shown in columns 2 and 3; with the former containing the stream power in the river should it contain no bends. Column 3 contains the stream power that should be added in the presence of a bend. In the example shown in Table 9.7 the stream power around the bends was calculated by making use of an equation developed by Chang (1992)—see Chap. 5.

Column 5 indicates the typical consistency of the earth material making up the channel boundary. In this particular case it is a loose, non-cohesive soil. The known properties of this soil are shown in columns 6, 8, 9, and 10 (representing respectively the mass strength number M_s, friction angle ϕ of the soil, the shear strength number K_d [i.e., $K_d = \tan(\phi)$], and the orientation number J_s, which is set equal to 1 by convention).

The value of the required block/particle size number K_b is calculated using either Eqs. (9.1) or (9.2). Column 11 contains the required erodibility index for the vegetated earth material, i.e., the product of columns 6, 7, 9, and 10. The anticipated threshold stream power that the vegetated earth material would offer is shown in column 12, and is determined by converting the erodibility index in column 11 to stream power using Annandale's erosion threshold relationship. Column 13 contains a check calculation to determine if the vegetated earth material will erode, i.e., it compares the total stream power shown in column 4 and the anticipated threshold stream power of the vegetated earth material in column 12 (i.e., the resistance that would be offered by the vegetated earth material). Column 14 contains the factor of safety (FOS) that was used to calculate the value of K_b (i.e., the value of F'). The last column, i.e., column 15, contains the necessary root depth (diameter).

The results of the calculated root depth (root bulb size) for plants that will protect the natural boundary material of the channel against the erosive capacity of the water are plotted in Fig. 9.11. This graph shows how the required root bulb sizes change as a function of river

TABLE 9.7 Example Calculation for Determining the Root Architecture Requirements for Vegetated Erosion Protection

1 River station	2 Stream power W/m^2	3 Bend power W/m^2	4 Total stream power W/m^2	5 Typical consistency	6 M_s	7 K_b	8 Friction angle (degree)	9 K_d	10 J_s	11 Erodibility index	12 Resistance (W/m^2)	13 Erosion (yes/no)	14 FOS	15 Necessary root depth (mm)
4125	748.8	316.0	1064.8	loose	0.04	203.716	30	0.577	1	4.70	3194.43	no	3.0	588
4025	474.6		474.6	loose	0.04	69.357	30	0.577	1	1.60	1423.78	no	3.0	411
3975	225.9		225.9	loose	0.04	25.778	30	0.577	1	0.60	677.74	no	3.0	295
3945	310.7		310.7	loose	0.04	39.427	30	0.577	1	0.91	932.11	no	3.0	340
3875	480.4		480.4	loose	0.04	70.497	30	0.577	1	1.63	1441.29	no	3.0	413
3775	212.2		212.2	loose	0.04	23.712	30	0.577	1	0.55	636.59	no	3.0	287
3725	97.5		97.5	loose	0.04	8.406	30	0.577	1	0.19	292.46	no	3.0	203
3695	130.2	254.5	384.7	loose	0.04	52.417	30	0.577	1	1.21	1154.06	no	3.0	374

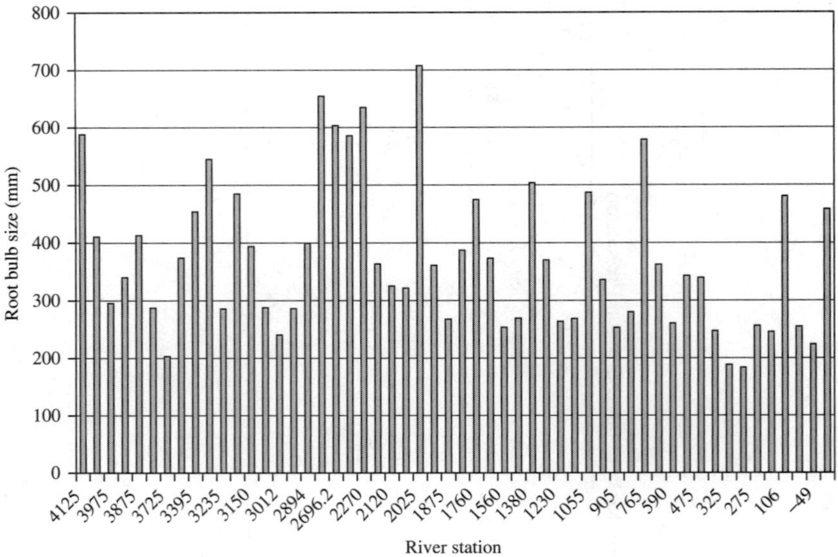

Figure 9.11 Graph showing required root bulb sizes as a function of river station.

station and provides the basis for preparing engineering specifications of the required plant material. Restoration ecologists use these specifications to select plant material.

The restoration ecologist will also inform the engineer of the maximum root bulb size that is available in the particular climate under consideration. Say, for example, that the restoration ecologist indicates that it is not possible to establish plants with root bulb sizes in excess of, say 400 mm. This information is then used to determine which parts of the river channel should be protected with harder protection, such as riprap. Alternatively, if a stream channel with soft protection only is required it will be necessary to redesign the geometric layout of the stream channel to reduce the erosive capacity of the water by making use of pre-forming techniques discussed earlier.

Rock anchoring without concrete lining

When using rock anchors to enhance rock properties to prevent scour the main objective is to increase the effective size of the rock blocks so that the erosive capacity of the water will not be able to remove individual blocks of rock. A secondary objective is to reduce the aperture width of discontinuities between rock blocks and minimize the effect of fluctuating water pressures within the discontinuities.

For example, if the rock under consideration has an *RQD* of 50 its erosion resistance can be increased by anchoring the rock in a manner that will result in its effective *RQD* increasing to, say, 100. This can most probably only be accomplished if the rock anchors are stressed over long enough distances, forcing the individual rock block elements to act as a single unit. This approach will only be effective if the mass strength of the rock is high enough to prevent it from failing in brittle fracture or fatigue.

The effectiveness of using rock anchors can be assessed using two optional and complementary approaches. The one approach is to use Annandale's (1995) erodibility index method, and the other is to use Bollaert's (2002) comprehensive scour method. When using the erodibility index method the two most important factors are the mass strength number (M_s) and the block size number (K_b) (see formulation of the erodibility index). The mass strength number represent the relative ability of the rock to resist brittle fracture and failure by fatigue, while the block size number represent the relative ability of the rock to resist scour due to the increase in effective weight of the rock blocks created by anchoring. Additionally the magnitude of the shear strength number (K_d) will also be affected due to the decrease in aperture width facilitated by the use of the rock anchors.

When using Bollaert's (2002) approach to assess the effectiveness of using rock anchors the potential for the rock to fail by brittle fracture or fatigue can be determined directly by comparing the fracture toughness of the rock to the stress intensity caused by the fluctuating pressures in close-ended fissures. If it is determined that the rock will not fail in either brittle fracture or fatigue, the length of the stressed rock anchors can be determined by conducting a dynamic impulsion analysis. This analysis provides an indication of the required size of rock blocks that should be created by stressing individual rock blocks together by anchoring. Once this is known, methods to bind the rock together with stressed anchors can be determined by considering the stratigraphy of the rock.

In summary, rock anchors (without a pool lining) can be used to resist rock scour if the mass strength of the rock is adequate to prevent breakup of the rock into smaller pieces by either brittle fracture or fatigue. Once it has been determined that the mass strength of the rock is adequate to resist breakup, the next part of the design entails development of an anchoring system that will increase the effective rock block size. The required anchoring length and pattern will bind individual rock blocks together to form effective rock blocks that are large and heavy enough to prevent removal by the erosive capacity of the jet.

Rock bolting design

In cases where scour is expected either from direct impingement of a jet onto a rock mass or from a jet falling into a plunge pool, rock bolting can be implemented to improve the ability of the rock mass to resist the erosive capacity of the water. Rock bolting increases the cohesive nature of the rock mass by bringing more rock blocks into contact with one another. This creates larger "effective" rock blocks that offer higher resistance to ejection.

For example, two rock blocks in a rock mass separated by a joint plane can be considered to act like a single rock block if properly connected by post-tensioned rock bolts. Post-tensioned rock bolts will only be effective in resisting the erosive capacity of water if scour occurs due to the ejection of the blocks. It will not be effective if the failure mode is by either brittle fracture or fatigue failure. It is therefore necessary to check for the potential occurrence of these two failure modes prior to commencing with the design of rock bolt protection. If the scour analysis indicates that the most likely failure mode is scour by ejection of rock blocks, rock bolting could be an effective way to protect the rock against scour.

Equations for quantifying the rock bolt tension and the spacing between rock bolts can be derived by referring to Figs. 9.12 and 9.13. The average pressure distribution on joint surfaces within the rock mass due to the presence of uniformly spaced tensioned rock bolts can be approximated as

$$P_{\text{average}} = \frac{T}{s^2}$$

where T is the tension in the rock bolts and s is the rock bolt spacing (Fig. 9.12).

This relationship assumes that the pressure associated with the rock bolts is applied uniformly across the face of the rock mass and is acting in

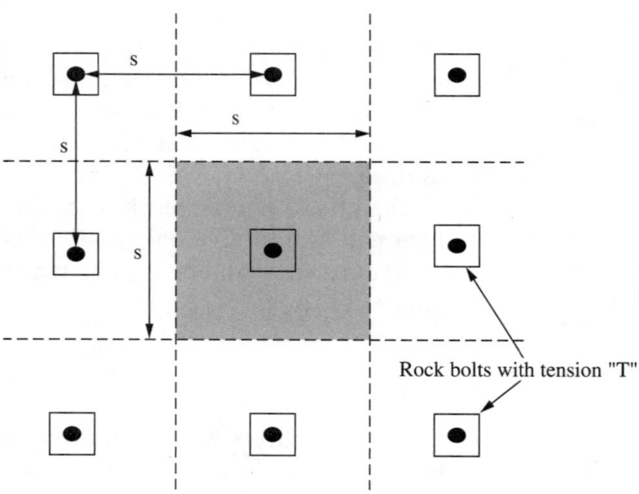

Figure 9.12 Rock bolt layout. The shaded region shows the distribution of P_{average} for one rock bolt. As indicated, a uniform distribution is assumed over the affected area.

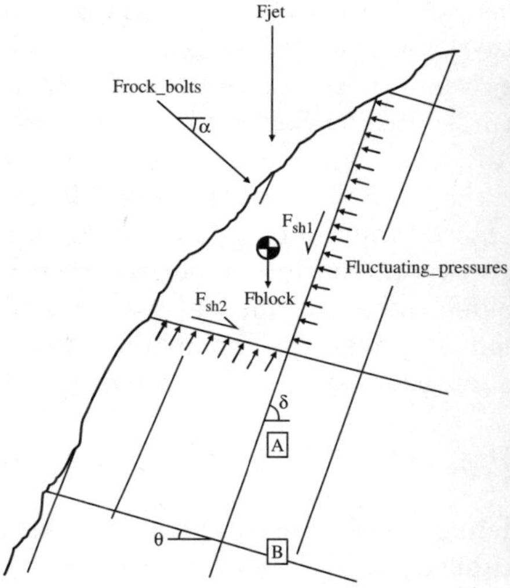

Figure 9.13 Rock block schematic for using the dynamic impulsion model to design rock anchoring.

the same orientation as that of the rock bolts. The vector component of the average pressure that is perpendicular to a joint plane forming a rock block may be multiplied by the surface area of the joint face of that rock block to quantify the resisting force acting on that particular joint plane.

$$F_{\text{rock\_bolts}} = P_{\text{average}}A = \frac{T}{s^2}A$$

where A is the total joint plane surface area projected at right angles to the resultant vector direction of the rock bolt force.

An equation for calculating the rock bolt force $F_{\text{rock\_bolts}}$ required to keep the rock in place when subjected to an impinging jet can be derived by following a procedure similar to that presented in Chap. 6 for calculating the magnitude of rock expulsion due to jet impingement (originally derived by Bollaert 2002). Consider the general arrangement of the abutment rock block shown in Fig. 9.13. Two joint planes are considered, A and B. The dip of joint plane A is δ and that of joint plane B is θ. Shear forces F_{sh1} and F_{sh2} respectively acts on the two joint planes when the rock block is set into motion. Fluctuating pressures induced by the impinging jet act at right angles on the two joint planes. The resultant force vector of the rock bolts is represented by the force $F_{\text{rock\_bolts}}$, which acts at an angle α to the horizontal plane. Using the approach in Chap. 6 the required rock bolt force for preventing block expulsion can be expressed as,

$$F_{\text{rock\_bolts}} \geq \frac{1}{\sin\alpha} \cdot \left[C_I \phi \gamma \frac{V_j^2}{2g} A - F_{\text{block}} - F_{\text{sh1}} \sin\delta - F_{\text{sh2}} \sin\theta \right]$$

The force F_{block} is the weight of the rock block. When the rock is submerged in water it is necessary to use the submerged weight of the rock block. If not, the actual weight of the rock block should be used. The coefficient C_I is quantified with methods presented by Bollaert (2002) for particular ratios of plunge pool depth to jet dimension—also see Chap. 6. Once the value of this coefficient and the jet impingement velocity V_j are known it is possible to calculate the required value of F_{rock_bolts}. Once known for a particular jet and rock configuration the other two equations presented in this subsection are used to determine the optimal spacing and tension requirements for each rock bolt. The design of the anchor should be such that its yield strength is not exceeded once fully loaded.

Hard Protection Design

Hard protection design entails hardening the surface of the earth material subject to attack by the erosive capacity of flowing water. Hardening can entail using materials such as riprap, grouted riprap, concrete, and commercial lining systems. It can also entail using rock bolts combined with a hard lining to strengthen a rock mass.

Concrete lining to protect against erosion has been used historically in the design and construction of flood control channels. It is also an option when designing protection in plunge pools, although the challenge to successfully engineer such systems is somewhat greater than when designing channel-lining systems.

The objective when designing concrete linings in systems where severe scour is anticipated, such as in plunge pools subject to plunging jets, is to enhance the properties of the natural rock on the flow boundary in a manner that will prevent scour from occurring. The design approach is therefore based on the understanding that was developed during the course of this book when the characteristics of rock scour were considered.

The most important rock characteristics relevant to scour of rock that were highlighted include the mass strength of the material, its block size, inter-block shear strength, and its shape and orientation. Enhancement of these properties should form the basis of design. It is advisable to consider the role of each of the rock properties when rock scours in order to conceive of design options that can be used to prevent degradation of the rock by the erosive capacity of water.

The mass strength of the rock principally relates to the ability of the fluctuating pressures in flowing water to break the rock into smaller pieces by either brittle fracture or fatigue. If the stress intensity introduced by the fluctuating pressures into the rock exceeds its fracture toughness the rock will fail in brittle fracture. In cases where the stress intensity caused by the fluctuating pressures does not exceed the fracture toughness of the rock it is possible for the rock to fail in fatigue if the fluctuating pressures are applied long enough.

Should rock fail in either brittle fracture or fatigue the effective block size of the rock is reduced and removal of the same is made easier. By devising protection systems that can prevent the breakup of rock into smaller pieces it is possible to protect it against the erosive capacity of the water and counter scour. This can e.g., be accomplished by covering the rock mass with a concrete lining with superior mass strength, or by preventing water pressures from entering the fissures in the rock mass. By preventing water from entering fissures it is no longer possible for fluctuating pressures to develop within the fissures. The potential for scour by brittle fracture or fatigue is therefore reduced.

The inter-block shear strength in a jointed rock mass is affected by the aperture size of the joints, their roughness properties and type of gouge. When fluctuating water pressures enter joints it is possible to remove individual blocks of rock by dynamic impulsion. A principal objective in the engineering of protection systems for jointed rock masses is therefore to prevent fluctuating pressures from entering in between the rock blocks. If it is not possible to prevent the development of fluctuating pressures in rock joints, an aim should at least be to reduce their magnitude and effect.

An effective way of countering the effects of joint properties on scour of rock is to cover them with a lining, such as concrete, or by infilling the joints with dental concrete. By doing this the effective size of the rock blocks, from a conceptual point of view, is increased. This is particularly true if the installation of a concrete lining is combined with anchoring. Post-tensioned anchors long enough to bind rock blocks together increase their effective size. By concurrently reducing the possibility of significant fluctuating pressure developing within rock joints and by effectively increasing rock block size the possibility of rock block removal by dynamic impulsion is reduced. Anchoring, if executed correctly, can also reduce the aperture sizes of joints, increasing friction forces and reducing the possibility of water entering the joints.

Lining a plunge pool without anchoring can be successful if the material used to line the pool, say, concrete, is strong enough to resist brittle fracture and fatigue failure, while concurrently being heavy enough to resist uplift. However, if this approach is followed reliance for protection against scour is placed mainly on the properties of the concrete. The properties of the rock underlying the concrete are not enhanced. When using post-tensioned anchored lining, coherence between rock and lining is facilitated. In the latter case the properties of the rock in the flow boundary is enhanced by making use of a combined system.

Using a concrete lining to protect a plunge pool against scour accomplishes two things; it can protect the plunge pool against the effects of rock failure that can occur due to brittle fracture or fatigue, and, when using post-tensioned rock anchors, it can increase the effective rock block size (i.e., it can help prevent removal of individual blocks of rock by dynamic impulsion due to the increased effective size of the rock blocks).

The design of an effective liner for plunge pool protection requires consideration of the following:

- The lining should be thick and strong enough to prevent failure by brittle fracture or fatigue, which could result from pressure fluctuations occurring in cracks within the concrete that develop in the lining over time due to either shrinkage or introduction of fluctuating pressures in the plunge pool.
- The water-stops sealing the concrete joints should be robust to prevent fluctuating water pressures from penetrating underneath the lining. Such pressures can lead to uplift of the lining.
- Fluctuating water pressures penetrating below the lining can potentially lead to failure of the rock underneath the lining should it penetrate the rock discontinuities. The anchoring of the lining should be such that it will prevent removal of rock fragments from underneath the lining.
- The tensile strength of the lining material should be high enough to resist failure by brittle fracture or fatigue. This can be accomplished by using steel reinforcement or other means of increasing the tensile strength of the concrete. The design should aim at reducing the possibility of crack formation, either due to age or vibration during spill events.
- A drainage system underneath the lining that will relieve pressures is of critical importance. However, the design of such a system should be executed in a manner that will prevent resonance of the fluctuating pressures within the drainage system. Careful attention to the geometric design of the drainage system is important.
- The anchors should be designed and post-tensioned to resist the maximum pressures that could occur underneath the lining.
- The anchors should be designed to resist failure by fatigue.
- The selection of anchor lengths should take account of the rock characteristics, particularly the number of joint sets and discontinuity spacing. One of the objectives of using anchors is to increase the effective rock block size, and thus the total weight of the combined lining/rock system.
- The rock anchors should be designed in a manner that will prevent punch failure, i.e., the detail where the anchors terminate on the surface of the concrete should be designed in a manner that will prevent failure by punch-out.

Gibson Dam

Gibson Dam, located 30 miles northwest of Augusta on the North Fork of the Sun River in Montana, USA, experienced very high reservoir inflows resulting from a storm that was sustained by upslope winds and unusually heavy moisture from the Gulf of Mexico. These conditions caused a rainstorm over an area of 160 km (100 miles) long on the eastern slope of the continental divide and produced rainfall for 30 h ranging between 200 and 400 mm (8 and 16 inches). The shallow soils of the Rocky Mountains and foothills were already saturated with spring snowmelt and contained very little capacity to retain any additional moisture. By 1400 h on Monday, June 8, 1964 overtopping commenced at Gibson Dam. Inflows into the reservoir reached an estimated maximum discharge of 1700 m^3/s (60,000 cfs) and remained at this rate for three hours. A high water mark in the spillway control house indicated that the dam overtopped by approximately 1 m (3.23 ft). The overtopping event lasted for 20 h.

Gibson Dam consists of a 60 m (199 ft) high thick concrete arch dam with a crest length of 292 m (960 ft). The dam crest is 4.6 m (15 ft) wide and the maximum base width is 35.7 m (117 ft). The foundation consists of crystalline limestone (known as the Madison Group) with regular beds striking normal to the river with an upstream dip of 75°.

During the overtopping event large volumes of water flowed over the dam crest, behaving like a huge waterfall with the maximum discharge over the parapet estimated at 510 m^3/s (18,000 cfs). The water flowed over the dam and along the abutments on both sides of the dam. Relatively minor damage occurred in the abutments. It is believed that some rock was removed from the right abutment, principally in the surface layers. No significant scour depths were experienced.

Although the rock strength and jointing appeared to be quite resistant to erosion during this event, a 0.9 to 1.5 m (3 to 5 foot) thick concrete overlay with anchor bolts was placed where the overtopping flow impinged on the right abutment and foundation (Fig. 9.14) to protect against even larger floods up to the probable maximum flood (PMF). The steeper left abutment was treated with rock bolts and a concrete cap in major surface fracture zones (Fig. 9.15). This modification was deemed prudent given the large degree of uncertainty associated with determining erodibility at the time when modifications were designed.

The ability of the rock to resist the erosive capacity of the water originating with the overtopping was estimated be making use of the erodibility index method. The erodibility index is computed as

$$K = M_s K_b K_d J_s$$

The value of each of the parameters can be obtained from tables in Chap. 4. M_s is a factor representing the mass (intact) strength of the foundation. The majority of the foundation rock at Gibson Dam is limestone and dolomite. The average UCS for this material from laboratory tests is 158 MPa. The rock can therefore, on average, be classified as very hard.

Some weaker intensely fractured beds of about 1.8 to 3.0 m thick are present, particularly on the left abutment. The mass strength of the rock

Figure 9.14 Concrete lining placed on right abutment of Gibson Dam after the occurrence of 1964 flood (Photo: Bureau of Reclamation, US Department of Interior).

Figure 9.15 Dental concrete in left abutment of Gibson Dam, placed after occurrence of 1964 flood (Photo: Bureau of Reclamation, US Department of Interior).

in these beds is estimated to be lower than the average, perhaps by a factor of 2 to 4. The UCS of the rock in the intensely fractured zones is estimated as 40 to 80 MPa. Laboratory testing performed on the concrete during original construction of the dam resulted in an average UCS of about 20 MPa. The values of M_s for these materials were set equal to their UCS.

K_b is an index related to the mean block size. It can be estimated as the rock quality designation (*RQD*) divided by the joint set number (J_n). The dam foundation limestone varies from thin beds a few centimeters thick to massive beds, 2.4 to 3.0 m thick. The rock was found to be broken by several fissures, which followed the bedding surfaces very closely. Another prominent joint set was mapped on each abutment, and there were other minor joints. The rock was therefore characterized as having two joint sets plus random, which corresponds to a joint set number of 2.24.

The *RQD* was not logged for holes drilled on the downstream right abutment. However, in general, the rock was recovered in long sticks with only a few fractured zones. Based on core recovery numbers and field observations, the average *RQD* is probably about 90–95 percent, with isolated areas ranging down to about 80 percent. This results in K_b values between about 35.7 and 42.4. ($K_b = RQD/J_n$).

The intensely fractured beds would have an *RQD* of about 17 percent based on field measurements. This corresponds to a K_b value of about 7.6.

The concrete was placed in 1.2 m lifts using large blocks generally encompassing the entire thickness of the dam. The contraction joints are widely spaced at 10.7 to 18.3 m (35 to 60 feet) and are keyed. Although some lift lines exhibit minor seepage at high reservoir elevations, the lifts were cleaned well and also keyed. The value of K_b for the concrete should be high, say about 80 or higher.

K_d represents the interblock frictional resistance. It can be estimated as the ratio between the joint roughness number and the joint alteration number (J_r/J_a), which is roughly equivalent to the tangent of the friction angle. Based on field observations, the limestone bedding planes are rough and planar, while the joints are very rough and irregular. Therefore, the joint roughness was assumed to be rough/planar ($J_r = 1.5$). The J_r for concrete would be described as "stepped" since all the joints are keyed, resulting in a value of about 4.0.

The majority of foundation joints were reported as calcite healed or clean and tight, increasing in tightness with depth from the surface [although one joint open up to 76 mm (3 inches) wide at the surface was observed on the right abutment]. This results in J_a values ranging between about 0.75 to 1.5. For this rock the value of $K_d = J_r/J_a$ ranges between 1 and 2.

For the intensely fractured rock beds, the joint roughness number would tend toward a value of 1.5 for rough/planar joint surfaces, and the joint alteration number could be as high as 2.0. This results in a K_d value of about 0.75.

For the concrete, the joints would be considered to be healed (lift lines) or tight and clean (contraction joints), resulting in joint alteration numbers J_a between 0.75 and 1.0. The K_d values for concrete therefore range between 4.0 and 5.3.

The relative ground structure number (J_s) represents the orientation of the discontinuities relative to the impinging water, and takes into account the block shapes (long and narrow or roughly cubic). The orientation of the beds is extremely regular, striking 5 to 8 degrees west of north (about cross-canyon) and dipping to the east at angles ranging from 70 to 86 degrees. The abutments give the appearance that the open bedding planes are spaced roughly twice as close as the open joints. Although the apparent dip of the bedding changes with respect to the plunging jet in relation to the curvature of the dam, an angle of 70 degrees against the flow (beds dip upstream) was assumed on the average. This results in a J_s value of about 0.9, which would also apply to the intensely fractured rock zones. For the concrete, a J_s value of 1.0 would be appropriate.

Using the values outlined above, the values of the erodibility index (K) for the various material types subjected to the erosive capacity of the water are:

Foundation Rock: 5100–12,000
Intensely Fractured Beds: 200–400
Concrete: 6400–8500

The stream power is initially low as flows on the sides of the dam flow over the top and impinge directly onto the upper abutments. The stream power increases as the fall height increases toward the channel when the reservoir reaches its peak. At maximum overtopping, the flow depth of 1.0 m (3.2 feet) over the crest corresponded to roughly 1.78 m^3/s/m (19.2 cfs/ft). Computations indicated the stream power ranged from low values of 43 kW/m^2 at the upper abutments [fall height of 2.7 m (9 feet)] up to 258 kW/m^2 near the central part of the dam [fall height of 54.9 m (180 feet)].

The results are plotted in Fig. 9.16. The exposed foundation rock and newly placed concrete are not expected to erode, based on these results. In general it was felt that the foundation rock performed well, indicating the results should fall well below the threshold for erosion as indicated. This is particularly true when it is recognized that the parameters used to estimate the erodibility index are based largely on the conditions remaining after the 1964 overtopping event. The exposed intensely fractured zones would be expected to erode, except perhaps near the crest. Although the observed amount of erosion in these zones was not excessive, this is believed to be consistent with the observed behavior.

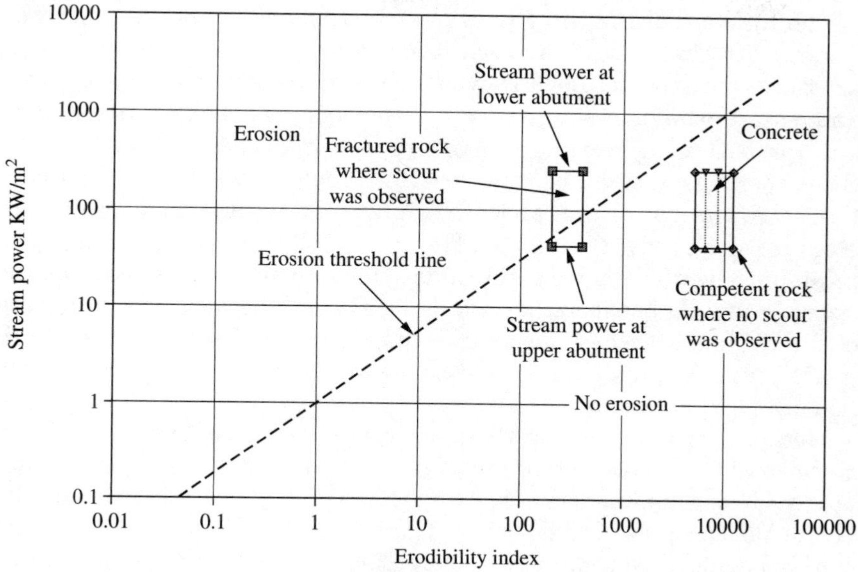

Figure 9.16 Relationship between estimated stream power and erodibility index of the rock at Gibson Dam for the overtopping event of 1964.

The results of this study support the conclusion that there probably was some erosion of weak rock (intensely fractured beds). There was some superficial erosion and scouring of loose material during the experienced overtopping, but not much, as judged from the condition of the foundation after the overtopping. The concrete should not be vulnerable to erosion in the future provided that it remains intact and does not experience degradation due to cracking, freeze-thaw action, or vandalism. The decision to protect the intensely fractured beds appears to be sound under any scenario. The areas of abutment rock most susceptible to erosion for higher overtopping flows have been protected.

Riprap

Protection of flow boundaries against the effects of scour by making use of riprap entails placing rock particles on top of a filter over the underlying soil. Successful protection of the underlying soils against the effect of scour when using riprap depends on the gradation and size of the riprap and provision of an appropriate filter. The riprap size must be such that it can resist removal by the erosive capacity of the water. In this regard it is important to place a range of rock sizes, adhering to a particular gradation. The interlocking effect of the graded material provides additional resistance to the erosive capacity of the water.

The function of the filter is to prevent removal of the fine material underneath the protection. Such removal occurs when the fluctuating pressures of the water flowing over the riprap penetrates below the rock and agitates the finer soil particles. If the openings between the individual riprap elements are large enough the fine material will move through these openings and will be removed. Such removal weakens the foundation below the riprap and can lead to failure of the riprap and, eventually, scour of the underlying material. The use of a filter, when designed appropriately, prevents removal of the underlying fine material by reducing the sizes of openings between the underlying soil and the water flowing on top of the riprap. Therefore, although agitation of the finer, underlying soil might still occur, it is no longer possible to remove the particles because no path for such removal exists.

Methods for determining the appropriate rock size, gradation of the riprap and formulation of filter designs are provided in engineering manuals such as the United States Army Corps of Engineers' Manual on the Hydraulic Design of Flood Control Channels (EM 1110–02–1601) which can be downloaded from their website. The required rock size can also be checked by making use of the erodibility index method, using the approach conventionally implemented when investigating scour of non-cohesive earth materials. Detailed design of riprap is therefore not dealt with here, but examples of such protection are provided.

Figure 9.17 shows the use of riprap in a confined area where it was not possible to restore a creek to its naturalized condition. The riprap, which adheres to a calculated gradation, is underlain by filter cloth.

Figure 9.17 Riprap protection of channel banks.

An example of the incorrect placement of riprap is shown in Fig. 9.18. The design of the protection for this particular channel entailed using large blocks of rock in the bottom of the channel and protecting the channel banks with grouted riprap (not shown in this figure). It should be noted that the large rock blocks in the channel bed were placed directly on top of the fine soil, without the presence of a filter underneath the rock. After the placement of the rock blocks the channel banks were protected with grouted riprap. It should be noted that although the rock blocks in the channel bed are large, they are roughly uniformly graded with no smaller rocks between the large rocks. The potential for providing additional strength by means of interlocking, therefore, does not exist.

Figure 9.19 shows the condition of the same channel after two years of operation. The photograph shows that the fine material underneath the large rock was washed out and lead to failure of the channel bed. This resulted in the grouted riprap suspended higher up on the channel banks. Figure 9.20 shows undermining of the riprap due to the lack of a filter between the riprap and the underlying fine soil.

The performance of this channel also points to one of the limitations of grouted riprap. Conventional riprap, without grouting, if placed correctly and with the right gradation and underlain by a filter, is flexible. Therefore, if minor settling or minor failures of the foundation below the riprap occur the hardened boundary will settle without severely jeopardizing its

Figure 9.18 Incorrect placement of riprap.

Figure 9.19 Condition of channel in Fig. 9.14 after two years of operation.

Figure 9.20 Undermining of riprap due to the lack of an appropriate filter between the large rock elements and the underlying fine soil.

integrity. When using grout, the riprap protection becomes rigid and no longer has the ability to settle and protect the underlying soil. Instead, as shown in Figure 9.19 the rigidity of the grouted riprap can lead to undercutting and eventual failure.

Figure 9.19 also illustrates the importance of toe-in, which entails anticipating the amount of scour that will occur in the channel bed and placing the channel bank protection down deep enough into the bed to prevent undercutting of the banks. Weaknesses of the design shown are the lack of a proper gradation of the riprap placed in the channel bed, no provision of a filter below the riprap, and no toe-in of the channel bank protection. The use of grouted riprap also diminishes the flexibility of the protection of the channel banks and will likely lead to failure by undercutting.

Accommodating Protection

Accommodating scour protection design essentially entails accepting that scour will occur and arranging infrastructure, infrastructure components, and property in a manner that will prevent adverse impacts resulting from scour. Figure 9.21 is a conceptual sketch of a bridge subject to scour, illustrating the accommodating protection design approach. The foundations of the pier and the abutments are placed deep enough below the river bed to prevent failure of the structure during design flood conditions. This approach is principally followed by the United States Federal Highway Administration (FHWA) in the guidelines provided for the design of bridges crossing rivers.

Plunge pools downstream of dams can for example also be designed in this manner. Although the original design intent for Kariba Dam in the Zambezi River on the border between Zambia and Zimbabwe, Africa, was

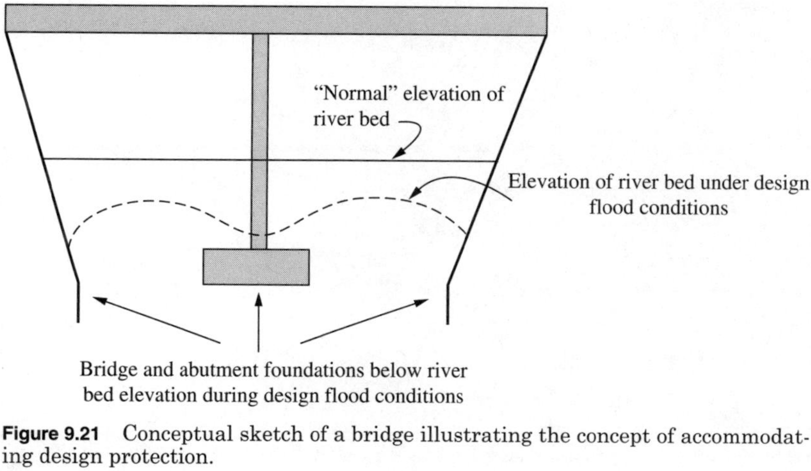

Figure 9.21 Conceptual sketch of a bridge illustrating the concept of accommodating design protection.

not to use an accommodating design approach to protect the dam against the effects of scour, the owners are currently forced to essentially accept this approach.

Kariba Dam, a 130 m high double curvature arch dam, was completed in 1959. The dam is founded on very hard to extremely hard gneiss rock, which has experienced significant scour over the years. The current depth of the scour hole is approximately 80 m, more than half the total height of the dam. Figure 9.22 shows the development of the plunge pool downstream of Kariba Dam over the period 1962 to 1982.

The spillway consists of six rectangular openings of 8.8 m by 9.1 m with a design capacity of 8400 m^3/s. Due to delays in the finalization of the hydroelectric power generation works the spillway operated frequently during the first five years, resulting in the removal of about 400,000 m^3 of rock and a scour hole that was approximately 50 m deep by 1966. Various attempts to stabilize the plunge pool include the placement of more than 20,000 sandbags, 344 m^3 of rockfill and 4170 sacks of cement in a single year. A very large flood, resulting in estimated release of 9444 m^3/s (i.e., greater than the design flood) occurred in 1981 and lasted for 15 days (ICOLD/CBDB 2002). This resulted in an additional increase in the plunge pool depth, to about 80 m deep.

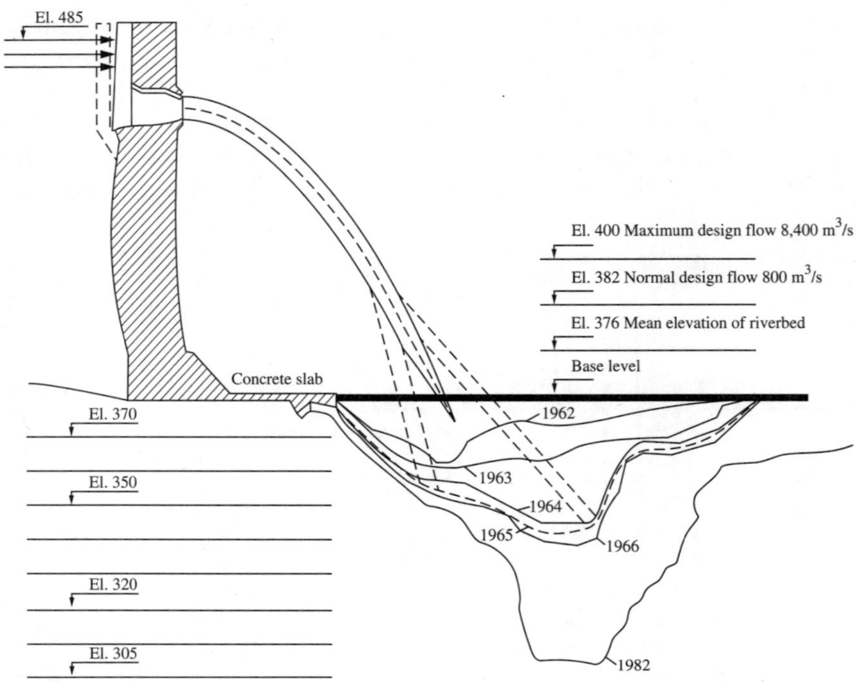

Figure 9.22 Scour pool development at Kariba Dam (ICOLD/CBDB 2002).

Although the owners are currently accepting the presence of the plunge pool, and are essentially following an accommodating design approach it is uncertain if this condition can be accommodated indefinitely. A more proactive approach to protecting the dam might be required in the near future.

Flow Modification

Flow modification entails implementing measures that result in a reduction of the erosive capacity of the water. This can include streamlining the flow to reduce its turbulence intensity in the flowing water, and thus its erosive capacity, or it can e.g., entail breaking up a plunging jet. An example of the effects of breaking up a plunging jet is provided in what follows.

A jet plunging into a pool leads to the development of average and fluctuating dynamic pressures that can cause the breakup and removal of rock and concrete. Attempts to minimize the erosive capacity of a jet should therefore aim at reducing the average and fluctuating dynamic pressures where it interacts with the rock on the boundaries of the plunge pool. This can be accomplished in two ways, by breaking up the jet prior to it plunging into the pool, or by constructing a plunge pool that is deep enough to adequately dissipate the erosive capacity of the jet. The former approach is discussed here.

The objective when breaking up a jet is to change its character from a coherent mass of water plunging through the air and into a pool (i.e., when the jet is intact) to a series of blobs of water falling onto and penetrating the water surface (when the jet is broken up). The maximum dynamic pressures caused by a coherent jet are greater than the maximum dynamic pressure caused by blobs of water falling into a pool. By combining research by Ervine et al. (1997), Castillo (1998), and Castillo (2004) it is possible to relate the average dynamic pressure coefficient (C_p) to the dimensionless plunge pool depth (Y/D) for varying dimensionless jet breakup lengths (L/L_b); where Y is the pool depth and D is the jet diameter or thickness, L is the total length of the jet; and L_b is the breakup length of the jet.

Figure 9.23 shows such relationships, indicating that if the jet breakup length is much smaller than the total length of the jet and the pool depth relative to the jet dimension is very large then the average dynamic pressure at the bottom of the plunge pool reduces to very low values. For example, if the breakup length of the jet is approximately one third of the total jet length (i.e., $L/L_b \approx 3$) and the dimensionless pool depth is about 15, then the average dynamic pressure coefficient is very close to zero.

The probability density functions of measurements at the bottom of a plunge pool that were made in physical hydraulic model studies to determine the impact of splitters on the average dynamic pressure resulting from a jet plunging into a pool are shown in Fig. 9.24. The jet causing the

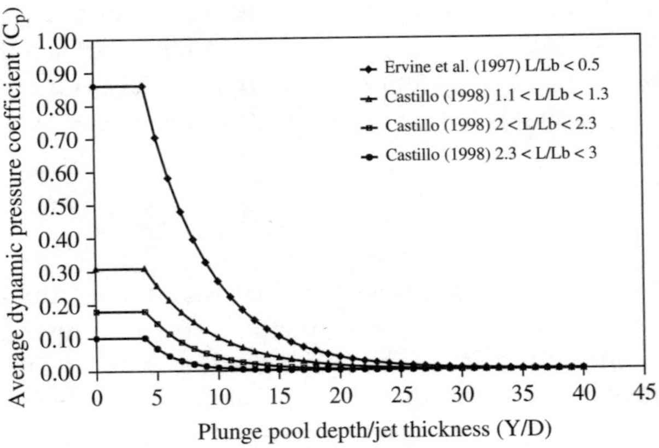

Figure 9.23 Average dynamic pressure as a function of pool depth to jet thickness and dimensionless breakup length developed from Ervine et al. (1987), Castillo (1998), and Castillo (2004).

dynamic pressures at the bottom of the plunge pool was discharged from the end of a spillway channel. The scenario without splitters consists of the jet merely discharging from the end of the spillway chute in a coherent fashion into the plunge pool. The scenario with splitters entailed incorporating splitters at the end of the spillway chute to break up the jet prior to plunging into the pool. (An example of the use of splitters can be found in the case study of Ricobayo Dam presented in Chap. 10.)

The graph on the left hand side of Fig. 9.24 shows the probability density function of the dynamic pressure in the plunge pool without the presence of splitters. The graph on the right shows the probability density function of dynamic pressures in the same plunge pool after introducing the splitters. Comparing the graphs indicates that the spread of dynamic

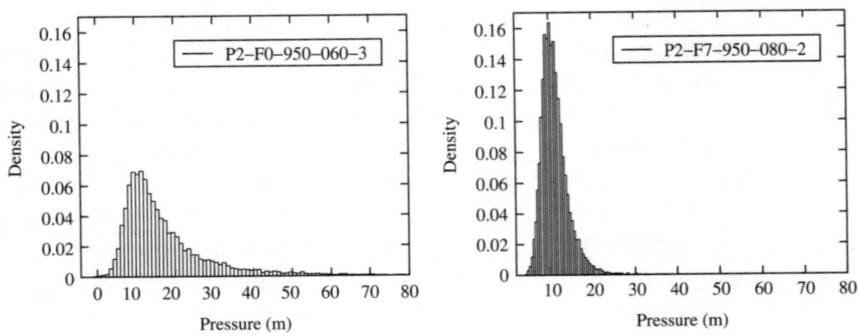

Figure 9.24 Probability density functions of pressures on the bottom of a plunge pool caused by a plunging jet without (left) and with (right) splitters (ETH, Zurich).

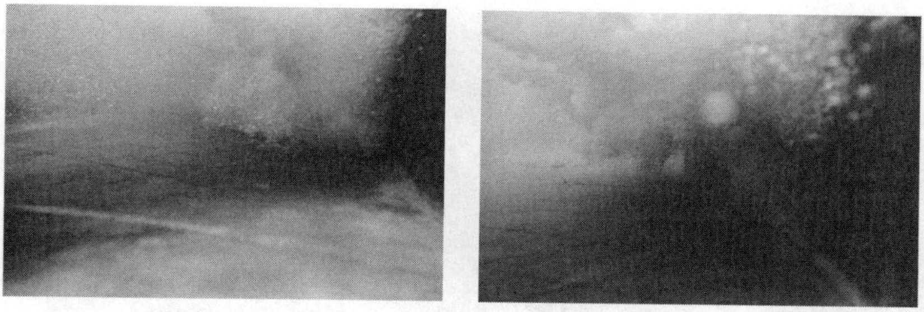

Figure 9.25 Comparison of the impact of plunging jets on the bottom of a plunge pool, without (left) and with (right) splitters (Photos: ETH, Zurich).

pressures on the plunge pool bottom is much greater for the case without splitters (i.e., the maximum pressures are much higher and pressures greater than the mode are also more persistent) than for the case with splitters, although the mode in both cases is approximately 10 m.

Photographs illustrating the difference in behavior of the jets within the plunge pool, without and with the splitters are shown in Fig. 9.25. The photograph on the left shows the interaction of the jet without the presence of splitters at the bottom of a plunge pool for a particular discharge. The photograph on the right shows the interaction of the jet at the bottom of the plunge pool for the same discharge, but with splitters installed at the end of the spillway chute. The jet is broken up prior to plunging into the pool in the latter case.

The interaction between the jet and the plunge pool boundary in the case without splitters is more pronounced, as indicated by the visual interpretation. In the case with splitters the air bubbles generated by the plunging jet in the pool often do not reach the bottom of the pool, indicating less intense interaction between the jet and the pool bottom.

Combined Approaches

The approaches to protecting infrastructure, property, and the public against the effects of scour can be combined in various ways. A few examples where combining protection approaches to safeguard against the effects of scour are presented in what follows. Other combinations can be conceived by the reader to develop unique project solutions.

One example illustrating the combination of methods to protect against the effects of scour can be found in the river restoration project that was implemented on the Eagle River adjacent to the town of Minturn, Colorado (Fig. 9.26). This is an example of a river with no significant disturbance to its sediment balance, which is an important consideration. This river was restored by making use of pre-forming, earth material enhancement and hardening.

Figure 9.26 Eagle River, Colorado. Left photo shows river prior to restoration. River is wide and featureless. Right photo shows river after restoration using pre-forming techniques, occasionally combined with hard and soft protection measures on the riverbanks. In addition to pools, runs and riffles, a flood plain has been formed on the left side of the photograph on the right.

In cases where a river's sediment balance is not disturbed pre-forming is implemented with sediment that is sized to move under design flow conditions. The reason for this is that sediment removed from the restored river during high flows will be replaced by sediment flowing in from upstream reaches. Balancing the in- and outflow of sediment results in a stable river.

The bed configuration of the Eagle River, i.e., the sizing of bed material that will move under design flow conditions and will be replaced by sediment discharging in from upstream is an example of pre-forming without bed hardening. (Compare this design approach with that of the Blue River, Colorado discussed in the section dealing with pre-forming and where bed hardening was concurrently used.)

However, in order to maintain the pre-formed design configuration of this river it was considered necessary to harden the river banks in some locations. This was done by placing rocks on the boundary in some locations that are resistant enough to prevent the river from changing its form. Additionally, the native earth materials of the river bank were enhanced in some locations by making use of vegetative erosion protection that was selected by making use of the techniques discussed under the section dealing with earth material modification.

Examples of optional solutions to protect the plunge pool and two water lines at Bull Run 2 Dam in Oregon, against the effects of scour are presented in Figs. 9.27 to 9.29. The optional approaches entail increasing the plunge pool water surface elevation to increase the energy dissipation of the plunging jet, pre-excavating the pool to increase the plunge pool depth and increase the energy dissipation of the jet without increasing the plunge pool water surface elevation, providing a post-tensioned concrete lining to protect the underlying rock against scour, and using combined implementation of riprap and a post-tensioned concrete lining systems.

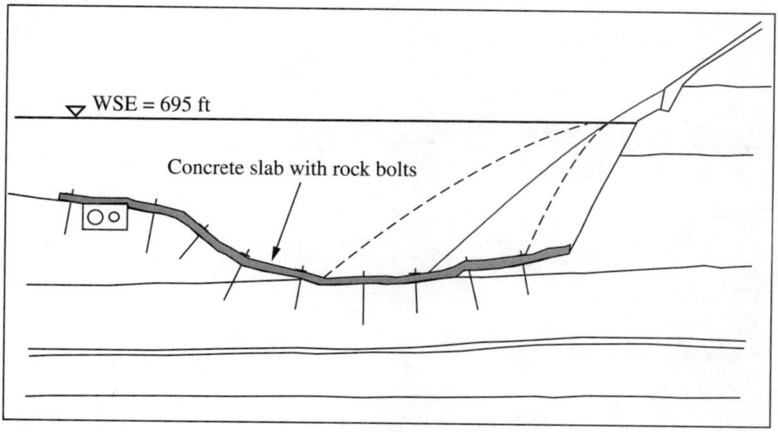

Figure 9.27 Protection of plunge pool by making use of a post-tensioned concrete lining and increasing plunge pool water surface elevation.

Another example of implementing combined solutions to protect against the effects of scour can be found at the plunge pool that developed downstream of the spillway chute at Bartlett Dam in Arizona. Bartlett Dam, constructed between 1936 and 1939 (Figure 9.26), is a multiple concrete arch dam with a height of 87 m located in the Verde River. The service spillway, which is located on the left abutment, was designed to pass a flood with a peak inflow of 4950 m^3/s through three 15.24 m by 15.24 m crawler-type gates into a spillway channel that is 52 m wide and 122 m long. The spillway channel is curved in plan and super-elevated, and contains a flip at its end. The exit channel of the spillway was in rock, almost flush with

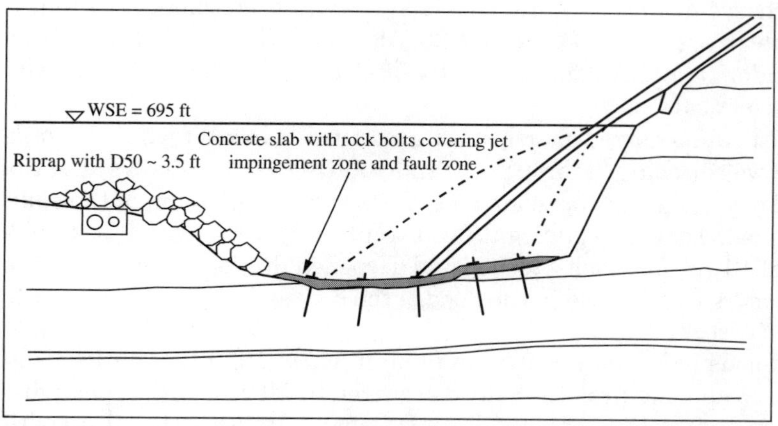

Figure 9.28 Protection of plunge pool by making use of a combination of post-tensioned concrete lining and riprap, and increasing plunge pool water surface elevation.

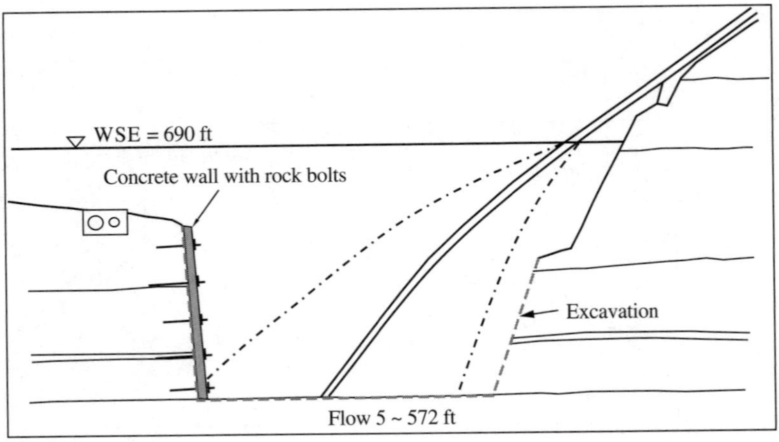

Figure 9.29 Protection of plunge pool by combining pre-forming techniques (excavation), concrete lining and post-tensioned rock anchors and retaining plunge pool water surface elevation at a lower level.

the exit of the spillway channel, and extended about 180 m to the river. The rock on the downstream side of the spillway channel has been eroded since and a drop of approximately 36 m currently exists at the downstream end of the flip bucket.

Two kinds of rock exist on the downstream side of the spillway channel, fine- and coarse grained granite. The fine-grained granite is slightly weathered to fresh, moderately hard to hard rock with an assumed UCS of about 20 MPa and a *RQD* ranging between 35 and 50. It has three plus random joint sets, that are intensely to moderately fractured. The dip of the fine-grained granite is approximately vertical.

The coarse-grained granite consists of weathered, moderately hard rock with an assumed UCS of 15 to 20 MPa and a *RQD* of 40. It has three plus random joint sets and is slightly fractured. The dip is also approximately vertical.

The fist spill occurred shortly after construction in 1941 when 925 m^3/s of water was discharged through the spillway (Fig. 9.30). Surveys after the event showed that a depression on the left side downstream of the spillway formed. Another significant flood occurred in 1965, with a peak discharge of 810 m^3/s, which widened and deepened the depression. Concrete with anchors were placed in the flood of the depression after this event to protect the rock.

Two floods occurred in 1978 and one in early 1979 that caused significant scour. The first 1978 flood occurred in March, with a peak discharge of 1980 m^3/s. In the middle of December 1978 another flood with a mean discharge of 1660 m^3/s occurred, followed by a flood in mid-January 1979 of 530 m^3/s (Fig. 9.31). This series of floods (principally

Figure 9.30 Bartlett Dam after construction in 1939 (left). First flood through spillway in 1941 (right) (Photo: Bureau of Reclamation, US Department of Interior).

the floods that occurred in 1978) caused significant scour, increasing the depth of the scour hole to about 36 m (Fig. 9.32).

After the 1978/79 floods a concrete cyclopean wall was constructed on the upstream side of the scour hole, i.e., just downstream of the end of the spillway channel. This wall was undermined during a flood the occurred in 1980 (peak discharge of 3030 m^3/s), which also resulted in deepening of the scour hole. The fix after this flood consisted of placing concrete on the plunge pool floor. A flood with a peak discharge of 3325 m^3/s that occurred in 1993 damaged the right side of the plunge pool, the downstream plunge pool wall and the downstream outlet channel.

Figure 9.31 High discharges through Bartlett Dam in 1978 lead to significant scour in granite, on the order of 36 m (Photo: Bureau of Reclamation, US Department of Interior).

Figure 9.32 A 36 m deep scour hole developed downstream of the spillway channel after the 1978 floods. Scour occurred in fine- and coarse-grained granite. Scale is given by two individuals and a dog in the bottom of the plunge pool. The end of the spillway channel is to the right of the photograph (only the gunite that was placed just downstream of the end of the spillway chute after the 1965 flood can be seen towards the top of the photograph) (Photo: Bureau of Reclamation, US Department of Interior).

The approach to protecting the dam against the adverse effects of scour was therefore to accept deepening of the plunge pool to a certain extent, while attempting to protect the upstream side of the scour hole by constructing a massive cyclopean concrete wall. This wall has been effective in protecting the spillway channel against continued back-cutting. However, it has been undermined and slightly damaged during flood events that occurred since construction of the cyclopean wall.

During a dam safety investigation that was conduct in 1992 an assessment of the erodibility of the rock downstream of the spillway channel was made by making use of the erodibility index method. Using the information about the coarse- and fine-grained granite it was determined that the erodibility index of the rock ranged between 90 and 100 for the coarse-grained granite, and between 30 and 85 for the fine-grained granite, depending on location. The erosion resistance of the rock was estimated at about 30 kW/m^2 for the coarse-grained granite and between 13 and 28 kW/m^2 for the fine-grained granite.

The erosive capacity of the water at the surface, prior to development of the scour hole and towards the end of the development of the scour hole was estimated at the impingement point of the jet and in the front- and back-rollers. The estimated stream power for the 1978 floods in the back-roller ranged between 600 and 700 kW/m^2 during the beginning of the flood event and between 1800 and 2300 kW/m^2 towards the end of the event. The stream power in the front-roller ranged between 3400 and 5000 kW/m^2 during the beginning of the event and between 5200 and 5900 kW/m^2 after formation of the scour hole; while the stream power at the impingement point was set between 4100 and 5600 kW/m^2 during the beginning of the event and between 7000 and 8200 kW/m^2 after formation of the scour hole.

The conclusion that can be made by comparing the resistance offered by the rock and the stream power that was available for scour indicates that the erodibility index method correctly indicates the potential for significant scour, as has been observed.

Chapter 10

Case Studies

Introduction

Chapter 1 introduces typical scour problems encountered in practice, while Chap. 2 provides a framework for civil engineering decision making. The information presented in Chap. 2 is considered of prime importance as it is frequently required of civil engineers to solve problems that are not necessarily well-understood and for which adequate data is often not available. The decision-making process summarized in that chapter outlines strategies for obtaining the maximum value from theoretical understanding and empirical knowledge for devising solutions to engineering problems.

The essential principles regarding the erosive capacity of water and the ability of earth materials to resist its effects are presented in Chaps. 3 to 6. Chapters 7 and 8 summarize techniques that can be used to calculate the extent and rate of scour in earth materials. Examples illustrating the application of the information in Chaps. 3 to 8 are presented in Chap. 9. This chapter provides case studies illustrating solutions to engineering problems that were dealt with in the past.

The first case study deals with the approach that was followed to protect the inverts of the diversion tunnels that were constructed to pass floods during construction of the San Roque Dam in the Philippines. This is followed by a case study dealing with the scour that was experienced at Ricobayo Dam, Spain, and the solutions that were developed to protect this dam against the effects of scour. The last case study presents the procedures followed to protect the piers of the Confederation Bridge in Canada, against scour of rock.

San Roque Dam Tunnels

The San Roque Dam, one of the largest hydroelectric, flood control, and irrigation projects in Asia, is located in a remote area approximately

Figure 10.1 San Roque Dam, Philippines (Photo: The Washington Group).

500 km from the Philippine capital, Manila (Fig. 10.1). It consists of a 650 ft high rock-fill dam, the twelfth highest of its kind in the world, and has a hydropower generating capacity of 345 MW.

Three diversion tunnels were built to protect the site against floods during the construction period (Fig. 10.2). The original intent was to line the tunnels with concrete to protect them against scour (Fig. 10.3). However, during the course of construction of the tunnels, after the walls

Figure 10.2 Inlet portals of two of the San Roque diversion tunnels (Photo: Rich Humphries).

Figure 10.3 View of one of the San Roque diversion tunnels from the inside (Photo: Rich Humphries).

and soffits of the tunnels were already lined, a decision was required to determine whether the floors of the tunnels should be lined with concrete as well. At that point in time the contractor was behind schedule. If the concrete lining could be omitted from the tunnel floors it would not only lead to cost savings but would also allow the contractor to get back on schedule.

It was therefore decided to analyze the erodibility of the rock on the tunnel floors and compare its strength with the anticipated erosive capacity of the water that would flow through the tunnels during the construction period. If the rock was found to be strong enough to resist the anticipated erosive capacity of the water a proposal would be made to the Independent Board of Review (IBR) to omit the concrete lining from the tunnel floors. If it was found that certain portions of the tunnel floor were not strong enough to resist the erosive capacity of the water but others were, only the weak reaches would be protected and the rest of the tunnel floor would remain unprotected.

Data

Tunnel geometry. All three tunnels were constructed around bends. The initial reaches were straight, with bends located at a distance of 230 m from the upstream portals. The bends extended from a distance of 230 to 410 m along the tunnels, at which locations widened sections (known as flares) were constructed to secure concrete plugs to block the tunnels at the end of the construction project. The width of the flared sections was 13.4 m. The tunnel orientations, and therefore the flow directions in the tunnels, were:

190 degrees azimuth at the entrance
250 degrees azimuth in the central section
255 degrees azimuth in the downstream section

Tunnels 1 and 2 were 10.4 m wide and 14.45 m high, while tunnel 3 was 3.6 m wide and 6 m high. The invert elevations for tunnels 1 and 2 were 115 m at the entrance and 104.55 m at the exit. Those of tunnel 3 were 110 m at the entrance and 104.55 at the exit. Tunnels 1, 2, and 3 were respectively 852.5, 888.3, and 816.8 m long.

Material properties. The engineering geologist on site classified the rock encountered during construction of the tunnels into six categories: Theses were identified as rock at the tunnel inlets and outlets, and categories I, II-high, II-low, and III rock. Category I was sound rock, whereas category II-high and II-low were intermediate quality rock. Category III rock was of poor quality. The pertinent parameters that were used to quantify the relative ability of the different rock types to resist the erosive capacity of water are presented in Table 10.1.

The rock formation has three joint sets J1, J2, and J3. The dip and dip direction of these are,

Joint set J1: 85/095
Joint set J2: 67/170
Joint set J3: 67/345

Table 10.2 illustrates calculation of the erodibility index for the rock encountered in the tunnel (see Chap. 4 for guidelines on calculating the erodibility index). The mass strength number is equal to the UCS of the rock, while the block size number is calculated as the *RQD* divided by the

TABLE 10.1 Pertinent Material Properties of Rock Encountered in the Tunnels at San Roque Dam

Parameter	U/S portal	D/S portal	Cat. I	Cat. II-high	Cat. II-low	Cat. III
Intact rock strength (UCS)	50 MPa	20 MPa	100 MPa	50 MPa	20 MPa	5 MPa
RQD	70	50	95	80	70	(rock consists of 50 mm gravel sized particles)
Number of joint sets	3 + random	3 + random	3 + random	3 + random	3 + random	3 + random
Joint Separation	2 mm	2 mm	1 mm	2 mm	2 mm	10 mm
Joint roughness	Planar, rough	Planar, rough	Planar, rough	Planar, rough	Planar, rough	Planar, rough
Joint filling	clay	clay	calcite	chlorite	calcite, clay, and chlorite	clay

joint set number. The shear strength number is equal to the value of the joint roughness number divided by the joint alteration number. A conservative approach was followed in the selection of the orientation number (J_s) by assuming a value of 0.49 for the shape factor of one.

The threshold stream power, i.e., the stream power that is required to scour the rock, is shown in the last column for each of the rock categories on site. This is quantified by determining the value of the stream power associated with the erodibility index on the threshold graph developed by Annandale (1995) (see Chaps. 6 and 7).

Quantification of the erodibility index and the threshold stream power for category III rock differs from the procedure followed for the other rock classes. The reason for this is that the rock is essentially fractured. The block size number is calculated using the equation,

$$K_b = 1000D^3$$

where D is the average diameter of the fractured rock (m). The rest of the numbers were determined in a way similar to the other rock classes.

Erosive capacity of water. Stream power was used to quantify the relative magnitude of the erosive capacity of water. Two kinds of flow were expected in the tunnels, i.e., open channel and pressure flow. The equations that were used to calculate stream power for open channel flow and pressure flow are summarized in Table 10.2.

TABLE 10.2 Quantification of the Erodibility Index Using the Geologic Information from Table 10.1

Tunnel section/ rock category	UCS (MPa)	M_s	RQD	No. joint sets	J_n	K_b	Joint opening (mm)	Joint surface	Gouge	J_r	J_a	K_d	Joint spacing ratio	J_s	Erodibility index, K	$P_{threshold}$ (KW/m^2)
D/S portal	20	20	50	3 + random	3.34	15.0	2	planar rough	clay	1.5	4	0.375	1	0.49	55	20
U/S portal	50	50	70	3 + random	3.34	21.0	2	planar rough	clay	1.5	8	0.188	1	0.49	96	31
Cat I	100	100	95	3 + random	3.34	28.4	1	planar rough	calcite	1.5	4	0.375	1	0.49	523	109
Cat II-high	50	50	80	3 + random	3.34	24.0	2	planar rough	chlorite	1.5	4	0.375	1	0.49	220	57
Cat II-low	20	20	70	3 + random	3.34	21.0	2	planar rough	calcite, clay, and chlorite	1.5	8	0.188	1	0.49	39	15
Cat III	5	5	0	3 + random		1.3E-01	10	planar rough	clay	1.5	13	0.115	1	0.49	4.E-02	2.E-01

Open channel flow. The largest portion of energy dissipation in open channel flow occurs on the bed of the channel. The power expended at the bed in open channel flow ($SP_{\text{open channel}}$), expressed in kW/m^2, can be calculated with Eq. (10.1).

$$SP_{\text{channel}} = \gamma q S_f \tag{10.1}$$

where γ = unit weight of water (kN/m^3)
q = unit discharge (m^3/s/m) = Q/W
Q = total discharge (m^3/s)
W = width of tunnel (m)
S_f = energy slope (assumed to be approximated by the estimated slope of the water surface)

A more accurate expression for calculating stream power in open channels is (see Chap. 5)

$$SP_{\text{channel}} = 7.853\rho\left(\sqrt{\frac{\tau}{\rho}}\right)^3$$

Equation (10.1) was used during execution of this investigation because the understanding represented by the equation referenced here was not available at that time.

Pressure flow. Streeter (1971) shows that the rate of energy dissipation of pressure flow in conduits (SP_{pressure}) is equal to the net power input, which can be expressed as

$$\text{Net power input} = Q\Delta p \tag{10.2}$$

where Δp is the pressure drop over distance L along the tunnel.

In the case of pressure flow, the major portion of the power is expended on the circumference of the tunnel. Therefore, the stream power per unit area on the tunnel circumference was conservatively expressed as

$$SP_{\text{pressure}} = \frac{Q\Delta p}{LP} \tag{10.3}$$

where P is the wetted perimeter of the tunnel over the distance L (m).

Equations that can be used to calculate the stream power under pressure flow conditions at the entrances to the tunnels, in its straight reaches and at bends, and at the flares where the proposed plugs were constructed at completion, were derived by using Eq. (10.3) as basis. These are summarized as follows:

Tunnel entrances:

$$SP_{\text{entrance}} = \frac{K_{\text{entrance}} \gamma Q^3}{2gPA^2L} \tag{10.4}$$

where K_{entrance} = entrance loss coefficient
g = acceleration due to gravity (9.82 m/s^2)
A = cross sectional flow area of the tunnel (m^2)

Straight reaches of tunnel:

$$SP_{\text{tunnel}} = \frac{\gamma f Q^3}{2gDPA^2} \tag{10.5}$$

where f is the Darcy–Weisbach friction factor = 0.0318 for tunnels 1 and 2 and is 0.053 for tunnel 3 and $D = 4A/P$ is the hydraulic radius of tunnel (m).

Bends:

$$SP_{\text{bend}} = \frac{\gamma(fL/D + K_{\text{bend}})Q^3}{2gLPA^2} \tag{10.6}$$

where K_{bend} is the bend loss coefficient = 0.066.

Flare at plug location:

$$SP_{\text{flare}} = \frac{\gamma(fL/D + K_{\text{flare}})Q^3}{2gLPA^2} \tag{10.7}$$

where K_{flare} is the energy loss coefficient at flare = 0.078.

Erosion assessment

The potential erosion of the rock in the tunnel floors were determined by comparing the available stream power provided by the flowing water to the threshold stream power of the different rock categories for each of the flow conditions. An example of the erosion assessment performed for the straight reaches in the three tunnels for category II-low rock is shown in Table 10.3. The threshold steam power of the rock (the stream power required to scour the rock) is shown at the top of the table, i.e., 15 kW/m^2. When the available stream power exceeds the required stream power the rock is deemed to be susceptible to scour. In cases where the required steam power of the rock exceeds the available stream power, the rock is deemed to be strong enough to resist the erosive capacity of the water.

TABLE 10.3 Scour Evaluation in Straight Reaches of Tunnels for Category II-Low Rock

Input parameters	Tunnel 1	Tunnel 2	Tunnel 3	Units
Darcy Weisbach, *f*	0.0318	0.0318	0.053	
Wetted perimeter	45.23	45.23	21.42	m
Area	138.67	138.67	32.14	m^2
Unit weight of water	9.82	9.82	9.82	kN/m^3
SP required to scour	15	15	15	kW/m^2
Accel. of gravity, *g*	9.82	9.82	9.82	m/s^2
Hydraulics radius, *D*	12.3	12.3	6.0	m
Width of channel, *W*	10.4	10.4	6	m
Length of tunnel, *L*	852.5	888.3	816.8	m

	Tunnel 1				Tunnel 2				Tunnel 3					
Headwater (m)	*Q*(1) (m^3/s)	Water slope	SP available (kW/m^2)	Scour (Y/N)	*Q*(2) (m^3/s)	Water slope	SP available (kW/m^2)	Scour (Y/N)	*Q*(3) (m^3/s)	Water slope	SP available (kW/m^2)	Scour (Y/N)	Total *Q* (m^3/s)	Comments
110	0				0				0				0	
114	0				0				124	0.016528	3.4	No	124.4207	Tunnel 3, open channel flow
115	0				0				175	0.019344	5.5	No	175.1159	Tunnel 3, open channel flow
120	600	0.015836	9.0	No	581.9895	0.0151976	8.35	No	152		1	No	1334.671	Tunnel 1 & 2, open channel flow
125	1580	0.018534	27.7	Yes	1532.1529	0.0177868	25.73	Yes	192		1	No	3303.987	Tunnel 1 & 2, open channel flow
130	1412		4	No	1395.0152		4	No	216		2	No	3022.614	
135	1547		6	No	1528.1625		5	No	236		3	No	3311.108	
140	1579		6	No	1560.454		6	No	241		3	No	3381.074	

(Continued)

TABLE 10.3 Scour Evaluation in Straight Reaches of Tunnels for Category II-Low Rock (*Continued*)

	Tunnel 1				Tunnel 2				Tunnel 3					
Headwater (m)	$Q(1)$ (m^3/s)	Water slope	SP available (kW/m^2)	Scour (Y/N)	$Q(2)$ (m^3/s)	Water slope	SP available (kW/m^2)	Scour (Y/N)	$Q(3)$ (m^3/s)	Water slope	SP available (kW/m^2)	Scour (Y/N)	Total Q (m^3/s)	Comments
141.8	1635		7	No	1615.5998		6	No	250		3	No	3500.56	10-year flood
150	1873		10	No	1850.6968		9	No	286		5	No	4009.95	
155	2007		12	No	1982.6891		12	No	306		6	No	4295.941	
156	2036		13	No	2011.9195		12	No	311		6	No	4359.276	
157	2066		13	No	2040.7313		13	No	315		6	No	4421.703	
158	2094		14	No	2069.1419		13	No	320		7	No	4483.261	
159	2119		14	No	2092.9877		14	No	323		7	No	4534.928	
160	2143		15	No	2116.7947		14	No	327		7	No	4586.512	
160.13	2147		15	No	2120.9278		14	No	328		7	No	4600.107	30-year flood
161	2167		15	No	2141.246		15	No	331		7	No	4738.491	
162	2191		16	Yes	2164.5224		15	No	334		7	No	4969.938	
163	2214		16	Yes	2187.551		16	Yes	338		8	No	5254.24	
164	2237		17	Yes	2210.3398		16	Yes	342		8	No	5581.198	
164.06	2239		17	Yes	2211.6601		16	Yes	342		8	No	5601.945	100-year flood
165	2259		17	Yes	2232.2422		17	Yes	345		8	No	5943.508	Coffer Dam Crest

Table 10.3 indicates that none of the category II-low rock in tunnel 3 was expected to scour, while this rock would scour in tunnels 1 and 2 under certain flow conditions. When rock that would scour was identified during construction it was protected against erosion by armoring those portions of the tunnel floor with concrete.

Summary of results

Table 10.4 summarizes the results of the scour assessment for tunnels 1 and 2, while the same for tunnel 3 is shown in Table 10.5. In general, it was concluded that scour is expected to occur at the entrance and exits to all tunnels. Away from the entrances and exits, category I and II-high rock are expected to be resistant to scour for all flow conditions, up to the 100-year flood, in all three tunnels.

Category II-low rock was not expected to scour in the straight and bend reaches of tunnel 3. However, the available stream power was very close to the required stream power to scour category II-low rock at the flares in this tunnel for flood events with recurrence intervals exceeding the 30-year event.

It was also found that scour of category II-low rock in tunnels 1 and 2 could commence at approximately 3300 m^3/s (total flow) during *open channel flow* conditions at the bends and straight reaches. However, when flows in these two tunnels change to pressure flow, the erosive capacity of the water at the tunnel inverts reduces. The reason for this apparent anomaly is that most of the energy in open channel flow is dissipated on the bed, whereas the energy dissipation under pressure flow conditions occurs over the entire circumference of the tunnel (see e.g., Streeter, 1971).

Category III rock is of poor quality and was expected to scour in all tunnels under all flow conditions.

Based on the scour analysis the following recommendations were made (also see Table 10.6).

- Scour protection was required at the entrances and exits to all three tunnels.
- Category I and II-high rock did not require scour protection in any of the tunnels.

TABLE 10.4 Scour Potential for Tunnels 1 and 2

Rock class	Entrance	Exit	Straight	Bend	Flare
U/S	Scour				
D/S		Scour			
I			No scour	No scour	No scour
II-high			No scour	No scour	No scour
II-low			Scour	Scour	Scour
III			Scour	Scour	Scour

TABLE 10.5 Scour Potential for Tunnel 3

Rock class	Entrance	Exit	Straight	Bend	Flare
U/S	Scour				
D/S		Scour			
I			No scour	No scour	No scour
II-high			No scour	No scour	No scour
II-low			No scour	No scour	No scour below the 30-year event, but scour expected for flow events exceeding the 30-year event.
III			Scour	Scour	Scour

- Category II-low rock required scour protection at all locations where it is encountered in tunnels 1 and 2.
- Category II-low rock did not require protection against scour in the straight and bend reaches of tunnel 3.

TABLE 10.6 Scour Protection Recommendations Developed for the Diversion Tunnels at San Roque Dam

Rock type	Tunnel 1	Tunnel 2	Tunnel 3
Entrance to tunnels	Scour protection required	Scour protection required	Scour protection required
Exit from tunnels	Scour protection required	Scour protection required	Scour protection required
Category I	No scour protection required in straight reach, bend or at flare.	No scour protection required in straight reach, bend or at flare.	No scour protection required in straight reach, bend or at flare.
Category II–high	No scour protection required in straight reach, bend or at flare.	No scour protection required in straight reach, bend or at flare.	No scour protection required in straight reach, bend or at flare.
Category II–low	Scour protection required at all locations	Scour protection required at all locations	No scour protection required in straight reaches and bend. Scour protection required at flare (plug location) for all flood magnitudes greater than 30-year recurrence interval.
Category III	Scour protection required at all locations	Scour protection required at all locations	Scour protection required at all locations

Figure 10.4 Flood discharge of about 3500 m^3/s flowing through the diversion tunnels. Tunnel 3 is completely submerged (Photo: Rich Humphries).

- Category II-low rock required protection for flood magnitudes exceeding the 30-year recurrence interval if encountered at the location of the flare in tunnel 3.
- Category III rock is weak and requires protection against scour at all locations where it was encountered in tunnels 1, 2, and 3.

Tunnel performance

During the dam construction period high flows were experienced on a number of occasions, which discharged through the tunnels (Fig. 10.4). The unlined tunnel floor resisted the scour as predicted by the scour analysis. This can be seen in Fig. 10.5, which shows the condition of the unlined tunnel floor after flooding. The figure also shows the condition of the shotcrete on the wall of the tunnel.

Ricobayo Dam

Ricobayo Dam consists of a 99 m high double-curvature arch dam that is located on the Esla River, which is a tributary to the Duero River in Spain. Construction of the dam and spillway commenced in 1929 and was completed in 1933. The power station at the base of the dam was completed in 1935.

The spillway, with a discharge capacity of 4650 m^3/s, is located on the left abutment and originally consisted of a 400 m long unlined chute with

Figure 10.5 Condition of rock on tunnel floor after flood occurrence–no significant scour (Photo: Rich Humphries).

a longitudinal slope of 0.45 percent that traversed across a rock outcrop before discharging over a cliff face and returning to the Esla River (Fig. 10.6). The upstream approach to the spillway is at elevation 670 m while the spillway crest is at an elevation of 674 m. Four gates with dimensions of 20.80 m × 10.60 m are located on the spillway, separated by intermediate piers that are each 3 m wide. The rock outcrop over which the spillway channel traversed consisted of open-jointed granite. An anticline and a fault ran across the spillway chute.

Figure 10.6 General Layout of Ricobayo Dam and spillway, also showing hydraulic model study of spillway in action. (1) Arch dam. (2) Hydropower plant. (3) Spillway chute.

Five separate scour events occurred along the spillway chute, commencing soon after commissioning of the dam. Each of the flood events occurred over a period of several months, usually from December to June of the next year. Figure 10.7 pictorially shows the progression of the scour as a function of time, from January 1934 to January 1939 (bottom portion of figure). A photograph showing the flood event of 1939 is not available. An additional scour event occurred in 1962.

Right from the beginning the scour of the granitic rock advanced rapidly upstream and lead to safety concerns. Attempts to minimize the scour and stabilize the rock were made after the flood events in 1934 and 1935. Additional protection was added in 1942 and after the flood event in 1962.

The vertical face of the drop and the right hand side of the pool that developed up to 1934 was protected with concrete (Fig. 10.8). Additional scour, approximately 25 m downwards, occurred at the base of the pool during the 1935 flood that reached a peak of 1000 m^3/s (Fig. 10.7).

After that flood event a concrete lip was added to the end of the spillway chute that would project the jet further away from the face of the drop. This structure failed during the course of the 1936 flood event, which had a peak discharge of 1280 m^3/s, as the pool deepened by another 30 m and regressive scour occurred along the face of the drop (Figs. 10.7 and 10.9). The flood of 1939 reached 3230 m^3/s, causing additional damage at the drop along the end of the spillway chute. The plunge pool did not scour any deeper during this event (Fig. 10.7).

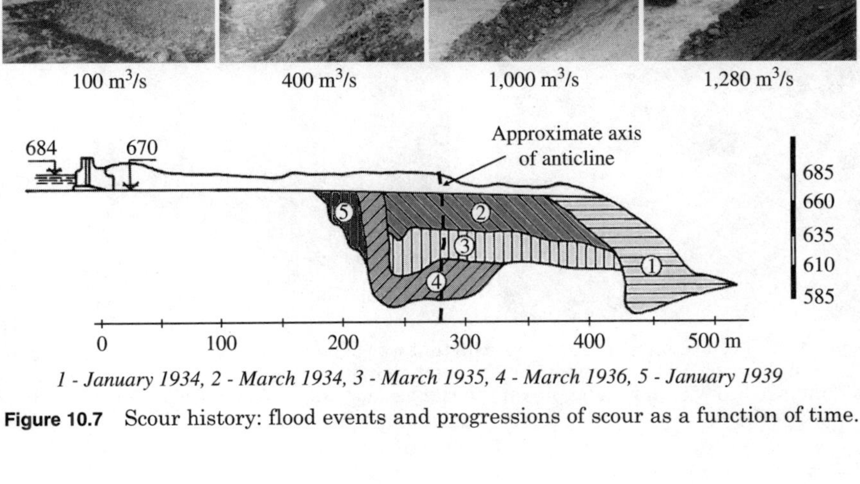

1 - January 1934, 2 - March 1934, 3 - March 1935, 4 - March 1936, 5 - January 1939

Figure 10.7 Scour history: flood events and progressions of scour as a function of time.

Figure 10.8 Concrete protection of the drop at the end of the spillway chute and the concrete protection to the right side of the plunge pool added in 1934.

In spite of the fact that the plunge pool did not become deeper a decision was made to design and construct measures to protect the pool and the drop face against future scour. These measures were implemented during the early 1940s. The protection measures that were implemented entailed installation of a concrete lining in the plunge pool, concrete protection of the spillway channel and additional protection of the drop at the end of the spillway channel.

The flood event that occurred in 1962 reached a peak of 4800 m^3/s and resulted in failure of the concrete plunge pool floor that was installed in 1942. The erosive capacity of the flood lead to the destruction of the

Figure 10.9 Performance of the concrete lip that was added to the end of the spillway chute during 1935. The photo on the right shows the flow over the spillway on January 8, 1936 and the photo on the right shows its performance on April 28, 1936 after failure.

Figure 10.10 Concrete lining of plunge pool and addition of splitters at the end of the spillway chute to break up the jet prior to plunging into the pool.

2 m thick concrete floor and removed individual concrete elements with masses of up to 50 t each. After that event hydraulic splitters were added to the end of the spillway channel to break up the jet prior to it plunging into the pool (Fig. 10.10). Since then floods with peak magnitudes ranging from 3000 to 3500 m^3/s occurred without causing additional damage (Fig. 10.11).

Local geology

The Ricobayo spillway is located within a granite massif known as the Ricobayo Batholith. The rock contains two systems of joint sets (A and B

Figure 10.11 Splitter performance during a flood in 1985.

in Fig. 10.12). Joint sets A and A`, comprising system A, are generally near vertically dipping with attitudes of S17 E, 80 SW and N23 E, 80 SE, respectively. Joint sets, B and B`, comprising system B, are more horizontally dipping with attitudes of S68 W, 30-40 NW and N68 E, 10-20 SE, respectively. Most likely, system B was originally one joint set but since its formation tectonic activity resulted in the formation of an anticline, resulting in differing dips to the north and south of the anticline axis (Fig. 10.12). The anticline trends N75 E and intersects the middle of the spillway at an angle of approximately 40°. Both joint sets (A and B) are relatively planar, but system B appears to be more persistent and continuous.

An opinion expressed by a geologist after the occurrence of scour was that the joints of the rock that failed most probably contained clay and that its presence contributed to the failure. However, observations during a site visit in 2005 indicated no clay gouge in the rock strata surrounding the plunge pool. The dry climate in the region is unlikely to lead to the development of clay from the weathered granite. The field characterization of the joints performed during the site visit in 2005 is that the joint separation of the rock is less than 5 mm (with a maximum separation of about 10 mm at some locations near the original surface). The 2005 site visit indicated that gouge may consist of rock flour, but no clay was observed.

For simplicity, the area to the south of the anticline containing joint set B` will be referred to as zone 1, while the area to the north of the

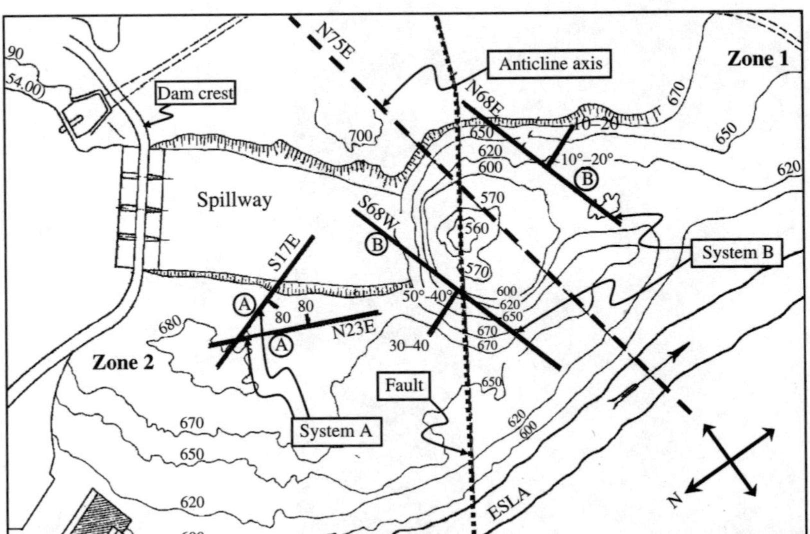

Figure 10.12 Structural geology of the Ricobayo spillway.

anticline containing joint set B will be referred to as zone 2. Additionally, a near vertical fault trends perpendicular to the spillway at S60 E. The fault trace shows mylonic texture, indicative of intense shearing. Rock quality within the fault zone is likely to be poor. Weathered bedrock sits atop competent bedrock in exposed areas.

Qualitative analysis of scour

The scour events at Ricobayo Dam provide excellent examples of how geology affects scour. Figure 10.7 (bottom portion of the figure) displays a cross-sectional view of the five scour events as well as the approximate location of the anticline axis. It is quite noticeable that to the right (south) of the anticline axis (i.e., zone 1) the scour progression is in the horizontal direction (events 2 and 3), while to the left (north) of the anticline axis (i.e., zone 2) the scour progression is in the vertical direction (event 4). The likely cause of this switch in scour direction is due to the difference in joint structure between zone 1 and zone 2. Figure 10.13 represents a cross-section of the spillway showing generalized joint patterns of systems A and B.

The dominant joint structure in zone 1, i.e., the dip direction of joint system B, which is aligned with the flow direction, is more conducive to scour than the dominant joint structure in zone 2. The persistent dominant orientation of joint system B in zone 2 is against the direction of flow. The latter system will therefore provide more resistance to the removal of rock blocks than the former (also see Chap. 4).

Scour event 2 (March 1934) consisted of two large landslides occurring within one day of each other, each removing about one-half of the total volume of material scoured during the course of that event. Evidence of failure along joint planes can be seen in Fig. 10.14, which

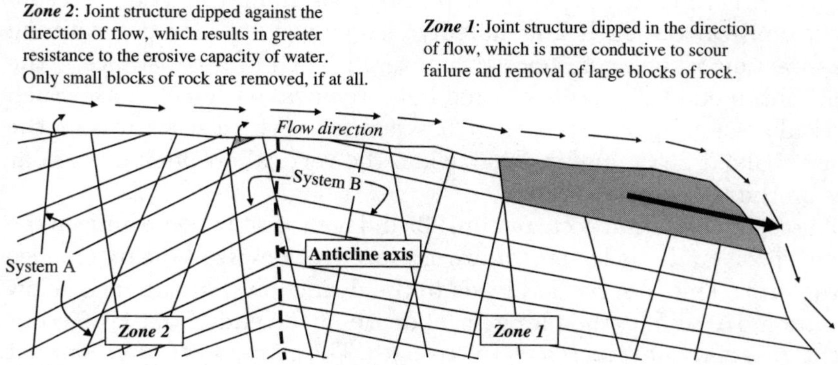

Figure 10.13 Cross-section showing joint patterns.

Figure 10.14 Failure along joint plane A`.

shows that almost the entire western rock face at the exit to the Elsa River, is defined by joint plane A`.

Once scour proceeded into zone 2, on the northern side of the anticline, the rapid upstream migration of scour reduced considerably. The orientation of joint set B, which dips against the direction of flow (compared to joint set B` that dips in the direction of flow), provides greater resistance to the removal of large blocks of rock by scour processes. Instead only smaller pieces of rock will be removed from the channel bed, if at all. The orientation of the scour therefore changed, from removal of rock blocks in an almost horizontal plane to vertical deepening of the plunge pool.

Additionally, the presence of the fault played a significant role in determining the orientation of the plunge pool that formed downstream of the spillway outlet. It undoubtedly also played a role in determining the direction of the scour, once encountered. The lower resistance of the rock contained in the fault resulted in its removal with more ease. Such vertical deepening occurred in 1935, when the vertical scour was on the order of about 25 m, and in 1936, when the vertical deepening by scour was on the order of about 30 m.

The scour event that occurred in 1939 did not result in deepening on the plunge pool, but it did lead to failure of a large rock mass at the upper edge of the drop. This observation leads to the deduction that the rock below elevation 570 m (i.e., the plunge pool depth at the end of the 1936 scour event) most probably increases in strength. The geologic information that is available does not provide an indication of how strong this rock might be, but its minimum strength can be inferred from the results of rock sour analysis presented further on.

Additional indications of the strength of the rock at the base of the plunge pool, and its ability to resist the erosive capacity of the water, can be determined by analyzing the flood event that occurred in 1962. This flood event lead to the destruction of the concrete lining that was constructed at the base of the plunge pool prior to that event, but did not result in deepening of the plunge pool. This means that the concrete that was placed on top of the rock was weaker than the underlying rock.

Quantitative Analysis of Scour

A quantitative analysis of scour at Ricobayo Dam was performed for Events 3, 4, 5, and 6 (i.e., 1935, 1936, 1939, and 1962). The maximum scour depth occurs when the erosive capacity of the jet is less than the ability of the rock to resist it. Calculated and observed scour depths were compared.

The erosive capacity of the jet plunging into the pool that formed diminishes as a function of plunge pool depth. Quantification of the erosive capacity of a jet at various elevations below the water surface of the plunge pool therefore becomes an important component of the analysis. When applying the erodibility index method the variation in the erosive capacity of the jet below the plunge pool water surface elevation is determined by multiplying the stream power of the jet per unit area at the water surface elevation of the pool with the change in total dynamic pressure coefficient (i.e., the sum of the average and fluctuating dynamic pressure coefficients) as a function of water depth. It has been shown in Chap. 5 that this is a conservative approach to quantifying the relative change in stream power originating from a jet impinging into a plunge pool.

Figure 10.15 is a schematic showing the calculation of the total dynamic pressure coefficient as a function of dimensionless plunge pool depth. When using Castillo's (1998) relationships for calculating the average dynamic pressure coefficient as a function of dimensionless plunge pool depth it is necessary to estimate the breakup length ratio of the jet. Once the latter is known a curve expressing the average dynamic pressure as a function of dimensionless depth is selected for quantifying the change in average dynamic pressure coefficient as a function of plunge pool depth [Fig. 10.15(a)].

The variation in fluctuating dynamic pressure coefficient as a function of dimensionless plunge pool depth is determined once the issuance turbulence intensity is known. Quantification of the issuance turbulence intensity allows selection of an appropriate curve relating the fluctuating dynamic pressure coefficient to dimensionless depth [Fig. 10.15(b)].

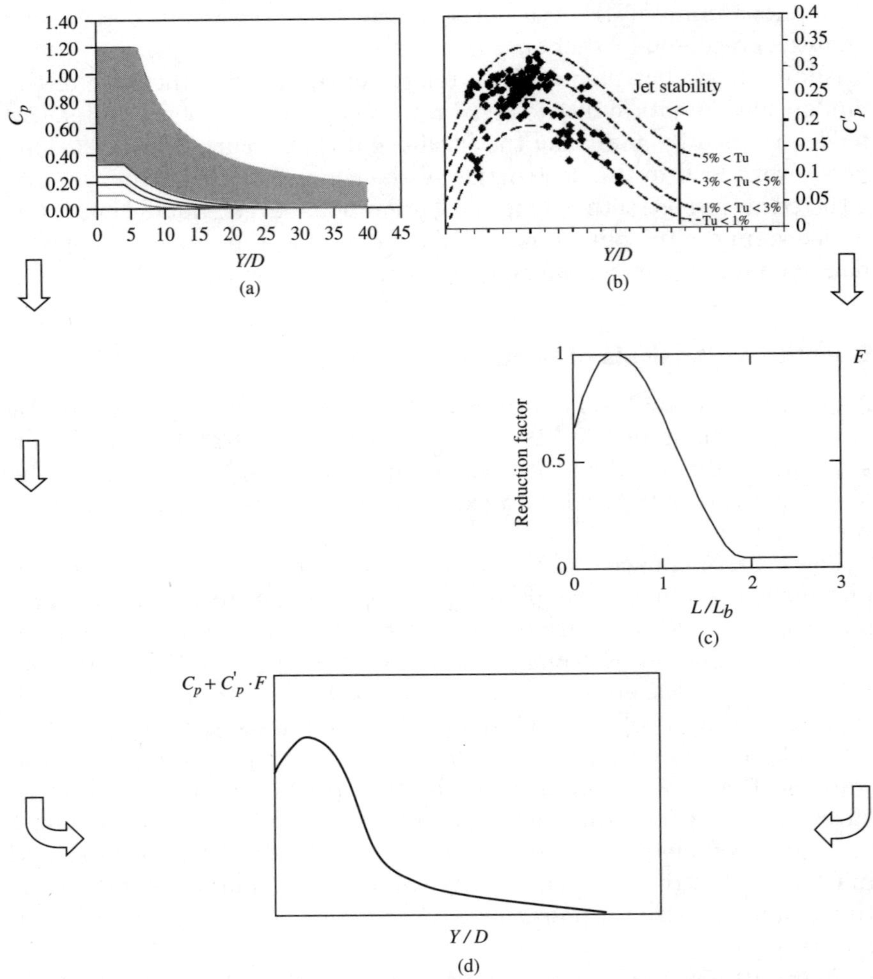

Figure 10.15 Development of total dynamic pressure coefficient. (a) Average dynamic pressure coefficient as a function of dimensionless plunge pool depth and jet breakup (Castillo 1998). (b) Bollaert's (2002) fluctuating dynamic pressure coefficient as a function of dimensionless plunge pool depth and turbulence intensity. (c) Fluctuating dynamic pressure reduction coefficient as a function of jet breakup length ratio. (d) Total dynamic pressure coefficient as a function of dimensionless plunge pool depth.

The effect of jet breakup on the fluctuating dynamic pressures in a plunge pool is estimated by assuming that the relationship between fluctuating dynamic pressures and jet breakup length ratio established by Ervine et al. (1997) can be made dimensionless and used for all dimensionless depths [Figs. 10.15(c) and 10.16]. An equation representing the change in the dimensionless fluctuating dynamic pressure

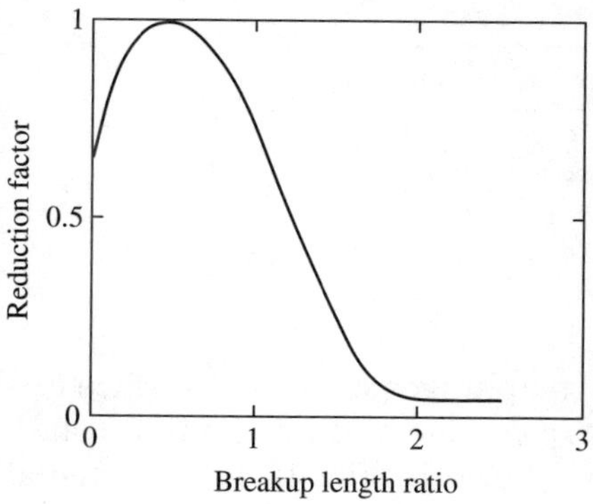

Figure 10.16 Fluctuating pressure reduction factor allowing for jet breakup.

reduction factor F and the breakup length ratio shown in Fig. 10.16 is expressed as follows:

$$F = 0.595L_{br}^3 - 2.075L_{br}^2 + 1.583L_{br} + 0.645 \qquad \text{for } L_{br} \leq 1.9 \quad (10.8)$$

$$F = 0.113 \qquad \text{otherwise}$$

where F = fluctuating dynamic pressure reduction factor
$L_{br} = L/L_b$ = jet breakup length ratio
L = total length of jet trajectory
L_b = jet breakup length

Once the average dynamic pressure coefficient (C_p), the fluctuating dynamic pressure coefficient (C_p') and the fluctuating dynamic pressure reduction factor (F) are known, the total dynamic pressure coefficient as a function of dimensionless water depth (Y/D) can be determined [Fig. 10.15(d)], i.e.,

$$C_t(Y/D) = C_p(Y/D) + C_p'(Y/D)F$$

Jet breakup

The role of jet breakup in determining the total dynamic pressure coefficient has been briefly summarized in the previous section. For rectangular nappe jets, such as the case at Ricobayo Dam, the breakup length of plunging jets may be calculated from an equation by Horeni (1956):

$$L_b = 6 \cdot q^{0.32}$$

where q is the discharge per unit width over the spillway ($m^3/s/m$).

Table 10.7 presents calculated jet breakup lengths at Ricobayo Dam for the scour events that were analyzed, indicating that the jets were

TABLE 10.7 Breakup Length for Flow at Ricobayo Dam

Date	Q (m^3/s)	q (m^2/s)	Jet breakup length (m)	L/L_b
1935	1000	24.7	17	4.2
1936	1280	31.6	18	3.9
1939	3230	79.8	24	2.9
1962	4800	118.5	28	2.5

broken up by the time they reached the plunge pool water surface elevation. The location where jet breakup commenced for the flow conditions that occurred in March 1936 is shown in Fig. 10.17. The visual identification of the commencement of jet breakup appears to correlate reasonably well with the calculated value (i.e., about one quarter of the total drop length).

Jet impact dimension

The jet impact dimension D used to calculate the dimensionless depth Y/D is the thickness of a rectangular jet, as is the case at Ricobayo. Castillo

Figure 10.17 Breakup of plunging jet at Ricobayo Dam.

(1998), who investigated rectangular jets, uses the overall thickness of the jet at impact as the indicator parameter (not the core thickness) i.e.,

$$D_j = \frac{q}{\sqrt{2gZ}} + 4\varphi\sqrt{h_o}\left[\sqrt{Z} - \sqrt{h_o}\right] \tag{10.9}$$

where $\varphi = 1.07T_u$

h_o = characteristic flow depth dimension over a free-flowing ogee spillway that is defined by Castillo as

$$h_o \approx 2D_i \approx 2\left(\frac{q}{C_d}\right)^{2/3}$$

$C_d = 2.1$ = discharge coefficient

This equation is applicable to nappe jets only, i.e., jets flowing through critical depth just prior to plunging. Experience has shown that an apparently reasonable estimate of the jet footprint for non-nappe jets is obtained by adding the jet expansion to the jet thickness at issuance, i.e.,

$$D_j = D_i + 4\varphi\sqrt{h_o}\left[\sqrt{Z} - \sqrt{h_o}\right] \tag{10.10}$$

This equation was used to calculate the overall jet thickness of the jet for Ricobayo Dam.

Jet stream power

The stream power of the jet per unit area at the point of impingement into the plunge pool, i.e., at the water surface elevation, is calculated as

$$P_{\text{jet}} = \frac{\gamma_w QZ}{A}$$

where A is the footprint area of the jet at impact = D_jW.

The variation of stream power as a function of plunge pool depth is determined by multiplying the stream power of the jet at the water surface elevation with the total dynamic pressure coefficient i.e.,

$$P_{\text{jet}}(Y/D) = C_t(Y/D)\frac{\gamma_w QZ}{A}$$

Scour threshold—erodibility index method

The threshold stream power of the rock at Ricobayo Dam was determined by making use of the erodibility index method. The fairly limited engineering geologic information that is available for Ricobayo Dam was combined with field observations that were made during a site visit in 2005 to estimate the scour threshold of the rock.

It was concluded that the rock strength possibly increases as a function of elevation below the water original ground surface. The principal reason for the increase in strength is considered to result from changes in the joint properties of the rock. Site observations indicated that the joint separation of the rock could possibly be on the order of about 5 mm (with a maximum of 10 mm) and that it contained no significant gouge. It is anticipated that the joint separation decreases as a function of elevation below the original ground surface, which leads to the anticipated increase in strength. However, no boring logs were available for accurately determining the change in rock strength as a function of elevation. A range of maximum and minimum rock strength was therefore determined using the erodibility index method. The erodibility index K for the rock was calculated using the following equation (Annandale 1995):

$$K = M_s K_b K_d J_s$$

The threshold stream power of the rock, expressed in kW/m^2, was calculated with the equation:

$$P_{\text{rock}} = K^{0.75}$$

[This equation expresses the relationship between the threshold stream power and the K on Annandale's (1995) scour threshold graph.]

Table 10.8 provides a summary of the values used for estimating the range of the threshold stream power of the rock mass at Ricobayo Dam.

TABLE 10.8 Erodibility Index Parameters and Rock Scour Threshold

Unit	UCS (Mpa)	M_s	RQD	J_n	K_b	J_r	J_a	K_d	J_s	Erodibility index (K)	Resisting power (kW/m^2)
Granite-zone 2-high	75	75	70	2.73	25.64	1.50	2	0.75	0.5	721.15	139.16
Granite-zone 2-low	75	75	70	2.73	25.64	1.50	4	0.38	0.5	360.58	82.75
Concrete (high)	35	35	100	1.5	66.67	3.00	1	0.58	1	1347.15	222.36
Concrete (low)	15	15	100	1.5	66.67	3.00	1	0.58	1	577.35	117.78

The upper rock strength is for rock with an assumed joint separation of 5 mm. The lower rock strength is for a rock separation of 10 mm, which is unlikely to penetrate to any significant depth below the original ground surface.

The estimated unconfined compressive strength of the rock, obtained during the site visit in 2005 by means of field observations, is set at about 75 MPa and the *RQD* at about 70 percent or higher. The dip and dip direction of the rock were obtained from the geologic report (see earlier discussions) and the number of joint sets was set at three. Table 10.8 shows the estimated range of threshold stream power of the rock in zone 2.

Table 10.8 also presents an estimate of the scour resistance of the concrete that was used to line the plunge pool floor. The strength of the concrete that was used is not known, but engineers at Iberdrola, which owns the dam, indicated that it could range somewhere between 15 MPa and 35 MPa. This range was therefore used to estimate the scour threshold of the concrete. The thickness of the concrete lining averaged about 2 m. The table indicates that the concrete lining is estimated to be more erosion resistant than the rock that scoured. However, as shown in the next subsection, the scour analysis indicates that the rock below elevation 570 m is most probably much stronger than the estimates shown in Table 10.8.

Scour extent

The extent of scour is determined by comparing the threshold stream power of the rock and the stream power of the jet below the water surface elevation. This comparison is provided in Fig. 10.18 for the scour that occurred during 1935.

The scour of rock during the 1935 event increased the pool depth from an elevation of about 640 m to 600 m, a maximum increase in depth of about 40 m. Comparison between the stream power of the jet in Fig. 10.18 and the maximum resistance offered by the rock indicates that the stream power of the jet is initially significantly greater than the threshold stream power of the rock. Once an elevation of about 622 m is reached the stream power of the jet gradually decreases until it is estimated to be lower than the maximum estimated strength of the rock at an elevation of approximately 602 m. The maximum observed scour depth is on the order of about 600 m. This indicates that the estimate of threshold stream power of the rock based on a 5 mm joint separation appears to be more realistic than the 10 mm assumption. The calculated and observed scour depths compare favorably for the 1935 event.

During the flood event of 1936 the rock scoured approximately another 30 m downwards. The comparison between the threshold steam power of the rock and that of the jet for the 1936 event, shown in Fig. 10.19, indicates favorable agreement between calculated and observed scour.

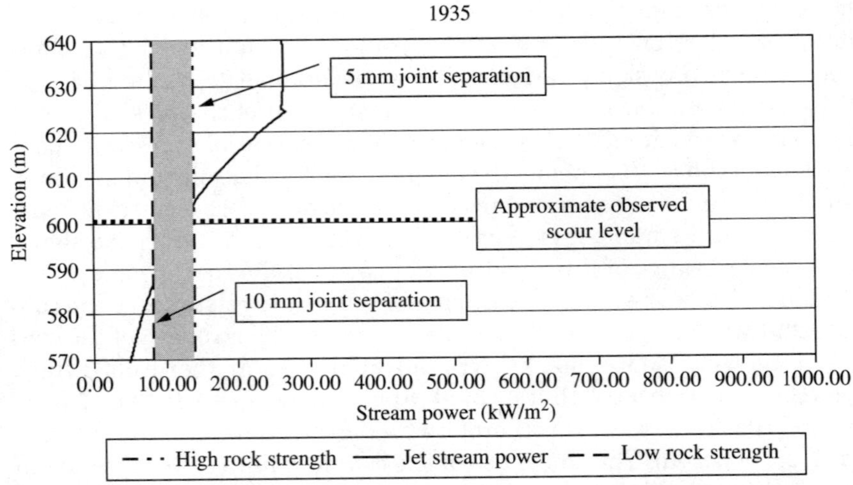

Figure 10.18 Comparison between jet stream power and scour threshold of rock for the flood event that occurred in 1935.

The stream power of the jet decreases below the scour threshold of the rock when using a joint separation of 5 mm to characterize the rock.

However, as indicated by the analysis of the 1939 and the 1962 flood events, the rock below an elevation of 570 m may be stronger than estimated with the surface data. The flood peak during the 1939 flood was on the order of about 3230 m^3/s. This is much greater than the peak discharge that occurred in 1936, which amounted to 1280 m^3/s. However, no additional downward scour was observed during this flood event.

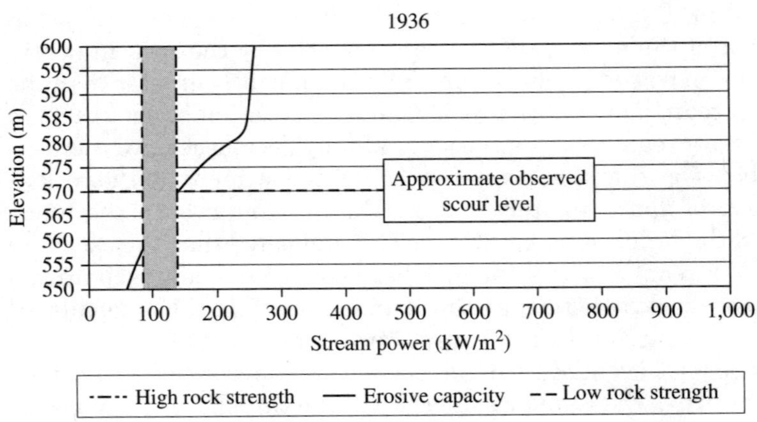

Figure 10.19 Comparison between jet stream power and scour threshold of rock for the flood that occurred in 1936.

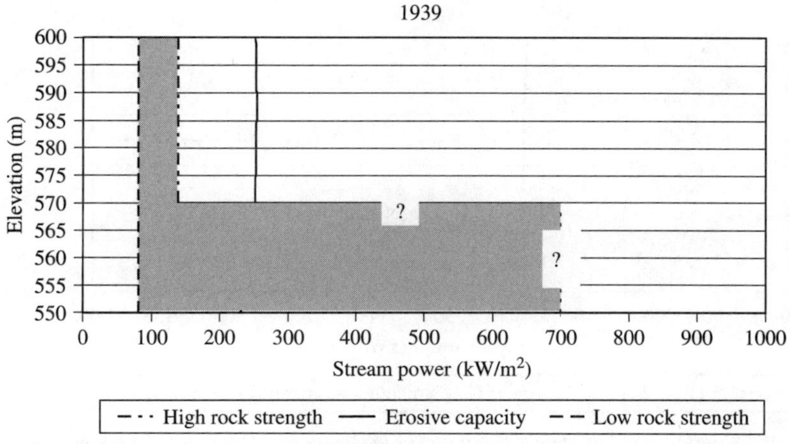

Figure 10.20 Comparison between stream power of the jet and the scour threshold of rock for the flood event of 1939.

The maximum scour depth of about 570 m was retained. Figure 10.20 shows that the jet stream power of the 1939 flood event was significantly larger than that which occurred during 1936. Therefore, if the rock did not experience any deepening the threshold stream power of the rock should be greater than about 260 kW/m^2, which is the magnitude of the stream power of the jet at elevation 570 m.

After the flood of 1939 the plunge pool was strengthened with a continuous concrete lining averaging about 2 m in thickness. The concrete lining was destroyed during the 1962 flood event, which had a peak discharge of about 4800 m^3/s. However, the plunge pool did not increase in depth, but remained at elevation 570 m. This means that the underlying rock must have greater resistance to scour than the concrete lining.

A comparison between the scour resistance of the rock and concrete, and the erosive capacity of the jet is shown in Fig. 10.21. This figure indicates that the concrete had a much lower scour resistance than the erosive capacity of the jet. The calculation thus confirms the observation that the concrete lining failed. It also indicates that the rock at this elevation must have a strength in excess of about 380 kW/m^2.

Summary

The scour analysis using the erodibility index method shows good agreement between observed and calculated scour of the 1935 and 1936 flood events. However, additional analysis indicates that the scour depth that was reached in 1936 might be due to a rock layer that might have much greater resistance that the estimated value that lead to the maximum scour in 1936.

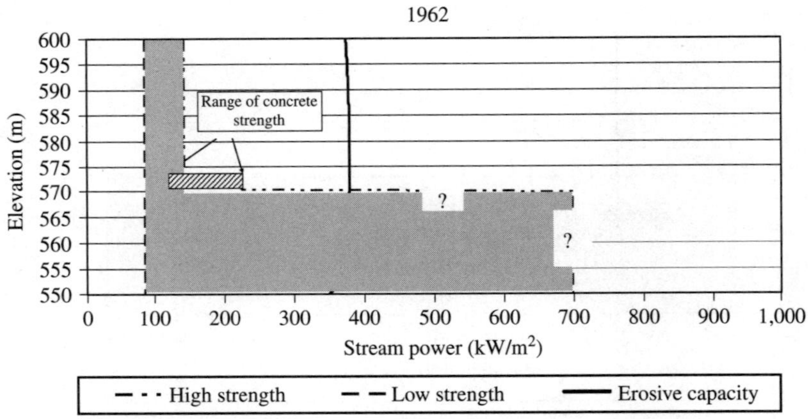

Figure 10.21 Comparison between stream power of the jet and the scour threshold of the concrete lining and rock for the flood event of 1962. The concrete lining was destroyed, but the rock remained intact.

The flood event in 1939 was much greater than that in 1936, but did not increase the depth of the plunge pool. This indicates a much higher strength of the rock at the bottom elevation of the pool (570 m). The analysis of the 1962 flood event also indicates that the destruction of the concrete lining that was placed on the bottom of the plunge pool after the 1939 flood event could have been predicted using the erodibility index method. It also confirms that the rock at the base of the plunge pool is much stronger than the rock that occurred above it prior to the scour events. The quantitative scour analysis that was performed on Ricobayo Dam indicates reasonable agreement between calculated and observed scour.

Confederation Bridge

C.D. Anglin[1] and R.B. Nairn[2]

Introduction

The 13 km long, $ 800 million Confederation Bridge crosses the Northumberland Strait and joins the provinces of New Brunswick and Prince Edward Island in Eastern Canada (Fig. 10.22). The bridge, developed under a finance-design-build-operate agreement between Strait

[1]Baird & Associates, 500-1145 Hunt Club Road, Ottawa, ON, Canada, K1V 0Y3 (danglin@baird.com)

[2]Baird & Associates, 200-627 Lyons Lane, Oakville, ON, Canada, L6J5Z7 (rnairn@baird.com)

Figure 10.22 The Confederation Bridge.

Crossing Bridge Limited (SCBL) and the Canadian Government, was constructed in 1994–97, and opened to traffic on June 1, 1997.

One of the engineering challenges associated with this $ 800 million project was the assessment of scour potential and scour protection requirements for the 65 bridge piers. The direct application of standard scour design techniques [such as those documented in TAC (1973) and FHWA (1993), available at the time of the original design investigations] was not possible due to the unique combination of complex flow conditions, complex pier base geometries and complex seabed conditions. In what follows an overview is provided of the multifaceted coastal engineering investigation undertaken to assess the scour potential and scour protection requirements for the Confederation Bridge, including the development of a new methodology to assess the potential for scour around bridge piers. This methodology was subsequently refined through detailed analysis of the results of a systematic scour monitoring program, as described further on.

Relevant site and project characteristics

At the crossing location, the Northumberland Strait is approximately 13 km wide, with water depths ranging from 0 to 30 m (typically in the order of 15 m). Under extreme design conditions, the bridge may be exposed to winds up to 120 km/h, waves of significant height (H_s) up to 4.5 m and peak period (T_p) up to 9 s, and currents up to 2.5 m/s (the latter generated by the combined effects of tides, surges and wave-driven longshore currents). Seabed conditions are highly variable, and consist of weathered mudstone, siltstone, and sandstone, which are sometimes overlain by glacial till.

The crossing consists of the main bridge (forty four 250 m long spans), the East approach (seven 93 m long spans) and the West approach

Figure 10.23 Schematic of main pier (left) and model of approach pier (right).

(fourteen 93 m long spans). The main bridge piers (P1 to P44) consist of an 8 m wide octagonal shaft cast integral with a conical ice shield (20 m base diameter) and supported by a conical pier base (22 m base diameter). The pier base rests either directly on the seabed or in dredged pits up to 14 m deep. Figure 10.23 (left image) provides a schematic illustration of a typical main pier. For the approach piers (E1 to E7 and W1 to W14), which are located in water depths less than 8 m, the conical pier base is also the ice shield, as shown or the right side of Fig. 10.23.

The individual bridge components, weighing up to 8000 t each, were precast on land and placed in the Strait using a heavy lift vessel (HLV Svanen) guided by a differential global positioning system (Fig. 10.24).

Key scour design issues

An extensive literature review was undertaken in an attempt to identify scour assessment techniques that could be applied to the Confederation Bridge. This included review of numerous technical papers and several

Figure 10.24 HLV Svanen placing the main bridge span.

design manuals related to scour around bridge piers, as well as numerous papers related to scour around coastal structures. In general, the bridge papers focused on the scour of non-cohesive sediments (i.e., sands and gravels) around rectangular or cylindrical piers in unidirectional flows, while the coastal structures papers focused on wave-induced scour of non-cohesive sediments along breakwaters, revetments, and seawalls (i.e., long, linear structures, without three-dimensional flow effects). In addition, a few papers were reviewed that dealt with erosion/scour of cohesive materials and weak rock by unidirectional flows. However, it was concluded that there were no acceptable techniques to define scour potential for the Confederation Bridge project because of the following unique features and complex conditions:

- combined waves and currents;
- conical pier bases, some located in dredged pits; and
- highly weathered and fractured bedrock seabed.

As such, it was necessary to develop a new methodology to assess scour potential for this project.

Development of new scour assessment methodology

Initial laboratory studies were undertaken at the Canadian Hydraulics Center in Ottawa, Canada, to characterize and quantify the erosion potential of the various seabed materials at the crossing location. This included flume tests to assess the erodibility of both core and slab samples of till, mudstone, siltstone, and sandstone. These test results showed considerable variability, as illustrated in Fig. 10.25 for the weakest (till and mudstone) samples.

Considering the rock materials (mudstone, siltstone, and sandstone), it was noted that the sample sizes were not sufficient to incorporate the highly variable bedding/fracture/joint patterns within the in situ rock mass, clearly an important parameter in the overall erosion process for these materials. Given the complex nature of the erosion process, and the associated variability in the test results, it was not possible to develop a reliable method to quantify the erosion process as a function of either shear stress or near bed velocity.

Subsequently, a promising new approach (Annandale, 1995) to estimate the erosion potential of "complex materials" was identified. In general terms, Annandale's (1995) approach relates the driving force for scour, as defined by the stream power parameter, P (which provides a measure of the rate of energy dissipation in the near bed flow), to the resistance to scour, as defined by the erodibility index, K (which provides

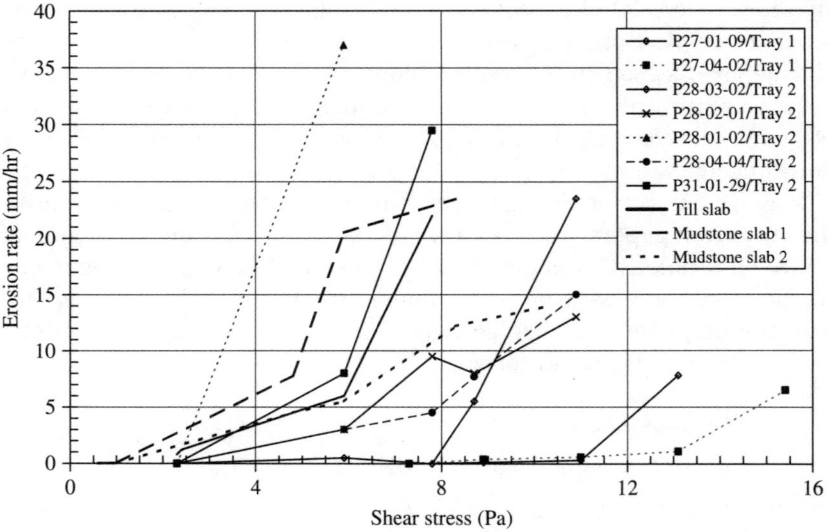

Figure 10.25 Erodibility test results for till and mudstone from Northumberland Strait.

a measure of the in situ strength of the material). Annandale's (1995) database, and his relationship between stream power and erodibility index (which defines the threshold for scour) is based on observations of erosion (or no erosion) in spillways downstream of dams. A subset of the Annandale database is presented in Fig. 10.26. This figure shows a log-log plot of

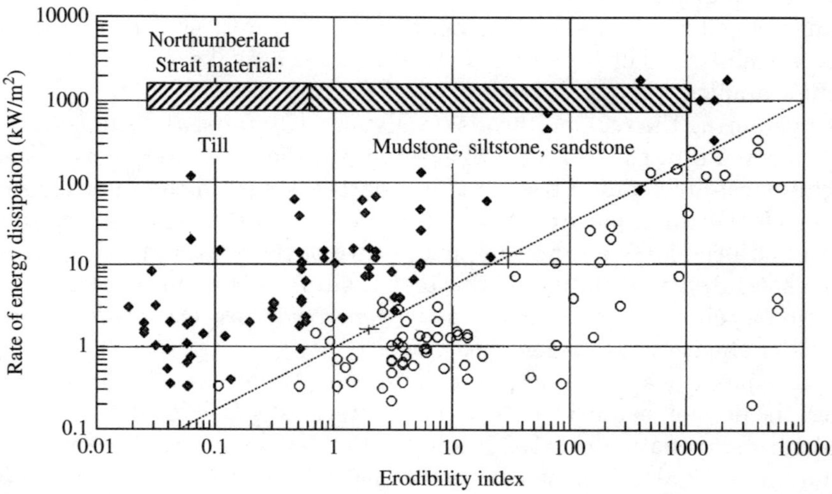

Figure 10.26 Erodibility of rock and other complex earth materials (Annandale 1995), with range of erodibility index for materials in the Northumberland Strait shown.

stream power versus erodibility index for approximately 150 field observations in materials ranging from cohesive sediments to hard, massive rock. The closed symbols represent events where erosion did occur, while the open symbols represent events where erosion did not occur. The sloping line is the estimated "erosion threshold" relationship between stream power and erodibility index.

As shown above, the estimated erodibility indices for the seabed materials encountered along the Confederation Bridge crossing alignment fall within the range of Annandale's database (the available geotechnical data and the calculations of K are summarized further on). However, in order to develop and apply this methodology to the Confederation Bridge project, it was necessary not only to evaluate the stream power (driving force for scour) and erodibility index (resistance to scour) for this project, but also to calibrate and verify the methodology for the assessment of scour potential around conical bridge piers exposed to combined waves and currents. These aspects of the investigation are described below.

Driving force for scour. Two general issues must be addressed with respect to quantifying the driving force for scour around the Confederation Bridge piers. First, the ambient flow conditions (waves and currents) at the crossing location must be defined, and second, the local influence of the bridge piers on these flow conditions must be defined.

Numerical modeling techniques were utilized to define the ambient flow conditions at the crossing location. Tidal and surge induced currents and water levels in the Strait were estimated on an hourly basis over a 23-year period (1973–95) using the MIKE21 hydrodynamic model of the Danish Hydraulic Institute. The model was driven by recorded water levels at either end of the Strait and the model predictions were successfully verified against available recorded current data (approximately 3 months) at the crossing location. The mean and large tidal ranges at the crossing location are approximately 1.5 m and 2.25 m respectively, with peak tidal currents (depth-averaged) in the order of 0.9 m/s (during large tides). Water level fluctuations and currents associated with storm surges can be similar in magnitude to the tidal effects.

A parametric wind-wave hindcast model was used to estimate hourly wave conditions at four locations along the crossing alignment for the same 23-year period. The wave predictions were validated against available recorded wave data (approximately 5 months) at the crossing location. Extreme wave heights were estimated using a peak over threshold (POT) extreme value model. For example, the 2- and 100-year significant wave heights (H_s, the average the highest one third of the waves; the maximum wave height, H_{max}, is typically 1.6 to 2 times H_s) near the middle of the crossing are in the order of 3 m and 4 m respectively. Shallow water processes, including refraction, shoaling, breaking, and wave-driven longshore currents, were estimated using the COSMOS coastal processes model (Southgate and Nairn, 1993).

The hindcast water level, current, and wave data were used to estimate the near bed velocity, shear stress, and stream power considering the combined effect of both waves and currents, on an hourly basis over the 23-year period of the environmental database. The wave orbital velocities were calculated using nonlinear wave theories selected based on guidance provided in USACE (1984), with Stokes second order theory (Stokes 2) used for deep water conditions, and Cnoidal theory used for shallow water conditions. The transition between the two wave theories was set at an Ursell number of 26 (USACE, 1984), corresponding to a water depth of approximately 8 to 12 m during storm wave conditions. The combined shear stress was calculated using the method of Myrhaug and Slaatelid (1990), as presented in Soulsby et al. (1993), while the combined velocity was calculated as the vector sum of the maximum wave orbital velocity and the depth-averaged tidal/surge current. Stream power was calculated as the product of the combined shear stress and the combined velocity. Given the importance of water depth on wave orbital motions, the calculations were repeated for a range in water depths representative of the 65 bridge pier locations.

Severe stream power events were extracted from the time series database and input to a POT extreme value model to estimate extreme events as a function of return period. The 100-year event was selected as the design condition to evaluate the requirement for, and the design of, scour protection.

The influence of the various pier shapes and dredged pit depths on the local flow conditions around the base of the piers were investigated for a range in water depths using a 1:70 scale model in a 1.2 m wide flume at the Canadian Hydraulics Center in Ottawa, Canada. Figure 10.27 presents a schematic diagram of the test configuration in the flume. The flume setup allowed the simulation of unidirectional currents with either

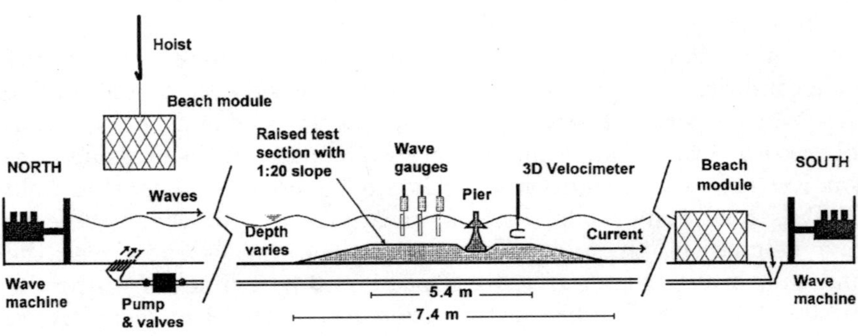

Figure 10.27 Schematic diagram of test configuration in flume.

"following" or "opposing" irregular wave conditions at a range in water depths.

The flow patterns around the base of the piers were defined with the aid of a laser-Doppler velocimeter, acoustic velocity meters, and flow visualization techniques. In addition, a unique "tracer test" was developed by Baird to quantify the local influence of the piers on seabed scour potential. More specifically, stream power magnification factors were developed for the various conditions encountered at the 65 bridge piers through a comparison of the stream power required to initiate the "scour" of tracer materials with and without the pier in place. Well-sorted coarse sands and fine gravels were used for the tracer materials, with median grain sizes (D_{50}) ranging from 1 to 5 mm. The tracer mat was placed to a thickness of two to three grains, and "scour" was defined as the complete removal of grains from any area resulting in an exposed patch on the flume floor. This was found to be a more repeatable "threshold condition" than the "initiation of motion" of individual grains for the irregular wave conditions in these tests.

In general, scour of the tracer mat initiated on either side of the piers as a result of the acceleration of flows in these areas. Following this, scour was usually experienced on the downstream/downwave side of the pier as a result of a wake vortex. Flow visualization techniques confirmed that a strong horseshoe vortex did not develop on the upstream/upwave side of the piers, probably due to the conical shape of the piers. Figure 10.28 shows photographs of the tracer mat around a model approach pier before and after a test. In this case, scour has developed on the downstream/downwave side of the pier; scour at the sides was prevented through the use of a coarser tracer material at the sides.

The resulting estimates of the stream power magnification factor (PMF) varied from approximately 1.6 for a deep water main pier placed in a deep pit, to approximately 6 for moderate depth main piers and shallow water approach piers placed directly on the seabed. The tracer test results for the main piers are summarized in Fig. 10.29.

Figure 10.28 Tracer mat around model approach pier before and after test.

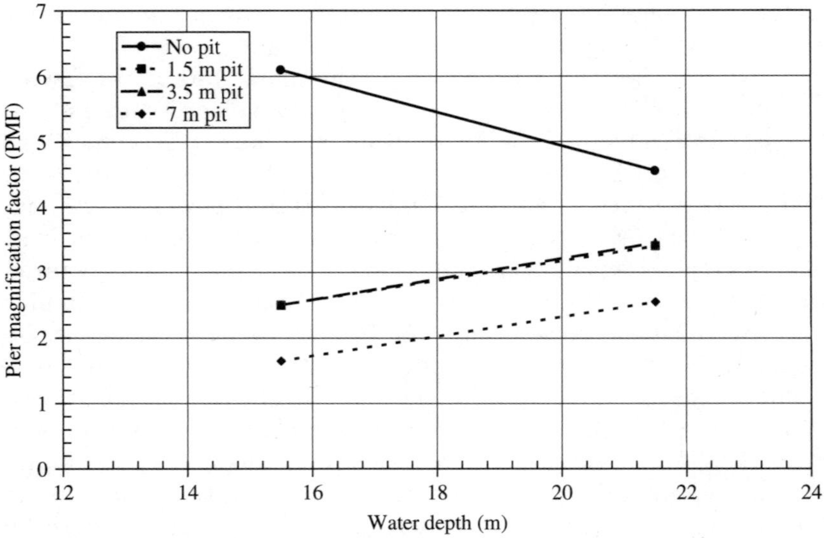

Figure 10.29 Main pier magnification factors estimated from tracer test results.

The PMF is proportional to the cube of velocity; hence, PMF's of 1.6 to 6 correspond to velocity magnification factors of approximately 1.2 to 1.8. The upper limit, for a conical pier base placed directly on the seabed, is somewhat larger than the values recommended in HEC-23 (FHWA, 2001b) for the application of the Isbash equation to design scour protection around round nose and rectangular piers (1.5 and 1.7 respectively). The reduction in PMF with deeper pits infers the existence of an "equilibrium" scour depth that could be estimated using such tracer tests (i.e., no further scour when PMF = 1.0).

Seabed resistance to scour. Estimating the erodibility of the highly weathered and variable seabed materials was one of the most challenging aspects of the project. As noted earlier, considerable variation was noted in the results of the erodibility flume tests. Further, the size of the test samples (both cores and slabs) was insufficient to incorporate the highly variable bedding/fracture/ joint patterns within the in situ rock mass. As such, it was not possible to describe or quantify the erosion process using only these test results. Ultimately, the empirical erodibility approach developed by Annandale (1995) for scour in spillways was adopted. In this approach, the erosion resistance of the material is quantified by the erodibility index, *K*, which accounts for the mass strength of the material, the typical block size, the inter-particle shear stress, and the orientation and shape of the layers of rock.

The erodibility index is calculated as the product of four dimensionless variables, all defined from standard borehole records, as summarized as following (see Chap. 6)

$$K = M_s K_b K_d J_s$$

As part of the geotechnical investigation undertaken to support the design of the bridge piers and foundations, between two and ten boreholes were drilled at each of the 65 bridge pier locations. A total of approximately 300 boreholes were drilled. Erodibility indices were calculated by a geotechnical engineer for each core run (approximately 0.3 m lengths) for each borehole. The erodibility indices showed considerable variability in both the horizontal and vertical dimensions, reflecting the highly variable nature of the materials on which the bridge piers are founded. This variability was a primary consideration in the incorporation of a factor of safety in the scour design methodology.

Calibration of methodology. The Confederation Bridge represents the first known application of Annandale's (1995) methodology to bridge piers, waves, and currents. As such, calibration and verification of the methodology was a key component of the investigation.

Fortuitously, observations of actual scour experienced around one of the East approach piers installed early in the project provided valuable information for calibration of the new methodology. Figure 10.30 provides

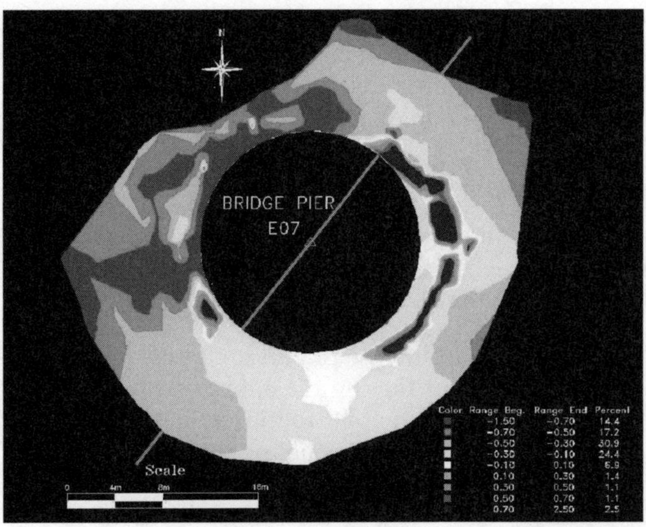

Figure 10.30 Scour in bedrock observed at approach pier E07 (November 1994).

an illustration of the measured scour around pier E07. The scour extended up to 5 m out from the base of the pier, with undermining of up to 1 m in and 1.5 m below the pier base. This scour was caused by a moderate storm event (return period of approximately five years) that occurred in November 1994. No significant scour was noted around several other East approach piers in place at that time.

The wave and current conditions during this event were hindcast using the numerical models, and the corresponding flow patterns around the piers were simulated in the physical model. Based on this information, along with the geotechnical data describing the seabed conditions in the vicinity of the East approach piers, comparisons of measured and predicted scour were used to calibrate the methodology. The key calibration parameters were the wave height (i.e., H_{avg}, H_s, $H_{1/10}$, H_{max}), the wave period (i.e., T_{avg}, T_p, T_{max}), the combined shear stress (i.e., τ_{mean}, τ_{max}), and the bottom roughness (k_s). The best comparison between estimated and observed scour at the single pier was obtained using H_{max}, T_{max}, τ_{max} and a bottom roughness of 0.3 m. The use of the maximum wave height, wave period, and shear stress can be qualitatively justified by the hypothesis that the erosion process for the weathered bedrock is a threshold process, and that once a rock fragment has been dislodged and removed from the surrounding matrix, the remaining material will be more susceptible to erosion. The lack of scour at the adjacent piers can be explained by the presence of stronger materials at these locations.

A qualitative confirmation of the methodology was also made through consideration of the morphological development of the seabed across the Strait. Through the application of the erodibility index approach, it was possible to explain the existence of glacial till over the underlying bedrock for areas with depths greater than about 13 m [i.e., in these areas the 100-year stream power event was less than the stream power required to erode the till material according to Annandale's (1995) relationship].

Requirement for scour protection

A pier by pier assessment was undertaken in order to define the requirement for scour protection at each of the 65 bridge piers. In general, this assessment included the following steps:

- for each borehole at each pier, define the maximum value of K in the "buffer zone" between the seabed (or the mass excavation level for piers placed in a dredged pit) and the pier founding elevation; these "local K values" represent the strength of the most erosion resistant material above the pier founding elevation (note that scour is allowed in the buffer zone, but not below the pier founding elevation);
- estimate the "design threshold K value" at each pier based on the local design stream power (100-year ambient stream power value

times pier magnification factor) and Annandale's (1995) scour threshold relationship;

- compare the "local K values" at each pier to the "design threshold K value," and recommend scour protection if the factor of safety (local K value/design threshold K value) is less than two to four (a higher factor of safety was used at piers with greater variability in seabed conditions, and/or where the tolerance for scour was lower).

Based on the results of this assessment, scour protection was recommended at 14 of the 65 bridge piers.

Modeling and design of scour protection. The design of the scour protection system was developed and optimized using a physical model investigation. These model tests were completed in the same wave flume as the tracer tests described earlier, again at a scale of 1:70. The scour protection tests were used to define the size of armor stone required to remain stable during the 100-year design wave and current conditions. The extent of the scour protection was defined based on the results of the tracer tests, which defined the "zone of influence" of the piers where ambient flow conditions were significantly affected by the presence of the pier. The recommended protection design consists of one or two layers of armor stone placed in a 10 m wide band around the base of the piers. The size of the armor stone is dependent on the water depth, with larger stones being required in shallower depths. Additional information on the modeling and design of the armor stone scour protection is provided in Anglin et al. (2002).

Construction

SCBL, the owner, chose to install scour protection at five of the 14 piers where Baird recommended protection. This decision was based on consideration of the cost of scour protection (approximately $ 0.5 million per pier) versus the risk of scour, recognizing the significant uncertainties and (possibly) conservative approach to the assessment of scour potential. Armor stone scour pads were installed at three approach piers (Fig. 10.31), while construction logistics led to the design and implementation of tremie concrete scour pads at two other approach piers [refer to Anglin et al. (2002) for additional information].

In response to SCBL's decision, Baird recommended a detailed and systematic scour monitoring program. This recommendation was accepted by SCBL; the resulting monitoring program is discussed following.

Scour monitoring program

An extensive long-term monitoring program was designed and implemented by Baird in order to assist SCBL in identifying any scour that

Figure 10.31 Armor stone scour pad at shallow water approach pier. (Photo: Boily)

might occur around the base of the bridge piers such that appropriate action could be taken before scour compromises the integrity of the structure (i.e., before scour extends beneath the founding elevation of any pier). The scour monitoring program is a critical component of the overall scour investigation for the following reasons:

- the Confederation Bridge represents the first known application of the new scour assessment methodology to bridge piers, waves, and currents;
- there are significant uncertainties associated with the estimation of the driving forces for scour and the seabed resistance to scour; and
- there is a desire to minimize seabed survey requirements around the pier bases.

In addition, and recognizing the limitations noted above, Baird recommended to SCBL that a systematic reassessment of the scour assessment and design methodologies be undertaken approximately five years after the bridge opened, with the seabed response over this time to be accurately quantified using appropriate survey techniques.

Initially, the 65 bridge piers were broken into priority classes based on the estimated risk of scour, with the highest priority class (nine AA piers) being those at which scour protection was recommended by Baird (FS < 4) but not implemented by SCBL. In addition, Baird developed and installed a near real-time wave and tide prediction system, using numerical models similar to those utilized in the original scour assessment and design study. This "SCOUR" software system is installed on the SCBL computer network in the bridge administration/operations building. The system is updated by SCBL staff on a biweekly basis (and immediately following severe storms) in order to define any requirement for action by bridge operations and maintenance staff. For example, a pier base inspection is "flagged" at specific piers if an event occurs which is more severe than any prior event, or if the factor of safety against scour is less than four, or if a certain period of time has elapsed since the last survey. Figure 10.32 presents a sample output report from the SCOUR system for one-half of the piers.

The output from the system is digitally archived to provide a historical database for each pier documenting the conditions (stream power) to which each pier has been exposed since the bridge opened in June, 1997.

Confederation Bridge Scour Monitoring Database

Update Date: 26JAN99 Update Period: 04JAN99 to 17JAN99

Pier	Priority Class or Protect. Type	Max Event Prior to Last Survey		Last Survey Date	Max Event Since Last Survey		Exceeded Time Int.	Exceeded Prior Event	FS Against Scour	Survey Required
E03	Arm 8	2388.0	10MAR98	15JUN98	2831.0	26NOV98	N	Y	Prot'd	Y
E04	AA	3847.0	10MAR98	23JUL98	4093.0	26NOV98	N	N	< 4	Y
E05	AA	1544.0	26NOV98	21DEC98	551.8	23DEC98	N	N	10 - 20	N
E06	AA	1826.0	26NOV98	21DEC98	624.5	23DEC98	N	N	4 - 10	N
E07	Trm10	1696.0	26NOV98	17DEC98	586.1	23DEC98	N	N	Prot'd	N
P01	B	233.4	10MAR98	17JUL98	298.0	26NOV98	N	N	20 - 100	N
P02	B	151.5	10MAR98	28JUL98	191.4	26NOV98	N	N	20 - 100	N
P03	A	170.4	10MAR98	28JUL98	215.7	26NOV98	N	N	10 - 20	N
P04	A	151.5	10MAR98	07JUL98	191.4	26NOV98	N	N	4 - 10	N
P05	B	135.4	10MAR98	07JUL98	170.8	26NOV98	N	N	10 - 20	N
P06	B	131.2	10MAR98	17JUL98	165.5	26NOV98	N	N	10 - 20	N
P07	A	129.2	10MAR98	10JUL98	163.0	26NOV98	N	N	10 - 20	N
P08	C	118.8	10MAR98	10JUL98	126.1	26NOV98	N	N	20 - 100	N
P09	A	110.7	10MAR98	22JUL98	117.2	26NOV98	N	N	10 - 20	N
P10	D	94.0	10MAR98	06JUL98	98.8	26NOV98	N	N	> 100	N
P11	D	83.5	10MAR98	02SEP98	87.4	26NOV98	N	N	> 100	N
P12	D	73.5	10MAR98	14SEP98	76.5	26NOV98	N	N	> 100	N
P13	D	63.3	10MAR98	06JUL98	65.6	26NOV98	N	N	20 - 100	N
P14	D	59.5	10MAR98	08JUL98	61.5	26NOV98	N	N	> 100	N
P15	C	55.4	10MAR98	09JUL98	57.1	26NOV98	N	N	20 - 100	N
P16	D	46.4	10MAR98	08JUL98	47.4	26NOV98	N	N	> 100	N
P17	C	58.1	10MAR98	24JUL98	60.0	26NOV98	N	N	20 - 100	N
P18	D	69.9	10MAR98	27JUL98	72.7	26NOV98	N	N	10 - 20	N
P19	AA	72.6	10MAR98	21JUL98	75.5	26NOV98	N	N	4 - 10	N
P20	D	54.7	10MAR98	09JUL98	56.4	26NOV98	N	N	> 100	N
P21	D	69.9	10MAR98	09JUL98	72.7	26NOV98	N	N	> 100	N
P22	B	64.0	10MAR98	09JUL98	67.6	26NOV98	N	N	20 - 100	N
P23	C	75.9	10MAR98	09JUL98	80.8	26NOV98	N	N	20 - 100	N
P24	D	68.8	10MAR98	13JUL98	72.9	26NOV98	N	N	> 100	N
P25	C	64.0	10MAR98	13JUL98	67.6	26NOV98	N	N	20 - 100	N
P26	D	50.0	10MAR98	14JUL98	52.2	26NOV98	N	N	> 100	N

* after Last Survey Date indicates accepted survey

Maximum Wave Conditions During Update Period

Pier	Date & Time	Max Hs(m)	Tp(s)	Approx. Return Period (yrs)
E04	16JAN99-02	1.38	5.5	n/a
P01	16JAN99-02	1.67	5.5	< 2
P14	16JAN99-02	1.69	5.6	< 2

Figure 10.32 Sample output report from SCOUR monitoring database system.

SCBL has followed this monitoring program since the bridge opened in 1997, and has undertaken at least one diver inspection around every pier, and numerous diver inspections around the high priority AA piers. Scour was detected during a diver inspection around one of the AA piers (P41) in 1998 (despite the fact that the scour hole was partially infilled with loose granular material). The scour reached a maximum depth of 1.6 m below the original seabed (approximately 1 m below the pier base elevation), and extended 12 to 15 m out from the pier base over a +/–100 degree sector. This pier was subsequently protected with an armor stone scour pad.

No significant scour has been observed around any of the other piers during diver inspections, despite the fact that the bridge has been exposed to several moderate storm events (the most severe of which had an estimated return period in the order of five years) for which the original scour assessment predicted scour at some of the AA piers. These results suggest that the original scour assessment may be conservative, as intended.

However, it is noted that diver inspections are qualitative, and can only identify significant changes or specific problems in localized areas (i.e., scour below the pier base). The diver inspections might not identify widespread, ongoing scour around a pier base. Further, it is interesting to note that P41 was not the most critical AA pier, and that scour was predicted to occur at other piers before it actually occurred at P41 (i.e., other AA piers had lower estimated factors of safety against scour). This information highlights the limitations and uncertainties in the original scour assessment methodologies, principally as a result of the large variation in (and limited characterization of) seabed material characteristics, but also the application of a new methodology for scour potential in complex materials subject to complex flow conditions.

Scour reassessment study

In response to these issues, as well as SCBL's interest in reducing their monitoring requirements/costs, a systematic reassessment of the scour assessment and design methodologies was recommended by Baird, and initiated by SCBL, in 2001. The first step in this process was the collection of detailed seabed surveys around the bridge piers. Multibeam sonar (MBS) seabed surveys were completed by the Federal Government (PWGSC) around 14 piers (including all nine AA piers) in the summer of 2001, with the remaining piers surveyed by PWGSC in the summer of 2002.

Detailed seabed surveys. The 2001–2002 MBS survey data provide an accurate description of the seabed surface around each pier approximately five years after construction. These surveys were overlain on the 1996 preconstruction and/or 1997 as built seabed surveys in order to

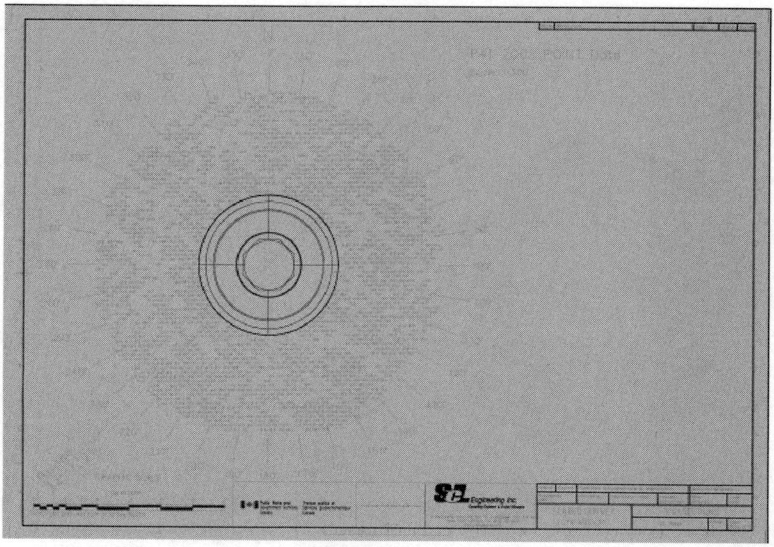

Figure 10.33 2002 survey of pier P41.

quantify changes in seabed elevation over this period. Detailed comparisons plots were developed at each pier, including:

- contour maps of seabed elevation for each survey;
- cross sections at 10 degree intervals; and
- contour maps of elevation change between 1996/97 and 2001/02.

Figures 10.33 to 10.35 provide examples of each of these presentations for P41.

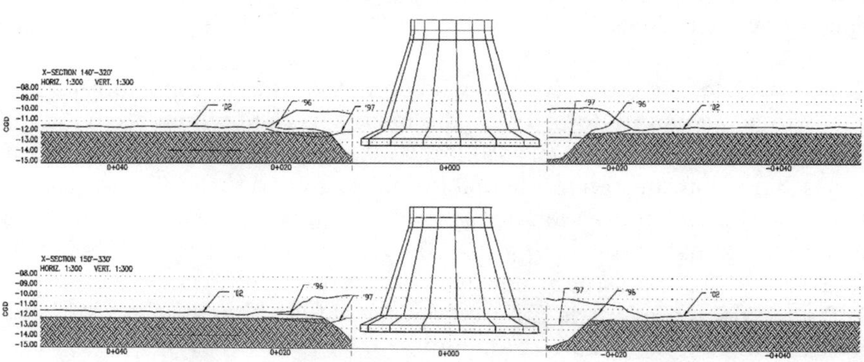

Figure 10.34 Selected survey cross section at pier P41.

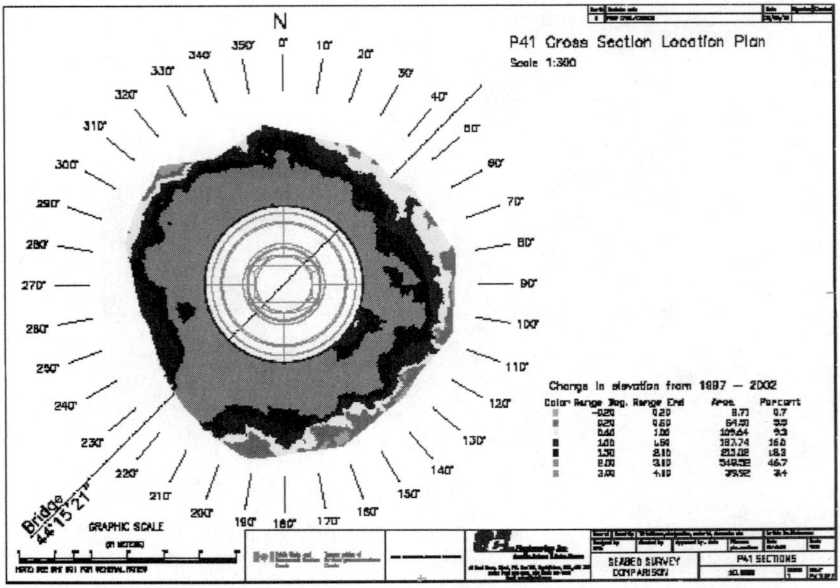

Figure 10.35 1997–2002 seabed survey comparison (elevation changes) at pier P41.

In this example, the 1996–97 cross sections show "infilling" of the dredged trench immediately adjacent to the pier base (at least some of this material would be the tremie concrete poured around the pier base after its placement), but erosion/scour in rock up to 10 m out from the pier base. The 2002 survey shows the armor stone scour pad that was subsequently constructed.

In general, the 2001–02 MBS surveys indicate infilling of the dredged pits, particularly in the deeper pits. In addition, these data do not show any signs of widespread erosion or general lowering of the overburden seabed beyond the dredged pits. Further, these data do not indicate any specific scour defects.

Storms to date. There have been several moderate storm events since the bridge opened in June 1997, the most severe of which occurred on November 26, 1998, with an estimated return period on the order of five years. The scour design methodology was based on the 100-year design event, with scour protection being recommended if the estimated factor of safety against scour during the 100-year design event was less than four (FS < 4). The stream power (driving force for scour) during the 100-year design event is approximately double that of the five year event, and generally more than double to which the Bridge has experienced till date.

Observed seabed response. Considering the diver inspections, the only "scour defect" identified through 2002 was at P41 (one of the AA piers), where erosion in bedrock, up to 1.6 m deep (and 1 m below the top of the tremie plug), was noted in July 1998. An armor stone scour pad was subsequently designed and constructed around this pier (completed in June 2000). Erosion (up to 2 m) of overburden was also noted around P38 in a July 1998 diver inspection. Aside from some flattening of the dredged pit side slopes, no other scour relevant issues have been identified by the ongoing diver inspections.

Considering the seabed surveys and comparisons, the 1996–1997 comparisons generally show erosion of the dredged pit side slopes (sometimes to a significant depth over a significant area), and sometimes show erosion at the base of the pit encroaching into the average bedrock elevation (as estimated from the available borehole data). In general, this (apparent) erosion has subsequently been buried by infilling documented in the 2001 and 2002 MBS monitoring surveys. There is some uncertainty regarding the 1997 survey data. However, it is possible that final trench clearing/cleaning operations performed after the 1996 surveys, as well as wave/current action, could have caused at least some of the changes suggested by the 1996–97 survey comparison.

FS < 4 trigger criteria and nearshore wave kinematics. As noted earlier, nonlinear wave theories were used to estimate the nearbed wave hydrodynamics. The application of Cnoidal wave theory in particular was anticipated to provide conservative estimates of wave orbital velocities in shallow water, but was retained in the scour assessment/design methodology to account for other (significant) uncertainties. As a result, the estimated stream power increases significantly where the calculation of wave kinematics changes from the deep water theory (Stokes 2) to the shallow water theory (Cnoidal). Cnoidal theory was utilized for all piers located in depths less than 10 m (all approach piers, and two main piers). The SCOUR database software system used the same assumptions/methodologies, so the stream power estimates in the monitoring database were consistent with those in the design database.

There has been much additional research into shallow water wave kinematics since 1996, including Hamm (1996), Dibajnia et al. (2001), and Grasmeijer and Ruessink (2003). The results of these studies provide general verification of the work of Isobe and Horikawa (1982), who proposed the use of an empirical correction factor with linear wave theory to estimate wave orbital velocities in shallow water. A review of the results of these studies was undertaken, including comparison to measurements of wave orbital velocities available from two physical model investigations, including those undertaken at CHC for the Confederation Bridge. These comparisons confirmed that Cnoidal wave theory significantly overestimates the nearbed wave orbital velocities in shallow water. Further, the results indicate that linear wave theory provides a reasonable estimate of maximum nearbed wave orbital velocities for the wave conditions and water depths of relevance to the shallow water piers of the Confederation Bridge. Figure 10.36 presents a comparison of estimated (with linear wave theory) and measured orbital velocities for the two physical model studies noted above. These data indicate that linear wave theory provides a slightly conservative estimate of the wave orbital velocities. This finding is generally consistent with the results of the previous research referred to above.

Linear wave theory versus physical model measurements. The use of linear (LIN) wave theory rather than Cnoidal (CN) wave theory in shallow water results in a significant decrease in the estimated orbital velocity (*U*) and stream power (*P*) under extreme wave conditions, as illustrated by the examples in Table 10.9. A comparison of Linear and Stokes 2 (ST2) theories is also presented for greater depths, and shows significantly less influence, progressively reducing in deeper water.

Based on this information, the stream power values in the design and monitoring databases are significantly overestimated for piers in shallow water (depths less than 10 m) during severe storm wave conditions. This is considered to be one of the primary reasons why the results of the scour monitoring program indicate that the $FS < 4$ trigger criteria appears to be conservative. Specifically, it is noted that the majority of the "scour critical" piers (i.e., unprotected AA piers with $FS < 4$ during the 100-year

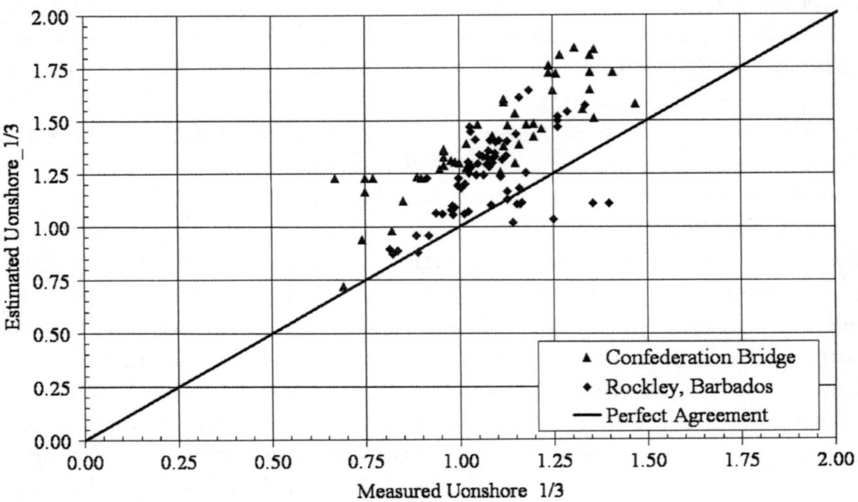

Figure 10.36 Wave orbital velocities.

TABLE 10.9 Impact of Wave Theory on Wave Orbital Velocity and Stream Power—Estimated 100-year Design Wave Conditions

Depth (m)	Non-linear wave theory	H_{max} (m)	T_{max} (s)	$U_{non\text{-}linear}/U_{linear}$	$P_{non\text{-}linear}/P_{linear}$
5.5	CN	5.3*	9.6	1.80	5.1
7.5	CN	6.8*	9.6	1.77	5.0
9	CN	7.2	9.6	1.73	4.8
12	CN	7.2	9.6	1.65	4.3
12	ST2	7.2	9.6	1.28	2.1
15	ST2	7.2	9.6	1.15	1.5
20	ST2	7.2	9.6	1.06	1.2
25	ST2	7.2	9.6	1.03	1.1
30	ST2	7.2	9.6	1.02	1.05

*H_{max} limited by water depth.

design event, as well as piers with scour protection in place) are located in shallow water.

Updated validation of scour design methodology. As noted earlier, the original validation of the scour design methodology was primarily based on the scour in bedrock observed at E07 following a storm event during construction. There was significant uncertainty in this validation, including:

- movement/destruction of the steel sheet pile cofferdam may have weakened the surrounding rock (and initiated the scour process);
- available borehole data may not accurately represent the actual rock characteristics in the scoured areas;
- potential inaccuracy in the wave and current conditions estimated during this event;
- range in PMFs estimated from model test results under various test conditions (i.e., with and without cofferdam, following and opposing waves).

The results of the monitoring program from 1997 to 2003, in particular the seabed response for those piers surrounded by exposed bedrock, provides additional data to support an updated validation of the scour design methodology. Of particular interest is the scour in bedrock at P41, noted in the May 1997 Mesotech survey as well as the July 1998 diver inspection, both prior to the $FS < 4$ criteria being reached at this pier. Several other piers have exposed bedrock that has not been subject to measurable erosion/scour since 1997. Additional investigations have been undertaken at these piers in order to estimate the factor of safety against scour during specific events known or believed to have caused scour (i.e., November 7–8, 1994 at E07, and December 30, 1996 at P41, the latter being the maximum event estimated between the 1996 and 1997 surveys at this pier), as well as the maximum event to date (June 1997 through December 2003) around the other piers where scour has not been observed. These calculations are summarized in Table 10.10 for both nonlinear (NL) and linear (LIN) wave theories.

These results demonstrate the significant increase in the estimated factor of safety against scour that occurs in response to the change in wave theory in shallow water from Cnoidal to linear (factor of four increase at Piers E04 to E07, and W13). The increase in estimated factor of safety against scour is much less at the deeper main piers (P01, P03, P19, and P41), where the original calculations for the events noted above were based on the deep water wave theory (Stokes 2).

The nonlinear and linear estimates of the factors of safety presented in Table 10.10 are plotted in Figs. 10.37 and 10.38 respectively.

These results suggest that linear wave theory provides a better estimate of the scour threshold for the Confederation Bridge than the nonlinear wave theories. Specifically, the nonlinear estimates (Fig. 10.37) show scour at $FS \sim 1$ at E07 and scour at $FS \sim 3$ at P41. Several of the other piers included in the analysis (the shallow water approach piers, where Cnoidal theory was utilized) show no scour at this same range in factor of safety. The linear estimates (Fig. 10.38) show scour at $FS \sim 3$ at E07 (considering the minimum estimated value, based on the uncertainty in the PMF at E07 during the November 7–8, 1994 event) and scour at $FS \sim 4$ at P41. The factors of safety at the other piers, where scour in rock has not yet occurred, is greater than four during the most severe event to date (i.e., between June 1997 to December 2003). Hence, a "factor of safety" (a misnomer, in this case) of four, based on linear wave

TABLE 10.10 Piers with Exposed Bedrock—Estimated Factor of Safety Against Scour—Events to Date

Pier	Depth (m GD)	Date of event of interest	Event causing scour		Maximum event since June 1997	
			NL	LIN	NL	LIN
E04	−3.1	Apr. 9/00	n/a	n/a	1.2	4.4
E05	−8.0	Nov. 26/98	n/a	n/a	4.6	18.2
E06	−7.4	Nov. 26/98	n/a	n/a	1.8	7.5
E07	−7.8	Nov. 7–8/94	0.6–1.9*	2.6–7.7*	protected	protected
P01	−12.5	Nov. 26/98	n/a	n/a	20.7	28.9
P03	−14.2	Nov. 26/98	n/a	n/a	4.5	5.7
P19	−20.5	Apr. 9/00	n/a	n/a	8.8	9.1
P41	−11.5	Dec. 30/96	3.0	3.9	protected	protected
W13	−6.5	Nov. 26/98	n/a	n/a	3.1	12.7

*E07-range is due to uncertainty related to pier magnification factor.

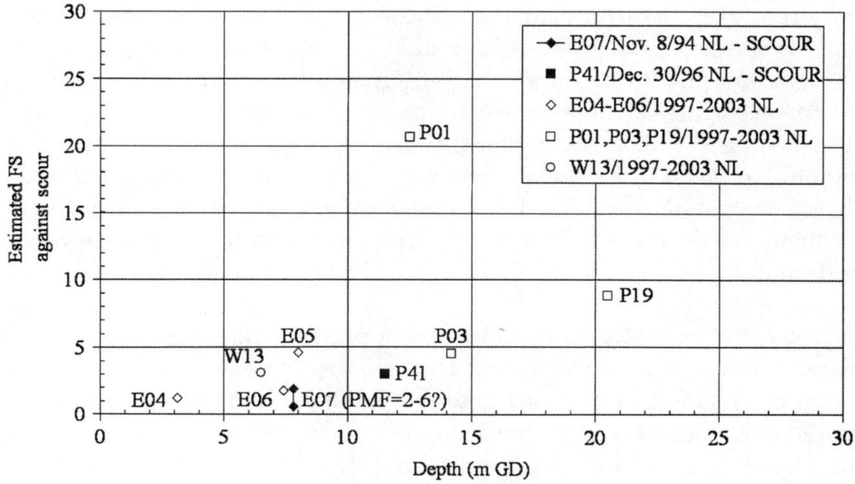

Figure 10.37 Estimated factor of safety against scour—events to date—nonlinear wave theory.

theory, appears to provide a reasonable estimate of the threshold for scour in bedrock for the Confederation Bridge.

As noted earlier, the original design investigations assumed a scour threshold at $FS = 4$ based on nonlinear wave theories. Considering the impact of the nonlinear and linear wave theories on stream power (refer to Table 10.1), it is apparent that the revised threshold criteria presented

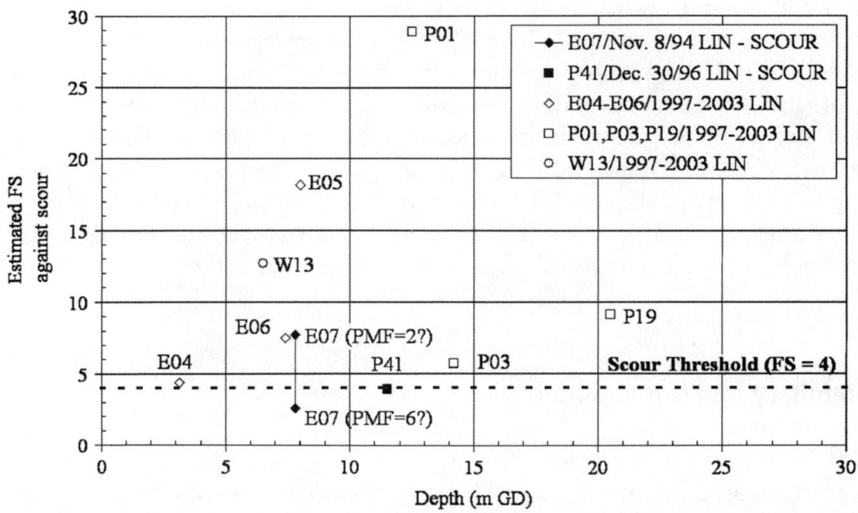

Figure 10.38 Estimated factor of safety against scour—Events to date—linear wave theory.

above (i.e., $FS < 4$ with linear wave theory) is similar to the original criteria for water depths greater than 20 m (i.e., the decrease in estimated stream power due to the change from ST2 to linear wave theory is less than 20 percent during extreme wave conditions). However, the revised threshold criteria is significantly higher (i.e., the seabed is more resistant than originally predicted) for the shallow water piers (depths less than 12 m), where the change from Cnoidal to linear wave theory results in a factor of four to five decrease in estimated stream power during extreme wave conditions.

Revised factors of safety and monitoring priorities. Based on the results presented above, it is concluded that the original scour assessment/design methodologies overestimate the potential for scour, particularly around the shallow water approach piers. The original design calculations were repeated for each pier using linear wave theory in order to provide a revised estimate of the factor of safety against scour during the 100-year design event. The monitoring priority class for each pier was revised to match these results.

Refined scour software/database system. An updated version of the SCOUR software/database system was developed to incorporate the results of the scour reassessment study. The most significant change in the system is the change from nonlinear to linear wave theory in the calculation of wave orbital velocities and stream power. Application of the updated version to the full monitoring period (June 1997 through December 2003) shows a significant reduction in the number of "$FS < 4$" triggers relative to the original version.

The updated version of the SCOUR software/database system was installed on SCBL's computer network in early 2004, including the change to linear wave theory and updated pier monitoring classifications. In addition, the updated version incorporates an increase in the maximum time intervals between inspections for all pier priority classes, which is deemed appropriate given the results of the monitoring program to date (i.e., good seabed response, with no scour defects) and the increased confidence in the assessment of scour potential gained through this scour reassessment study.

Summary and Conclusions

Summary

A multifaceted coastal engineering investigation was completed to support the assessment of scour and design of scour protection around the Confederation Bridge piers. This investigation led to the development of a new methodology to assess scour potential around bridge piers which can not only address complex flow conditions, pier geometries, and foundation materials but can also be applied to less complicated scour design problems. This methodology, derived from the empirical erodibility approach of Annandale (1995), has been calibrated for the Confederation Bridge project on the basis of the seabed response (scour or no scour) measured around four approach piers in place during a moderate storm event which occurred early in the construction period.

An extensive long-term monitoring program has been implemented to quantify the exposure of the bridge piers to potential scour events, to identify and address any scour that does occur, and to provide the information necessary to verify and improve the new scour assessment methodology. In general, the results indicate that the original scour assessment methodologies were conservative, particularly for piers in water depths of less than 10 m. Specific improvements to the scour assessment/design methodologies have been developed, tested, and verified. The refined approach provides an improved estimate of scour potential for the Confederation Bridge. Further, the results of the scour monitoring program support a reduction in the scope of the program, specifically a reduction in the frequency of seabed inspections.

It is noted that the Annandale (1995) approach is referenced in the most recent version of HEC-18 (FHWA, 2001a) in Appendix M—Scour competence of rock. The Confederation Bridge project represents the first known application of this method to bridge scour assessment and design.

Conclusions

- The Confederation Bridge has been exposed to a number of moderate storm events (estimated return period up to five years), but not to an extreme event. The estimated stream power (driving force for scour) during the 100-year design event is approximately double that to which the bridge has been exposed to date.
- There have been no significant seabed changes to the seabed around the approach piers since construction.
- Significant infilling of the dredged pits has occurred around the majority of the main piers since construction.
- Aside from the erosion/scour in bedrock at P41, which has been addressed through the construction of scour protection (as recommended in the original design study), there have been no "scour defects" identified by the scour monitoring program.
- The original scour assessment/design methodologies significantly overestimate the driving force for scour in shallow water.

- Specific improvements to the scour assessment/design methodologies have been developed. In particular, the primary cause of the conservatism in shallow water has been identified (nonlinear wave theory), and a refined approach (using linear wave theory) has been developed, tested, and verified.
- The refined approach provides an improved estimate of scour potential for the Confederation Bridge, as demonstrated by its ability to predict the scour in bedrock that occurred around E07 in the fall of 1994 and around P41 in the winter of 1996–97, as well the absence of scour to date around other piers with exposed bedrock.
- The refined approach will result in a significant reduction in future seabed inspection/survey requirements for the Confederation Bridge.

References

Anglin, C.D., Itamunoala, F., and Millen, G. 2002. Riprap as a Permanent Scour Protection Measure, *Proceedings of First International Conference on Scour of Foundations,* Texas A&M University, College Station, TX.

Annandale, G.W., Murthy, R., and Beckwith, G. 2004. Simulation of Dam Breach Initiated by Cracks and Earth Fissures in Semi-Arid and Arid Environments, *Proceedings of Dam Safety '04,* Association of State Dam Safety Officials, Phoenix, AZ, September 26–30.

Annandale, G.W., Bollaert, E., and Schleiss, A. 2002. Theoretical and Practical Aspects of Rock Scour Prediction, *First International Conference on Scour of Foundations,* Texas A&M University, College Station, TX.

Annandale, G.W., Wittler, R.J., Ruff, J.F., and Lewis, T.M. 1998. Prototype Validation of Erodibility Index for Scour in Fractured Rock Media, *ASCE Conference on Water Resources,* Memphis, TN.

Annandale, G.W. 1995. Erodibility, *Journal of Hydraulic Research,* Vol. 33, No. 4, pp. 471–494.

Annandale, G.W. 1993. Risk Analysis of Bridge Failure, *Proceedings of Hydraulic Engineering '93,* ASCE, San Francisco, CA.

ASTM D-2938. Standard Test Method for Unconfined Compressive Strength of Intact Rock Core Specimens, ASTM International, www.astm.org

ASTM D-2573. Standard Test for Field Vane Shear Test in Cohesive Soil, ASTM International, www.astm.org

ASTM D-4648. Standard Test Method for Laboratory Miniature Vane Shear Test for Saturated Fine-grained Clayey Soil, ASTM International, www.astm.org

ASTM D-2166. Standard Test Method for Unconfined Compressive Strength for Cohesive Soils, ASTM International, www.astm.org

ASTM D-1586. Standard Test Method for Penetration Test and Split Barrel Sampling of Soils, ASTM International, www.astm.org

Atkinson, B.K. (ed.). 1987. *Fracture Mechanics of Rock,* Academic Press, London.

Bakhmeteff, B.A. and Allan, W. 1946. The Mechanism of Energy Loss in Fluid Friction, *Trans. ASCE,* Vol. 111, pp. 1043–1102.

Beckwith, G.H. 2002. Background: Engineering Geology, Embankment Details, Observed Cracking and Special Studies Program, *Workshop on Cracking of Homogenous Embankments in Semi-Arid Environments,* Maricopa County Flood Control District, Phoenix, Arizona, January 29 and 30.

Baron, T. 1949. Technical Report, University of Illinois, No. 4.

Bell, F.G. 1992. *Engineering Properties of Soils and Rocks,* 3d ed., Ch. 7, Butterworth Heinemann, Oxford, 345 pp.

Benoist, J.M. and Cox, G.D. 1998. Failure of the Centennial Narrows Dam, *Proceedings of ASDSO Dam Safety Conference,* Las Vegas, NV.

Bieniawski, Z.T. 1988. The Rock Mass Rating (RMR) System (Geomechanical Classification) in Engineering Practice. In *Rock Classification Systems for Engineering Purposes,* ASTM STP-984, L. Kirkaldie, (ed.), American Society for Testing and Materials, Philadelphia, PA, pp. 17–34.

Blaisdell, F.W., Clayton, L.A., and Hebaus, G.G. 1981. Ultimate Dimension of Local Scour. *Journal of Hydraulics Division, ASCE,* Vol. 107(HY3), pp. 327–337.

Bollaert, E. 2002. Transient Water Pressures in Joints and Formation of Rock Scour due to High-Velocity Jet Impact, Communication No. 13. Laboratory of Hydraulic Constructions, Ecole Polytechnique Federale de Lausanne, Switzerland.

Booij, R. and Hofland, B. 2004. Measurement of the Mechanisms of Stone Entrainment, *Proceedings of Second International Conference on Scour and Erosion,* Singapore.

Briaud, J.-L., Ting, F.C.K., Chen, H.C., Gudavalli, R., Perugu, S., and Wei, G. 1999. SRICOS: Prediction of Scour Rate in Coheisve Soils at Bridge Pers, *Journal of Geotechnical and Geoenvironmental Engineering,* Vol. 125, No. 4, pp. 237–246.

Briaud, J.-L., Ting, F., Chen, H.C., Cao, Y., Han, S.-W., and Kwak, K. 2001. Erosion Function Apparatus for Scour Rate Predictions, *Journal of Geotechnical and Geoenvironmental Engineering,* Vol. 127, No.2, pp. 105–113.

Briaud, J.-L. 2004. Personal Communication.

Bureau of Reclamation. 1998. Case History Summary, Gibson Dam, Internal Report, Denver, CO.

Carollo, F.G., Ferro, V., and Termini, D. 2002. Flow Velocity Measurement in Vegetated Channels, *Journal of Hydraulic Engineering*, Vol. 128, No. 7, pp. 664–673.

Castillo, L.G. 1998. Revisión de las formulaciones de presión en los disipadores de energía en presas bóveda y corrección del coeficiente de presión dinámica. Personal Communication. Unpublished.

Castillo, L.G. 2004. Personal Communication.

Chamani, M.R. and Beirami, M.K. 2002. Flow Characteristics at Drops, *Journal of Hydraulic Engineering, ASCE,* Vol. 128, No. 8, pp. 788–791.

Chang, H.H. 1988. *Fluvial Processes in River Engineering*, Wiley, New York.

Chien, N. and Wan, Z. 1999. *Mechanics of Sediment Transport* [Translated under the guidance of John S. McNown], ASCE Press, Reston, VA, pp. 249–256.

Colebrook, C.F. and White, C.M. 1937. Experiments with Fluid Friction in Roughened Pipes, *Proceedings of the Royal Society, London,* Vol. A161, pp. 367–381.

Cox, J. 2002. Picacho Reservoir, *Workshop on Cracking of Homogenous Embankments in Semi-Arid Environments,* Maricopa County Flood Control District, Phoenix, AZ, January 29 and 30.

Croad, R.N. 1981. *Physics of Erosion of Cohesive Soils*, PhD Thesis, Department of Civil Engineering, University of Auckland, Auckland, New Zealand.

Deere, D.U. and Deere, D.W. 1988, The rock quality designation (RQD) index in practice. In *Rock Classification Systems for Engineering Purposes.* ASTM STP-984, L. Kirkaldie (ed.), American Society for Testing and Materials, Philadelphia, PA, pp. 91–101.

Dibajnia, M., Moriya, T., and Watanabe, A. 2001. A Representative Wave Model for Estimation of Nearshore Local Transport Rate, *Coastal Engineering Journal,* Vol. 43, No. 1, pp. 1–38.

Einstein, H.A. and Li, H. 1956. The Viscous Sub-layer along a Smooth Boundary, *Journal of Engineering Mechanics Division, Proceedings of ASCE,* Vol. 82, No. EM2, p. 27.

Emmerling, R. 1973. The Instantaneous of the Wall Pressure under a Turbulent Boundary Layer Flow, Max-Planck-Institut fur Stromungsforschung, Report No. 9.

Ervine, D.A. and Falvey, H.T. 1987. Behavior of Turbulent Jets in the Atmosphere and in Plunge Pools, *Proceedings of the Institution of Civil Engineers,* Part 2, Vol. 83, pp. 295–314.

Ervine, D.A., Falvey, H.T., and Withers, W. 1997. Pressure Fluctuations on Plunge Pool Floors, *Journal of Hydraulic Research,* Vol. 35, No. 2, pp. 257–279.

Ervine, D.A. 1998. Air Entrainment in Hydraulic Structures: A Review, *Proceedings of the Institution of Civil Engineers, Wat., Marit. & Energy,* Vol. 130, pp. 142–153.

Ewalds, H.I. and Wanhill, R.J.H. 1986. *Fracture Mechanics,* Delftse Uitgevers Maatschappij, The Netherlands.

Federal Highway Administration, National Highway Institute. 2001a. Evaluating Scour at Bridges, Hydraulic Engineering Circular No. 18, 4th ed.

Federal Highway Administration, National Highway Institute. 2001b. Bridge Scour and Stream Instability Countermeasures, Hydraulic Engineering Circular No. 23, 2d ed.

FERC. 2005. Collection of documents on findings on Silver Lake can be found at http://www.ferc.gov/industries/hydropower/safety/silver-lake.asp

Fiorotto, V. and Salandin, P. 2000. Design of Anchored Slabs in Spillway Stilling Basins, *Journal of Hydraulic Engineering, ASCE,* Vol. 126, No. 7, pp. 502–512.

Fiorotto, V. and Rinaldo, A. 1992a. Fluctuating Uplift and Lining Design in Spillway Stilling Basins, *Journal of Hydraulic Engineering, ASCE,* Vol. 118, HY4.

Fiorotto, V. and Rinaldo, A. 1992b. Turbulent Pressure Fluctuations under Hydraulic Jumps, *Journal of Hydraulic Research, IAHR,* Vol. 30, No. 4.
Flood Control District of Maricopa County. 2002. *Workshop on Cracking of Homogeneous Embankments in Semi-Arid Environments,* Proceedings, sponsored by the Flood Control District of Maricopa County, Phoenix, AZ, January 29 and 30.
Geological Society of London. 1977. The Description of Rock Masses for Engineering Purposes. Report by the Geological Society Engineering Group Working Party. *Quarterly Journal of Engineering Geology,* Vol. 10, pp. 355–388.
Grim, R.E. 1968. *Clay Mineralogy,* 2d ed. McGraw-Hill, New York, 596 pp.
Goodman, R.E. and Shi, G.H. 1985. *Block Theory and Its Application to Rock Engineering,* PrenticeHall, London.
Grasmeijer, B.T. and Ruessink, B.G. 2003, Modeling of Waves and Currents in the Nearshore—Parametric vs. Probabilistic Approach, *Coastal Engineering,* Vol. 49, pp. 185–207.
Hamm, L. 1996. Computation of the Near-Bottom Kinematics of Shoaling Waves, *25th International Conference on Coastal Engineering,* Orlando, FL.
Hanson, G.J. and Cook, K.R. 2004. Apparatus, Test Procedures, and Analytical Methods to Measure Soil Erodibility In Situ, Applied Engineering in Agriculture, *American Society of Agricultural Engineers,* Vol. 20, No. 4, pp. 455–462.
Hanson, G.J. and Simon, A. 2001. Erodibility of Cohesive Streambeds in the Loess Area of the Midwestern USA, *Hydrological Processes,* Vol. 15, pp. 23–38.
Henderson, F.M. 1966. *Open Channel Flow,* MacMillan, New York.
Hinze, J.D. 1975. *Turbulence*, 2d ed. McGraw-Hill, New York.
Hoffmans, G.J.C.M. and Verheij, H.J. 1997. *Scour Manual*, Balkema, Rotterdam.
Hofland, B., Battjes, J.A., and Booij, R. 2005. Measurement of Fluctuating Pressures on Coarse Bed Material, *Journal of Hydraulic Engineering, ASCE,* Vol. 131, No. 9, pp. 700–781.
Horeni, P. 1956. Desintegration of a Free Jet of Water in Air, *Sesit 93,* Praha, Pokbaba.
Hutchison, D.L. 1972. Physics of Cohesive Soils, PhD Thesis, Department of Civil Engineering, University of Auckland, Auckland, New Zealand.
International Society for Rock Mechanics 1981. *Rock Characterization, Testing, and Monitoring: Suggested Methods*, E.T. Brown (ed.), Pergamon Press, London, 211 pp.
Isobe, M. and Horikawa, K. 1982. Study on Water Particle Velocities of Shoaling and Breaking Waves, *Coastal Engineering in Japan*, Vol. 25, pp. 109–123.
Jennings, J.E.B., Brink, A.B.A., and Williams, A.A.B. 1973. Revised Guide to Soil Profiling for Civil Engineering Purposes in Southern Africa, *The Civil Engineer in South Africa*, Vol. 15, No. 1, pp. 3–12.
Kim, H.T., Kline, S.J., and Reynolds, W.C. 1971. The Production of Turbulence Near a Smooth Wall in a Turbulent Boundary Layer, *Journal of Fluid Mechanics*, Vol. 50, Pt. 1, pp. 133–160.
Kirkaldie, L. (ed.). 1988. *Rock Classification Systems for Engineering Purposes*, American Society for Testing and Materials, ASTM STP-984, Philadelphia, PA.
Kirsten, H.A.D. 1982 A Classification System for Excavation in Natural Materials, *The Civil Engineer in South Africa*, pp. 292–308, July (discussion in Vol. 25, No.5, May, 1983).
Kirsten, H.A.D., Moore, J.S., Kirsten, L.H., and Temple, D.M. 1996. Erodibility Criterion for Auxiliary Spillways of Dams, *ASAE International Meeting,* Phoenix, Arizona, Paper No. 962099.
Knight, D.W. and Patel, H.S. 1985. Boundary Shear in Smooth Rectangular Ducts, *Journal of Hydraulic Engineering, ASCE,* Vol. 111, No. 1, pp. 29–47.
Korhonen, K.H., Gardenmeister, R., and Saari, K. 1971. On the Diggability Classificaiton of Soils, State Inst. Tech. 1 Res. Pub. 163, Helsinki, Finland.
Mantz, P.A. 1973. Cohesionless, Fine Graded, Flaked Sediment Transport by Water, *Nature, Physical Science*, Vol. 246, pp. 14–16.
Mason, P.J. and Arumugam, K. 1985. Free Jet Scour below Dams and Flip Buckets, *Journal of Hydraulic Engineering*, ASCE, Vol. 111, No. 2, pp. 220–235.
May, R.W.P. and Willoughby, I.R. 1991. Impact Pressure in Plunge Pool Basins due to Vertical Falling Jets, Report SR 242, HR Wallingford, UK.

Melville, B.W. 2002. Local Scour Depths at Bridge Foundations: New Zealand Methodology, *Proceedings of First International Conference on Scour of Foundations*, Texas A&M University, College Station, TX.

Melville, B.W. and Coleman, S.E. 2000. *Bridge Scour*, Water Resources Publications, Denver, CO.

Moore, W.J. 1972. *Physical Chemistry*, 5th ed. Longman, London.

Moore, W.L. 1941. Energy Loss at the Base of a Free Overfall: Transactions, *ASCE*, Vol. 108, pp. 1343–1359, with discussion by M.P. White, p. 1361, H. Rouse, p. 1381, and others.

Moore, W.L. and Masch, F.D. 1962. Experiments on the Scour Resistance of Cohesive Sediments, *Journal of Geophysical Research*, Vol. 67, No. 4, pp. 1437–1449.

Myrhaug, D. and Slaatelid, O.H. 1990. A Rational Approach to Wave-Current Friction Coefficients for Rough, Smooth and Transitional Turbulent Flow, *Coastal Engineering*, Vol.14, No. 3, pp. 265–294.

Nairn, R.B. and Anglin, C.D. 2002. Confederation Bridge—New Scour Design Methodology for Complex Materials, *Proceedings of First International Conference on Scour of Foundations*, Texas A&M University, College Station, TX.

National Transportation Safety Board (NTSB). 1988. Collapse of New York Thruway (1–90) Bridge over the Schoharie Creek, near Amsterdam, New York, April 5, 1987, Highway Accident Report: NTSB/HAR-88/02, Washington, DC.

Odgaard, A.J. 1986. Meander Flow Model I: Development. *Journal of Hydraulic Engineering,* Vol. 112, pp.1117–1136.

Odgaard, A.J. 1986. Meander Flow Model II: Applications. *Journal of Hydraulic Engineering,* Vol.112, pp.1137–1150.

Offen, G.R. and Kline, S.J. 1974. Combined Dye-Streak and Hydrogen-Bubble Visual Observations of a Turbulent Boundary Layer, *Journal of Fluid Mechanics*, Vol. 62, Pt. 2, pp. 223–239.

Offen, G.R. and Kline, S.J. 1975. A Proposed Model of the Bursting Processes in Turbulent Boundary Layers, *Journal of Fluid Mechanics*, Vol. 70, Pt. 2, pp. 209–228.

Pagan-Ortiz, J.E. 2002. Impact of the Federal Highway Administration's Scour Evaluation Program in the United States of North America's Highway Bridges, *Proceedings of First International Conference on Scour of Foundations*, Texas A&M University, College Station, TX.

Paris, P.C., Gomez, M.P., and Anderson, W.E. 1961. *Trend Engineering*, Vol. 13, pp. 9–14.

Partheniades, E. 1965. Erosion and Deposition of Cohesive Sediments, *Proceedings of ASCE*, Vol. 91, No. HY1, pp. 105–139.

Raudkivi, A.J. and Hutchison, D.L. 1974. Erosion of Kaolinite Clay by Flowing Water, *Proceedings of Royal Society of London*, A. 337, pp. 537–554.

Raudkivi, A.J. 1976. *Loose Boundary Hydraulics*, 2d ed. Pergamon Press, New York.

Raudkivi, A.J. 1998. *Loose Boundary Hydraulics*, 4th ed. Balkema, Rotterdam, the Netherlands.

Rao, M.N. 1971. An Erosion Study of Halloysite Clay, Report No. 75, Department of Civil Engineering, University of Auckland, New Zealand.

Rhodes, D.G. and Knight, D.W. 1994. Distribution of Shear Force on Boundary of Smooth Rectangular Duct, *Journal of Hydraulic Engineering, ASCE,* Vol. 120, No. 7, pp. 787–807.

Richardson, E.V. and Davis, S.R. 2001. *Evaluating Scour at Bridges*, 4th ed. FHWA NHI 01-001 HEC-18, National Technical Information Service, Springfield, VA.

Richardson, E.V. and Davis, S.R. 2001. *Evaluating Scour at Bridges*, Publication No. FHWA NHI 01-001, Hydraulic Engineering Circular No. 18, U.S. Department of Transportation, Washington, DC.

Rooseboom, A., Basson, M.S., Loots, C.H., and Wiggett, J.H. 2005. Handleiding vir Paddreinering (Road Drainage Manual) National Transport Commission, Pretoria 1981 Revised 2005.

Rucker, M.L. 2004. Percolation Theory Approach to Quantify Geo-Material Density-Modulus Relationships, *9th ASCE Specialty Conference on Probabilistic Mechanics and Structural Reliability*, Albuquerque, New Mexico, July 26–28, available on the Web at: http://www.esc.sandia.gov/PMCpostconf.html (accessed 28 April 2005).

Sahimi, M. 1994. *Applications in Percolation Theory*, Taylor & Francis, Bristol, PA.

Sande, van de E. and Smith, J.M. 1973. Surface Entrainment of Air by High Velocity Water Jets, *Chemical Engineering Sciences*, Vol. 28, pp. 1161–1168.
Schlichting, H. and Gersten, K. 2000. *Boundary Layer Theory*, 8th ed. (Revised and Enlarged), Springer-Verlag, New York.
Shi, G-H. 1982. A Geometric Method for Stability Analysis of Discontinuous Rocks, *Scienta Sinica*, Vol. XXV, No. 1, January.
Shi, G-H. 1988. Discontinuous Deformation Analysis: A New Numerical Model for the Statics an Dynamics of Block Systems, PhD Thesis, Department of Civil Engineering, University of California, Berkeley, 378 pp.
Shields, A. 1936. *Application of Similarity Principles, and Turbulence Research to Bed-Load Movement*, California Institute of Technology, Pasadena, CA (translated from German).
Smith, S.P. and Annandale, G.W. 1997. Scour in Erodible Bedrock II: Erosive Power at Bridge Piers, *Proceedings of North American Water and Environmental Congress*, Anaheim, CA.
Soulsby, R.L., Hamm, L., Klopman, G., Myrhaug, D., Simons, R.R., and Thomas, G.D. 1993. Wave-Current Interaction Within and Outside the Bottom Boundary Layer, *Coastal Engineering*, Vol. 21, No.1, pp. 41–69.
Southgate, H.N. and Nairn, R.B. 1993. Deterministic Profile Modelling of Nearshore Processes—Part 1—Waves and Currents, *Coastal Engineering*, Vol.19, No.1, pp. 27–56.
Streeter, V. L. 1971. *Fluid Mechanics*, 5th ed. McGraw-Hill, New York, pp. 238–241.
Tan, S.K. 1983. Erosion of Cohesive Soils, PhD Thesis, Department of Civil Engineering, University of Auckland, Auckland, New Zealand.
Temple, D.M. and Moore, J.S. 1994. Headcut Advance Prediction for Earth Spillways, *Proceedings of ASAE International Winter Meeting*, Atlanta, Georgia, Paper No. 942540.
Transportation Association of Canada (TAC). 1973. *Guide to Bridge Hydraulics*.
Van Schalkwyk, A., Jordaan, J.M., and Dooge, N. 1995. The Erodibility of Different Rock Formations under Varying Flow Conditions (in Afrikaans), Water Research Commission, Report No. 302/1/95, Pretoria, South Africa.
Vick, S.G. 2002. *Degrees of Belief*, ASCE Press, Reston, VA.
Wan, C.F. and Fell, R. 2002. Investigation of Internal Erosion and Piping of Soils in Embankment Dams by the Slot Erosion Test and the Hole Erosion Test, UNICIV Report R-412, July, The University of New South Wales, Sydney, Australia.
White, S.J. 1970. Plane Bed Threshold of Fine Grained Sediments, *Nature*, Vol. 228, No. 5267, pp. 152–153.
Willmarth, W.W. and Lu, S.S. 1972. Structure of the Reynolds Stress near the Wall, *Journal of Fluid Mechanics*,Vol. 55, pp. 65–92.
Wittler R.J., Annandale G. W., Ruff J. F., and Abt S. R.1998. Prototype Validation of Erodibility Index for Scour in Granular Media, *ASCE Conference on Water Resources*, Tennessee,TN.
Yang, C.T. 1996. *Sediment Transport: Theory and Practice*, McGraw-Hill, New York.
Yang, C.T. 1973. Incipient Motion and Sediment Transport, *Journal of Hydraulics Division, ASCE,* Vol. 99, No. HY 10, Proceeding Paper 10067, pp. 1805–1826.
Yildiz, D. and Üzücek, E. 1994. Prediction of Scour Depth from Free Falling Flip Bucket Jets, *International Water Power and Dam Construction*, November.

Symbols

a	Average thickness of a soil clump; length of a close-ended fissure; initial length of a close-ended fissure; crack length; characteristic dimension of a fissure (usually the extent of a fissure into a rock mass from the surface of the rock)
a, b	Dimensions of a rectangular conduit cross section; the side dimensions of a rectangular tube
A	Cross-sectional area of flow; surface area of a block impacted by a impulsive force; value of the semitransverse and semi-conjugate axis of a hyperbola; cross-sectional area of tube; cross-sectional flow area in a tunnel; footprint area of a jet; footprint area of a jet at impact = $D_j \cdot W$; cross-sectional area of a fissure = $A = Dw_f$
A_1, A_2	Coefficients of clay rate in erosion equation
A_i	Flow area of a jet at a desired depth below water surface elevation
A_t	Cross-sectional area of flow in a test section for HET test (it is assumed that the flow through the specimen around the drilled hole is negligible)
B	Width of a rectangular jet
c	Pressure wave celerity
c_{air}	Wave celerity of air (340 m/s)
c_{liq}	Pressure wave celerity of liquid (assume 1000 m/s for water)
C	A roughness coefficient known as the Chezy coefficient; material property for fatigue failure calculations
C, m	Rock material parameters that can be determined by experiment
C_d	Discharge coefficient (= 2.1); diffusion constant (= 6.3)
C_f'	Friction coefficient defined by Hanson and Cook (= 0.00416)
C_{pa}	Mean dynamic pressure coefficient
C_{pa}'	Fluctuating dynamic pressure coefficient
$C_p\left(\frac{Y}{D}\right)$	Average dynamic pressure coefficient as a function of $\frac{Y}{D}$, which is assumed to be same as $C_{sp}\left(\frac{Y}{D}\right)$

d	Diameter of a soil particle; representative particle diameter; flow depth;
d_o	Nozzle diameter; orifice diameter
$d_{\min}$	Minimum particle size
$d_{\max}$	Maximum particle size that remains in an armor layer
dR_t	Incremental change in the radius of a hole as it erodes
$d\phi_t$	(Differential) change in hole diameter at time t
D	Hydraulic diameter; mean block diameter; characteristic particle diameter of a bed material; diameter of root bulb bounded by fine fibrous roots; maximum flow depth
$D = \frac{4A}{P}$	Hydraulic radius of tunnel
D_i	Thickness of jet
e	Width of a conduit (in this case a very thick boundary of a rock fissure or joint, assumed to be equal to infinity)
E	Modulus of elasticity of a conduit (such as rock in case of scour)
E_a	Activation energy
$E_{\Delta t}$	Kinetic energy imparted to a particle over a short pulse period Δt
EI, K	Erodibility Index
ΔE	Energy head loss over a jump
f	Factor that accounts for the shape of a close-ended fissure; function that accounts for the geometry of a rock block and its crack extension, the loading conditions and edge effects; Darcy–friction factor; Darcy-Weisbach friction factor = 0.0318 for tunnels 1 and 2 and = 0.053 for tunnel 3
$f(K)$	Expression that defines the relative ability of earth or engineered earth materials to resist the erosive capacity of water
f'_L	Friction factor for laminar flow
f'_T	Friction factor for turbulent flow
F	Fluctuating dynamic pressure reduction factor; fluid modules
$Fr_1 = \frac{V_1}{\sqrt{g \cdot y_1}}$	Froude number of flow upstream of a jump
Fr_i	Issuance Froude number
$F_{\Delta t}$	Net impulse on a particle during the period Δt
F_{s1}, F_{s2}	Instantaneous shear forces generated on the sides of a block of rock during the pulse period Δt
F_{up}	Total upward impulse caused by transient pressure in a joint

F_{down}	Total downward impulse caused by fluctuating pressures on top of a rock block
g	Acceleration due to gravity
h	Differential head measurement
h_o	Overflow depth over a free-flowing ogee spillway
h_f	Total head loss through a fissure
$h'_f = Qh_f$	Energy loss flux
Δh_t	Pressure head difference over a test section determined from a manometers at time t
H	Depth to a point in question from the original ground surface
$I_{\Delta t\_impulse}$	Net impulse on a block of rock
j	Incremental section number along a bend
J	Actual length of a jet; distance from the orifice to the centerline depth of scour
J^*	Dimensionless scour term
J^*_i	Dimensionless scour term at J_i/J_e
J_e	Distance between the orifice and the ground surface when equilibrium scour conditions are reached
J_i	Distance between the orifice and the ground surface; initial distance from orifice to soil surface
J_p	Length of the core of a jet $= K_jD_j$; potential core length of a jet $= C_d d_o$
J_s	Orientation and shape number
J_x, J_y, J_z	Average spacing of joint sets measured in three mutually perpendicular directions
k	Rate constant
k'	Magnification coefficient[1]
k_1	A constant
k_d	Rate of erosion coefficient; erodibility coefficient
k'_d	Erosion rate coefficient associated with stream power
K	Energy loss coefficient; Erodibility Index (*EI*); stress intensity
K_1	Coefficient allowing for the nonhydrostatic head in the zone just downstream of a knick point (usually assumed 1)

[1]The usual range of value is given by $3 \le k' \le 18$. The value most often used to express extreme pressure fluctuation in open-channel flow is $k' = 18$.

K_2	Coefficient allowing for the effects of air resistance on a jet trajectory
K_b	Block size number
K_d	Interparticle or interblock shear strength number
K_j	Empirically determined factor = 6.3 for most jets
K_{bend}	Bend loss coefficient (= 0.066)
$K_{entrance}$	Entrance loss coefficient (= 0.42)
K_{flare}	Energy loss coefficient at flare (= 0.078)
K_{liq}	Bulk modulus of a liquid
ΔK	Difference between the maximum and minimum stress intensity factors at the tip of a crack; range of stress intensities introduced to a material by the fluctuating pressures
l	Distance between two pressure measurement locations
L	Fissure length; effective length over which energy is dissipated; length of an open-ended discontinuity around a rock block in a plunge pool boundary; length of transition zone 2 (in absence of more detailed information, assume a unit length); distance of crack growth required for a material to fail; variable length of a close-ended fissure (it changes at the crack grows during the process of fatigue failure); length of the test section; total length of jet trajectory
L_b	Breakup length of a jet
$L_{br} = L/L_b$	Jet breakup length ratio
L_f	The thickness of the material layer (i.e., $L_f - a$ = distance through which a crack must grow in order for fatigue failure to occur)
m	Exponent that varies between 0.10 and 0.25; mass of a sediment particle; mass of a rock block; mass of free air per unit volume of water
M_s	Mass strength number
n	A roughness coefficient known as Manning's n
n_B	Number of bonds per unit area
N	Number of cycles of a fluctuating pressure that will lead to fatigue failure
p	Absolute pressure of a fluid
p_o	Frequency function of the original grain size distribution
p_0	Pressure at the entrance to a close-ended fissure, i.e., at the surface of a rock
p_1, p_2	Pressure measurements upstream and downstream of a sample section

p_{jet}	Stream power per unit area
$p_{jet}\left(\frac{Y}{D}\right)$	Average stream power per unit area as a function of $\frac{Y}{D}$
$p_{\max}$	Maximum pressure at the closed end of a fissure
p_{pool}	Stream power per unit area at a particular depth below water surface elevation of a plunge pool
Δp	Pressure drop over distance L along a tunnel
P	Stream power per unit area; relative magnitude of the erosive capacity of water; wetted perimeter of a tube;
$P = (2D + 2w_f)$	Wetted perimeter of a fissure
P_a	Available stream power in rivers upstream of a bridge pier
P_c	Critical stream power that will result in incipient motion; threshold stream power
P_e	Effective stream power
P_t	Wetted perimeter at time t; stream power per unit area at time t
P_u	Net uplift pressure over an entity with open-ended joints impacted by a jet
P_R	Power required to scour earth material with Erodibility Index values less than 0.1
$P_{applied}$	Applied stream power per unit volume of water (also known as turbulence production)
$P_{available}$	Available stream power
P_{jet}	Total stream power of a jet
$P_{\max}$	Total dynamic pressure
q	Discharge per unit width over a spillway; unit discharge = Q/W; unit flow
q_a	Unit flow of air
Q	Total discharge through a fissure
r	Radial axis
r_1, r_2	Inner and outer radii of a flow region; inner and outer radii of the space between the sample and the outer edge of the section in the CFD containing a fluid
r_c	Radius of the center line of a bend
R	Universal gas constant; hydraulic radius $=A/P$
$R_h = \frac{4A}{P}$	Hydraulic radius of an enclosed conduit
$\mathrm{Re}_* = \frac{u_* d}{V}$	Particle Reynolds number
s	Energy slope of flow in a river; energy slope in an open-channel flow

s_f	Energy slope
s_t	Hydraulic gradient over a section at time t
$[SF^*]$	Activated complex
S	Soil modules
S_f	Average energy slope of water discharging over a knick point
S_u	Undrained shear strength of soft and very soft alluvial deposits
$SP_{knickpoint}$	Stream power over a knick point
t	Time of data reading
t_m	Measured time
T	Absolute temperature measured in Kelvin; tensile strength of a rock
T'	Temperature in degrees Celsius
T_B	Period of turbulent bursts
$\overline{T}_B$	Average turbulent burst period
T_r	A reference time $= J_e/(k_d\tau_c)$
u	Tangential flow velocity at a particular depth z
$\overline{u}$	Average flow velocity; flow velocity in a CFD device
u_0	Flow velocity at water surface
$u_* = \sqrt{\frac{\tau}{\rho}}$	Shear velocity
U	Average flow velocity
$U_o = \sqrt{2gh}$	Flow velocity at the exit from a nozzle; velocity of a jet at an orifice (origin); average flow velocity in a rectangular pipe; average flow velocity through a hole
UCS	Unconfined compressive strength (of a rock)
v	Velocity of water; transverse flow velocity; kinematic viscosity of water; kinematic viscosity of water
V	Total volume
V_0	Minimum plunging velocity for aeration to start (1 m/s)
V_a	Volume of air
V_b	Volume of a block of rock
V_g	Velocity of water over a knick point
V_j	Jet velocity at water surface of a plunge pool
V_w	Volume of water
V_z	Jet velocity at an elevation z
V_{cr}	Critical average flow velocity, beyond which a particle will move

$V_{\Delta t}$	Average velocity achieved by a particle over the period Δt; average velocity attained by a mass of rock during the time period Δt
V_{air}	Volume of air
V_{liq}	Volume of a liquid
W	Width of a tunnel
W_g	Submerged weight of a block of rock
x	Variable distance along a fissure, from the opening to the close-ended side; horizontal distance
x, y, z	Joint set spacing[2]
y	Flow depth in a wide channel; distance from the boundary; flow depth in an open channel
y_1	Upstream water depth; downstream depth[3]
z	Vertical axis; vertical distance
z_b	Vertical height of a rock block, assuming it is prismatic
Z'	Partition functions per unit volume
β	Free air content
γ	Unit weight of water
γ_s	Unit weight of a sediment
δ	Thickness of a wall layer
δ_b	Boundary layer thickness (which is equal to the depth of flow in a fully developed, turbulent open-channel flow)
$\varepsilon, \varepsilon_r$	Rate of erosion
θ	Issuance angle; Shields parameter
ϑ^*	Frequency of passage of activated complexes over the energy barrier (i.e., in excess of the activation energy)
ξ_1	Dimensions in the direction of flow
ρ	Fluid density
ρ_d	Dry density of a soil sample
ρ_s	Mass density of a soil
ρ_r	Mass density of a rock
ρ_{air}	Density of air (1.29 kg/m^3)
ρ_{liq}	Density of liquid (1000 kg/m^3 for water)
σ_i	Confining stresses in a rock

[2]This can be determined by a Fixed Line Survey (see, e.g., International Society for Rock Mechanics 1981, Geological Society of London 1977, Bell 1992).

[3]It can be shown that the total energy loss can be expressed in dimensionless form solely as a function of drop height and critical depth at the drop.

σ_{water}	Water pressure in a close-ended crack; stress introduced by turbulent, fluctuating pressure in a close-ended fissure
τ	Shear stress; shear stress applied by a testing equipment; tractive shear stress ($=\rho gys$ in case of open-channel flow); total boundary shear stress; shear due to drag calculated in EFA and HET devices
τ_c	Critical shear stress (Pa)
τ_e	Effective stress (Pa)
τ_i	Maximum shear stress at the interface between a jet and ground surface
τ_t	Turbulent shear at the boundary; turbulent boundary shear stress measured directly with CFD device
$\overline{\tau}_w$	Mean shear stress on a wall
$\overline{\tau}_b$	Mean shear stress on the top and bottom
τ_{time}	Shear stress in a sample at time t
ϕ	Angle of friction of soil; residual friction angle of a granular earth material; kinetic energy velocity coefficient (often assumed to be 1)
ω	Settling velocity of a sediment particle

Index

Page numbers followed by italic *f*, *t*, or *i* indicate figures, tables, or illustrations, respectively.

ABOUT THE AUTHOR

George W. Annandale, D.Ing., P.E., is an internationally known expert on scour and president of Engineering and Hydrosystems, Inc. A civil engineer with 30 years' experience, he has worked on projects involving fluvial hydraulics, sediment transport, scour and sedimentation, and hydrology and hydraulics. He is the developer of the Erodibility Index Method used to assess scour on major projects around the world, including the San Roque Dam in the Philippines, the Karahnjukar Hydroelectric Project in Iceland, and the new Woodrow Wilson Bridge in the United States. His method has also been incorporated into federal and state guidelines, including those of the Federal Energy Regulatory Commission, U.S. Bureau of Reclamation, Colorado Department of Transportation, and Federal Highway Administration. Dr. Annandale has worked on projects on five continents and speaks English, German, and Afrikaans. He is author or contributing author to six books and close to 100 papers. He lives in Denver, Colorado.